# Matroids: A Geometric Introduction

Matroid theory is a vibrant area of research that provides a unified way to understand graph theory, linear algebra and combinatorics via finite geometry. This book provides the first comprehensive introduction to the field, which will appeal to undergraduate students and to any mathematician interested in the geometric approach to matroids.

Written in a friendly, fun-to-read style and developed from the authors' own undergraduate courses, the book is ideal for students. Beginning with a basic introduction to matroids, the book quickly familiarizes the reader with the breadth of the subject, and specific examples are used to illustrate the theory and to help students see matroids as more than just generalizations of graphs. Over 300 exercises are included, with many hints so students can test their understanding of the material covered. The authors have also included several projects and open-ended research problems for independent study.

GARY GORDON is a Professor in the Mathematics Department at Lafayette College, Pennsylvania.

JENNIFER MCNULTY is a Professor in the Department of Mathematical Sciences at the University of Montana, Missoula.

# Matroids:
# A Geometric Introduction

GARY GORDON
*Lafayette College, Pennsylvania*

JENNIFER MCNULTY
*University of Montana, Missoula*

CAMBRIDGE
UNIVERSITY PRESS

# CAMBRIDGE
## UNIVERSITY PRESS

University Printing House, Cambridge CB2 8BS, United Kingdom

Published in the United States of America by Cambridge University Press, New York

Cambridge University Press is part of the University of Cambridge.

It furthers the University's mission by disseminating knowledge in the pursuit of education, learning and research at the highest international levels of excellence.

www.cambridge.org
Information on this title: www.cambridge.org/9780521145688

© Gary Gordon and Jennifer McNulty 2012

First published 2012

*A catalogue record for this publication is available from the British Library*

*Library of Congress Cataloguing in Publication data*
Gordon, Gary, 1956–
Matroids : a geometric introduction / Gary Gordon, Jennifer McNulty.
p. cm.
Includes bibliographical references and index.
ISBN 978-0-521-76724-8 (hard back)
1. Matroids. 2. Combinatorial geometry. I. McNulty, Jennifer. II. Title.
QA166.6.G67 2012
511.6 – dc23    2012009109

ISBN 978-0-521-76724-8 Hardback
ISBN 978-0-521-14568-8 Paperback

The first author thanks Liz McMahon for much love and support (and some very good ideas) throughout this project, and Rebecca and Hannah Gordon for their love, inspiration and general wonderfulness.

The second author wants to thank Jim Tattersall for his encouragement and introduction to the wonders of mathematics, her parents for their love and support, and her sons, Ben and Nate, for their patience, understanding and flexibility (especially in moving to PA for a year).

This book is dedicated to Thomas Brylawski (1944–2008) who taught both authors matroid theory. His enthusiasm, energy and humor were inspiring to his students, colleagues and friends.

# Contents

# Preface

**Matroids – a quick prehistory**

Matroid theory is an active area of mathematics that uses ideas from abstract and linear algebra, geometry, combinatorics and graph theory. The study of matroids thus offers students a unique opportunity to synthesize several different areas within mathematics typically studied at the undergraduate level. Furthermore, matroids are an active area of research; *Mathematical Reviews* lists some 2000 publications with the word "matroid" in the title, with more than a third of these appearing in the last decade.

Why have we written this book? Our motivation is direct: There is no comprehensive text written for undergraduates on this topic. There are several more advanced treatments of the subject, suitable for graduate students or researchers, but most of these are difficult for undergraduates to read. To paraphrase an old joke, this text seeks to fill this "much-needed gap."

This text introduces matroids by emphasizing geometry, focusing especially on geometric (*affine*) dependence. Interpreting this approach for finite subsets of a vector space, points in Euclidean space or the edges of a graph gives a matroid spin to linear algebra, discrete geometry and graph theory. We believe the geometric approach, which both authors learned from their common Ph.D. advisor, Thomas Brylawski, to be the most natural, useful and powerful in understanding the subject.

The common thread that ties the various classes of matroids together is the abstract notion of independence. This unifying idea is due to Hassler Whitney, who defined matroids in his foundational paper [42] in 1935. The field developed slowly in the 1940s and 1950s, attracting the attention of Garrett Birkhoff [3], who studied the flats of a matroid from a lattice-theoretic viewpoint, Saunders MacLane [23], who related matroids to projective geometry, and Richard Rado [27], who made important connections between the transversals of a bipartite graph and matroids. Most significantly, William Tutte [35] anticipated much of the modern approach to the field in 1958, where he characterized binary and regular matroids in terms of *excluded minors*. Finding excluded minors for classes of matroids remains one of the deepest and most vibrant areas of the field today.

Matroids can be defined in a surprisingly large number of *cryptomorphic* (= "secretly isomorphic") ways, and using different characterizations can be quite useful for different kinds of problems (it can also be a bit daunting for the beginner). The fact that there are so many different ways to define a matroid is a striking feature of the subject. Gian-Carlo Rota, a mathematician who taught several of the top researchers in the field, remarked that "It is as if one were to condense all trends of present-day mathematics onto a single finite structure, a feat that anyone would *a priori* deem impossible, were it not for the mere fact that matroids do exist."

### Using this text

Since matroids impinge on several subjects studied at the undergraduate level, we believe this book could serve as an ideal text for a capstone course. Such a course might incorporate and unify abstract algebra, linear algebra, combinatorics and geometry. The text could also serve as a special topics course in linear algebra, geometry or combinatorics. In addition, the text could form the basis of an independent study course for undergraduates or beginning graduate students.

Although the study of any mathematical subject requires the reader to understand the conventions and details of mathematical proofs, we have tried to make this process as gentle as possible. Proofs are often accompanied by examples that illustrate the main ideas in the proof. Examples abound throughout, and the reader is given ample opportunity to understand the theory by working some 300 exercises.

While no specific background is assumed, a reader who has studied some linear algebra and understands the basics of mathematical reasoning should be ready to read this book. Some exposure to abstract algebra, combinatorics or graph theory might be useful, but is not a requirement. We have included introductory material on graphs, matrices and linear algebra, finite fields, and finite geometry. Some of this material appears as a subsection of a chapter, some as a section, and we devote an entire chapter (Chapter 5) to finite geometry, a topic not often taught to undergraduates. Depending on the background and preparation of a class, an instructor might choose to review material by lecture, by student presentations, or by independent study. Of course, any instructor may opt to select chapters to match the students' backgrounds.

A quick note about style: We have tried to make this book somewhat "chatty." This means we have included some jokes and lots of ancillary material on related areas of mathematics. Writing a mathematics text in a conversational style is a tricky business; if the jokes are overdone (or too corny, or not corny enough, or . . . ), then this may turn readers away.

On the other hand, both authors have taught enough undergraduates to appreciate a text that shows some of its author's personality. We hope the reader will be tolerant with this approach, and understand what we've tried to achieve.

## Note to instructors

Chapter 1 is designed to pique the student's interest through lots of examples. In this introductory material, we try very hard to develop some geometric intuition, and we do so without proofs. We recommend instructors spend enough time on this chapter to develop this intuition in their students.

Chapter 2 is long,[1] and it gives many of the equivalent (cryptomorphic) definitions of a matroid; it is an important chapter and one that the students typically find challenging. We recommend a thorough treatment (complete with proofs) of Sections 2.1–2.3. Moving quickly through the remaining sections of the chapter is recommended, with instructors choosing how much or how little to cover. While it is amazing (to us) that matroids can be defined in so many different ways, most of our students have hidden their amazement very well. Consequently, it's easy to get bogged down with this material. Moving on to matroid operations (Chapter 3) when your students have had their fill of cryptomorphisms is a wise approach.

Chapters 3, 4, 5 and 6 are the heart of the text. Chapter 3 introduces matroid operations and is needed for all subsequent chapters. This includes the fundamental operations of deletion, contraction and duality, all of which are of central importance. We have included applications to graphs and matrices here – the entire notion of duality is presented as a generalization of duals for planar graphs. Chapter 4 treats graphs carefully, providing matroid proofs for many theorems of graph theory. We also make a general argument for generalization: proving a theorem about matroids gives you, *gratis*, theorems about graphs, vector spaces, and all the other classes of matroids we care about.

Chapter 5 gives an extensive treatment of affine and projective geometry, and it also includes a quick and dirty introduction to finite fields and vector spaces. This is beautiful, classical material, and it has somehow slipped through the cracks of modern undergraduate education. Chapter 6 builds on the finite geometry of Chapter 5 with a treatment of matroid representation questions. This chapter gives careful proofs of non-representability for several classical examples, including the Desargues' and Pappus' configurations.

Chapters 7, 8 and 9 include various special topics students often find appealing. Chapter 7 treats two important classes of matroids in some depth: transversal matroids (based on matchings in bipartite graphs)

---

[1] Paraphrasing Pascal, we would have written a shorter chapter if we had had more time.

and hyperplane arrangements in Euclidean space. An instructor could choose to cover either of these topics, or both (or neither). Chapter 8 is the deepest chapter of this text, and it treats the problems of excluded minors. We have attempted to give as much example-based proof as possible here, and this is especially true in our proof of Tutte's Theorem on excluded minors for binary matroids. Chapter 9 develops some of the theory of the Tutte polynomial, a two-variable matroid invariant that is also a very active area of current research.

It is not possible in such a text to be comprehensive, of course. The topics we have selected reflect our taste (and the tastes of our students), but the common theme is the geometric approach to the subject. The field has grown tremendously in the past dozen years, with major advances from a variety of researchers. But there are also opportunities for beginners to make contributions to the subject. We have included several projects that could serve as jumping off points for ambitious students.

A typical one-semester course might include the introductory material in Chapter 1, the first three sections of Chapter 2 and all of Chapter 3, followed by three or four additional chapters, ending perhaps with student projects. For instance, one could follow the material in the first three chapters by covering Chapters 4, 5 and 9 (for a gentle introduction), Chapters 4, 5, 6 and 8 (for a course focusing on matroid operations and minors), or Chapters 4, 5, 6 and 7 (for a thorough treatment of matroid classes). The flowchart below shows the interdependencies of the chapters.

Figure 1. Interdependence between chapters.

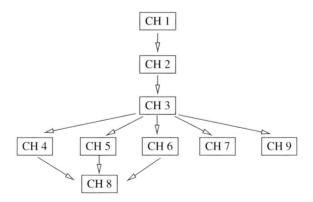

## A note to students

This book is meant to be read actively! That means having paper and pencil out while you're reading, with an example or two in mind. Then try your example on each new theorem, definition or concept you encounter in the text. While the theory can be rather abstract, the concrete examples should help fix the ideas in your mind. We've included lots of examples in the text to help you get used to this style of reading.

The exercises are meant to be worked through carefully. There are plenty of computational problems, but there are also a large number of problems where we ask you to show or prove something. Some of these are routine proofs that follow closely a proof in the text, but many of them require some new idea to complete. We have given extensive hints on the harder problems, and some of the more important exercises appear later in the text as propositions, with proofs.

A note about the jokes. The footnotes contain lots of remarks that might be uttered in a lecture, but usually don't survive the editing process in writing a book. If you like these comments, feel free to contact the authors to express your appreciation; if you don't, you should not necessarily feel compelled to tell us.

A general comment on reading mathematics: draw lots of pictures and don't get bogged down in terminology, definitions and details. Most professional mathematicians read technical material in stages, first skimming them for the main ideas, then going back and working through the details at one level (still skipping some technical details), then, finally, going back again to get all of the details down. Most of the time, there is a nicely chosen example (usually a picture) being updated as they read.

Our goal in writing this book is to show you both the beauty of matroids as well as the interconnectedness of mathematics. We use ideas from many different branches of mathematics in this book, and (we hope) this approach will strengthen your overall understanding of abstract mathematics. We hope you will learn much, enjoy the book, and laugh at our jokes.

## Acknowledgements

No book can be written without the help, encouragement and advice of others. This text is no exception. We are indebted to Joe Bonin, Garth Isaak, Dylan Mayhew, Liz McMahon, Nancy Neudaur, James Oxley, Lorenzo Traldi and Geoff Whittle for suggestions on topics, tone and a variety of other issues. We also thank Simeon Ball, Jim Geelen and Charles Semple for pointing out recent results and Gordon Royle for contributing data on matroid connectivity from his matroid database.

Roger Astley of Cambridge University Press was a reassuring presence in this project; we thank him for his help and calm demeanor.

Many undergraduate and graduate students also helped us by providing feedback to preliminary drafts of the text and the exercises. These include Chencong Bao, Stephanie Bell, Sean Burke, Bidur Dahal, Hannah Gordon, Rebecca Gordon, Peter McGrath, Joseph Mouselike, Joseph Oldenburg, Demitri Plessas, Liam Rafferty, Mary Riegel, Jordan Rooklyn, Share Russell, Hannah Stanton and Lahna VonEpps.

Finally, Thomas Brylawski's (1944–2008) spirit guided us throughout the project.

# 1

# A tour of matroids

## 1.1 Motivation

Matroids were introduced by Hassler Whitney, in a seminal[1] paper [42] that anticipated much of the early development of the subject.[2] Since its birth, the theory of matroids has undergone enormous growth, and it remains one of the most active research areas in mathematics. We'll give you more historical nuggets about the development of matroid theory along the way (often in footnotes), but not now.

Right now, as a tease, here's a question for you: How are each of the things[3] in Figure 1.1 related to each other, and how are they related to the following matrix?

$$B = \begin{array}{c} \begin{array}{ccccc} a & b & c & d & e \end{array} \\ \left[ \begin{array}{ccccc} 0 & 0 & 0 & 1 & 1 \\ 0 & 1 & 1 & 0 & 0 \\ 1 & 1 & 0 & 1 & 0 \end{array} \right]. \end{array}$$

This chapter should help you to answer this open-ended question, with matroids playing the unifying role. In particular, you should be able to find each of these pictures – and lots of matrices – in this and subsequent chapters.

Typically, when you study mathematics, especially in the undergraduate curriculum, you study topics separately, in individual courses like *linear algebra, abstract algebra, discrete math, combinatorics, geometry, graph theory*, and so on. While it's not illegal to think about a topic from *abstract algebra* (algebraically closed fields, say) while you're

---

[1] There will be plenty of footnotes in the text. You can ignore them, or just glance at them to decide if reading them is worth the effort. There is no reason we footnoted this word – it's just to get you used to looking down.

[2] Hassler Whitney (1907–1989) was an outstanding mathematician with wide interests; his fundamental work in algebraic topology, differential geometry and differential topology earned him the Wolf Prize in 1983. He invented matroid theory in the 1930s after working in graph theory. Whitney was an avid and accomplished rock climber, and his grandson wrote about his famous unprotected ascent in 1929 of the Cannon Cliff in NH, now named the Whitney–Gilman Ridge.

[3] This is not a technical term.

Figure 1.1. Four pictures.

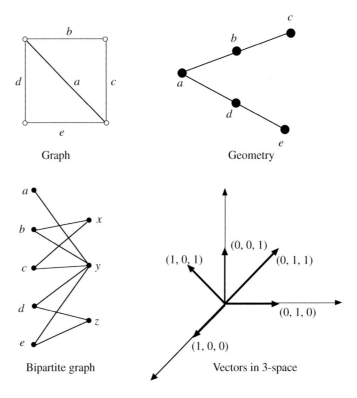

sitting in your *complex analysis* course, the textbooks for these courses and the way most undergraduate courses are designed do not encourage you to do so.

Our main point is this: much of the beauty and the richness of mathematics comes from the many connections between the various branches of mathematics. We believe the study of matroids is especially well suited for this purpose.[4] Figure 1.1 shows four objects that represent the same matroid-dependence structure: a graph, a point–line incidence geometry, a bipartite graph and an arrangement of vectors. Furthermore, they are all equivalent – as matroids – to the column dependences of a matrix, giving us a connection to linear algebra, as well.

> Matroid theory uses linear and abstract algebra, graph theory, combinatorics and finite geometry.

You shouldn't understand any of the details yet – we haven't given any. This chapter is devoted to introducing you to matroids by giving you lots of examples as they appear in several different areas of mathematics.

---

[4] Well, of course we believe this – we've written a book about them.

After reading this chapter, you should be able to understand what a matroid is from the viewpoint of graphs, linear algebra and geometry.

This text emphasizes the geometric approach popularized by Gian-Carlo Rota.[5] We learned Rota's approach from his student Thomas Brylawski, the Ph.D. advisor for both authors.[6]

Rota used the term "combinatorial pregeometry" instead of the term matroid, but this is sufficiently awkward to have not caught on. Rota had very strong feelings about terminology, believing the word "matroid" to be "ineffably cacophonic." In 1971, he and Kelly wrote:

> Several other terms have been used in place of geometry, by the successive discoverers of the notion; stylistically, these range from the pathetic to the grotesque. The only surviving one is "matroid," still used in pockets of the tradition-bound British Commonwealth. [19]

## 1.2 Introduction to matroids

We will tell you what a matroid is very soon – we promise – but we begin with two examples.

**Example 1.1.** Let $A$ be the following matrix.

$$A = \begin{array}{cccc} a & b & c & d \\ \left[ \begin{array}{cccc} 1 & 0 & 1 & 1 \\ 0 & 1 & 1 & 2 \end{array} \right]. \end{array}$$

We care about the four columns[7] $a = (1, 0), b = (0, 1), c = (1, 1)$ and $d = (1, 2)$. More to the point, we are interested in those subsets of columns that are *linearly independent* and those that are *linearly dependent*. (Remember that a set of vectors is *linearly dependent* if some non-trivial *linear combination*[8] of the vectors is the zero vector.)

Now in our matrix $A$, every pair of vectors forms a linearly independent (= not linearly dependent) subset of $\mathbb{R}^2$, but any subset of three of these vectors forms a linearly dependent set because the vectors all live in $\mathbb{R}^2$. (It is always true that when the number of vectors is larger than the dimension of your space, the vectors are linearly dependent.) Of course, the entire set of four vectors is linearly dependent.

How can we describe the linearly dependent subsets of $\{a, b, c, d\}$? Here's a surefire way that should appeal to the computer

---

[5] Gian-Carlo Rota (1932–1999) was an eloquent mathematician and philosopher who worked in combinatorics, but also made deep contributions to invariant theory and analysis. He received the Steele Prize in 1988 for his paper *On the foundations of combinatorial theory* [29] which is credited as "...the single paper most responsible for the revolution that incorporated combinatorics into the mainstream of modern mathematics."

[6] This last paragraph really belongs in the Preface, but scientists have proven that nobody ever reads the Preface.

[7] For convenience, we will sometimes write column vectors horizontally, like so: $(1, 0)$.

[8] This is the problem with definitions – they rely on *other* definitions.

Figure 1.2. Four vectors.

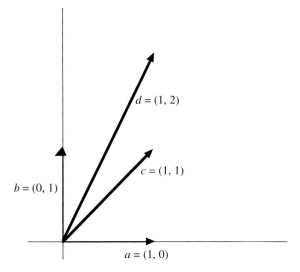

programmer: list them all. In this case, it's easy: $\{a, b, c\}$, $\{a, b, d\}$, $\{a, c, d\}$, $\{b, c, d\}$, $\{a, b, c, d\}$. You should convince yourself that this is not a good approach in general.

While there are many ways to "describe" these sets, we focus on a geometric way that will be central to the rest of this text (you might want to pay attention now). We will draw a picture that represents the linear dependence and independence of the subsets of the four columns from the example. The procedure has three easy steps.

---

**Rank 2 matroid drawing procedure from a matrix**

- Step 1: Draw the vectors in the plane. See Figure 1.2.
- Step 2: Draw a line in a "free" position – this means we want a line that is not parallel to any of our vectors. Now extend or shrink (and reverse, if necessary) each vector to see where it would hit this free line. See Figure 1.3.[9]
- Step 3: Finally, to get a picture of the column vector dependences corresponding to this matrix, just keep the line and discard the original vectors. See Figure 1.4 for a picture of the resulting "matroid," which just consists of four collinear points.

---

Here are two important things that you might notice.

(1) The length of a vector doesn't matter; for example, the picture in Figure 1.4 would be the same if we replaced $(1, 1)$ by $(2, 2)$.

---

[9] Hey – isn't this really two steps?

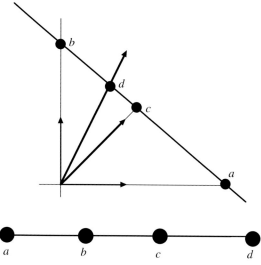

Figure 1.3. Our four vectors and a "free" line.

Figure 1.4. A picture of the matroid for the column dependences of the matrix $A$.

(2) We could replace a vector by its negative without changing our picture in Figure 1.4; for instance, replacing $(1, 2)$ by $(-1, -2)$ wouldn't change which subsets of vectors were dependent.

**Example 1.2.** Let's do another example. This time, let $B$ be the following matrix:

$$B = \begin{array}{c} \begin{array}{ccccc} a & b & c & d & e \end{array} \\ \left[ \begin{array}{ccccc} 0 & 0 & 0 & 1 & 1 \\ 0 & 1 & 1 & 0 & 0 \\ 1 & 1 & 0 & 1 & 0 \end{array} \right] \end{array}.$$

As before, we wish to draw a picture that represents the column dependences in the matrix. This time, we'll think of this as projecting the vectors onto a free plane, so we use the same drawing procedure as before, substituting "plane" for "line" in step 2:

---

### Rank 3 matroid drawing procedure from a matrix

- As before, draw the column vectors – or, better yet, make a three-dimensional model of the vectors using some nice model building kit (or, if no such kit is available, use toothpicks and gumdrops).
- Next, find a plane $P$ that is "free" with respect to your set – that means a plane that is not parallel to any of your vectors. See Figure 1.5.
- Finally, extend or shrink each of your vectors (or their negatives) until the extended or shrunken vector meets your plane $P$. These points in the plane will be the picture of your column vector dependences. See Figure 1.6. Congratulations!

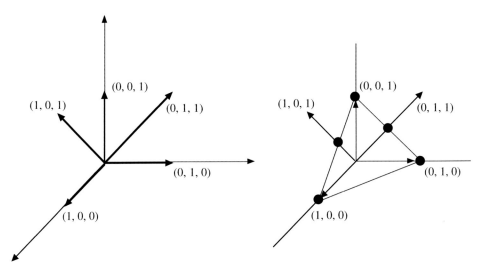

Figure 1.5. Projecting vectors in $\mathbb{R}^3$ onto the plane $x + y + z = 1$.

Figure 1.6. The matroid corresponding to the column vectors of the matrix $B$ with the two-point lines omitted.

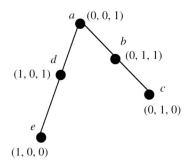

One more important comment about the picture in Figure 1.6 – by convention, we don't draw line segments connecting two points if they are the only two points on that line. For example, the points $b$ and $d$ form a two-point line, but we don't draw this line. The main reason for this is that no one else draws these lines,[10] but adding these lines would also increase the clutter in the picture (see Figure 1.7). There are four two-point lines in this matroid: $\{b, d\}$, $\{b, e\}$, $\{c, d\}$ and $\{c, e\}$. Remember – even though we haven't drawn two-point lines, they're still there.

Our drawing procedure amounts to a way of reducing dimension: in our first example, the rank of the matrix $A$ was 2, but the corresponding picture – the four-point line – is a one-dimensional object. (Recall the *rank of a matrix* is the dimension of its row space or its column space – these two subspaces have the same dimension.) In the second example, the matrix rank for $B$ is 3, but our matroid dependence picture was

---

[10] We know this sends a mixed message about peer pressure, but that's life.

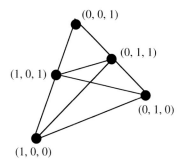

Figure 1.7. The matroid corresponding to the column vectors of the matrix *B*, with two-point lines drawn.

two-dimensional. So, in each case, we have "rank = dimension + 1" for the matroid dependence pictures we will draw.

It's now time for our first definition of a matroid[11] – you'll see several equivalent definitions in Chapter 2.

**Definition 1.3.** Let *E* be a finite set and let $\mathcal{I}$ be a family of subsets of *E*. Then the family $\mathcal{I}$ forms the *independent sets of a matroid M* if:     Matroid definition

(I1) $\mathcal{I} \neq \emptyset$;     Non-triviality

(I2) if $J \in \mathcal{I}$ and $I \subseteq J$, then $I \in \mathcal{I}$;     Closed under subsets

(I3) if $I, J \in \mathcal{I}$ with $|I| < |J|$, then there is some element $x \in J - I$     Augmentation
      with $I \cup \{x\} \in \mathcal{I}$.

The set *E* is called the *ground set* of the matroid. In our example, *E*     Ground set
was a set of vectors, but in another important example, *E* will be the
edges of a graph. The *rank* of a matroid, which is written $r(M)$, is just     Rank
the size of the largest independent set. The matroid associated to the
matrix *A* in Example 1.1 has rank 2, and the matroid associated to *B* in
Example 1.2 has rank 3. Most of the examples in this chapter have rank 3.
Also, the matroid rank equals the matrix rank – that seems fortuitous.     Matroid rank = matrix rank
Bet you didn't see that one coming.

This definition was first formally stated by Whitney [42]. He noticed
that the independence properties (I1), (I2) and (I3) were enjoyed by
linearly independent subsets of a *vector space*, and he wanted to understand how much (or how little) of the special features of vectors depend
on the field of coefficients (more precisely, how much of linear algebra
is independent of coordinates).

We'll prove that finite sets of vectors are examples of matroids in Theorem 6.1 in Chapter 6. In that chapter, we concentrate on the connections
between matroids and matrices.

**Theorem 6.1.** *Let E be the columns of a matrix A with entries in
a field* $\mathbb{F}$, *and let* $\mathcal{I}$ *be the collection of all subsets of E that are linearly independent. Then* $M = (E, \mathcal{I})$ *is a matroid, that is* $\mathcal{I}$ *satisfies the*     Matrices give matroids
*independent set axioms (I1), (I2) and (I3).*

---

[11] Your job: Read the definition and turn it into sentences you can understand.

**Definition 1.4.** A matroid whose ground set $E$ is a set of vectors is called a *representable matroid*.

Not all matroids are representable, and we'll study these matroids in some detail in Chapter 6. For now, note that it is very easy to see why properties (I1) and (I2) are satisfied by subsets of linearly independent vectors. It's a bit more work to check that (I3) always holds – this is really a fact from linear algebra. We defer the proof of Theorem 6.1 to Chapter 6.

A not-matroid

**Example 1.5.** It helps to understand a definition by looking at examples where it is *not* satisfied. As an easy example, suppose $E = \{a, b, c, d\}$ and you are given the following subsets: $\emptyset, a, b, c, d, ab, cd$. If you don't mind, for the sake of brevity, we will write $ab$ instead of $\{a, b\}$, $cd$ instead of $\{c, d\}$, and so on.[12]

Could these subsets be the independent sets of some matroid? (Pause to think!) The answer is no. While the subsets satisfy (I1) and (I2), axiom (I3) is violated: $c$ and $ab$ both independent requires either $ac$ or $bc$ to be independent.

By the way, it's possible to add some new sets to $\mathcal{I}$ and satisfy (I3). Of course, we could simply add all subsets of $E$ to $\mathcal{I}$, but this is overkill. You should check that adding the two subsets $ad$ and $bc$ to $\mathcal{I}$ works.

**Example 1.6.** This time, let $E = \{e_1, e_2, \ldots, e_n\}$. Let $k \leq n$ and define $\mathcal{I}$ to be all subsets of $E$ with $k$ or fewer elements. (For example, the matroid from Example 1.1 has this property for $n = 4$ and $k = 2$.) Then Uniform matroids $\mathcal{I}$ satisfies (I1), (I2) and (I3). This is called the *uniform* matroid, and it's denoted $U_{k,n}$. You will see it frequently in this text; see Exercise 4.

Boolean algebra

The matroid $U_{n,n}$ is called the *Boolean algebra*, and we denote it by $B_n$. Every subset is independent, and this clearly satisfies (I1), (I2) and (I3).[13]

Before finishing this section, we give two more matrix examples. In particular, given the column vectors of a matrix, could we skip all the vector drawing and jump straight to the matroid picture somehow? Well, could we?

The answer is yes, if we're a tiny bit clever.[14] We use the following picture drawing rules to go directly from our matrix $A$ to a matroid picture:

[12] You really have no choice here.
[13] George Boole (1815–1864) was the most famous logician of his day. Boole's daughter Alicia Boole Stott made important contributions to higher-dimensional geometry, for instance, proving there are precisely six regular solids in four dimensions, and constructing physical models of them.
[14] It's even easier if we're very clever.

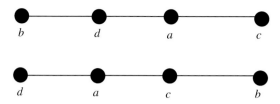

Figure 1.8. Two different labelings for the column dependences of *A*.

- each column vector will be represented by a point.
- if three vectors $u$, $v$ and $w$ are linearly dependent, then the corresponding three points will be collinear.

Only using these two rules frees us up a little. For instance, we can now represent the column dependences of

$$A = \begin{matrix} & a & b & c & d \\ & \begin{bmatrix} 1 & 0 & 1 & 1 \\ 0 & 1 & 1 & 2 \end{bmatrix} \end{matrix}$$

with either of the pictures in Figure 1.8. There are plenty[15] of other labelings of the four-point line that work.

Here's why the drawing procedure works in this example:

Three vectors are linearly dependent $\Leftrightarrow$
the vectors are coplanar $\Leftrightarrow$
the three corresponding points in our matroid picture are collinear.

This is easy to see for vectors in $\mathbb{R}^2$: two vectors in the plane are linearly independent as long as they point in different (not opposite) directions. But, if they point in different directions, then they meet our free line in distinct points, so they're independent in the matroid.

What if a pair of vectors is linearly dependent? Two dependent vectors will result in a pair of "multiple points" in the matroid, i.e., a dependent set of size 2. It is even possible for our matroid to have dependent *singletons*. Both of these pathologies occur in the next example.

**Example 1.7.** Let $C$ be the following matrix.

$$C = \begin{matrix} & a & b & c & d & e & f & g \\ & \begin{bmatrix} 1 & 0 & 1 & -1 & 0 & 2 & 0 \\ 1 & -1 & 0 & 0 & 1 & 0 & 0 \\ 0 & 1 & 1 & 2 & 2 & -4 & 0 \end{bmatrix} \end{matrix}.$$

Since $C$ is a rank 3 matrix, we expect our matroid picture to be planar. Two features of this matrix we have not seen before deserve some attention.

- $g$, which corresponds to the zero vector, is a dependent set of size 1.
- The pair $df$ is a dependent set of size 2.

---

[15] 4! We're excited about this!

Figure 1.9. A picture of the
matroid on the columns of $C$.

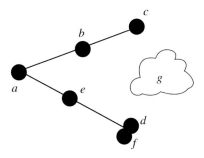

Other dependent sets should be more familiar. For instance, note that
columns $a$, $b$ and $c$ are linearly dependent:

$$(1, 1, 0) + (0, -1, 1) - (1, 0, 1) = (0, 0, 0).$$

From the linear dependences in the matrix, we get the following dependence story in the matroid:

(1)  $g$ is a dependent set.
(2)  $df$ is a dependent set; all other pairs of points taken from the set
     $\{a, b, c, d, e, f\}$ are linearly independent.
(3)  $abc$, $ade$ and $aef$ are dependent; all other triples of points taken
     from the set $\{a, b, c, d, e, f\}$ that don't contain both $d$ and $f$ are
     linearly independent.
(4)  Any set of four or more points is dependent.

The picture in Figure 1.9 gives all of the dependence information that
the matrix did (if we interpret four or more coplanar points as a dependent
set). For example, the three points $a$, $c$ and $e$ correspond to a linearly
independent set of vectors (since the three points are *not* collinear).
Note the two three-point lines $abc$ and $ade$ correspond to the two linear
dependences $1 \cdot a + 1 \cdot b - 1 \cdot c = 0$ and $1 \cdot a + 1 \cdot d - 1 \cdot e = 0$. As
usual, we don't draw two-point lines, like $ce$ or the line through $b$ and
the double point $df$.

How do we represent the double point $df$? And what is going on
with that weird cloud-like object that seems to have swallowed $g$? Well,
we[16] have a fundamental problem with trying to represent dependent
sets of size 1 or 2 geometrically. For multiple points, this problem isn't
too serious; we simply overlap our big black disks suggestively.

For the dependent set of size 1, we have a more fundamental problem.
It's not really possible to draw this as a point in any reasonable way, so

Loop = dependent singleton

we indicate this by enclosing $g$ in a cloud.[17] By the way, a dependent
singleton is called a *loop*, and you'll see loops throughout the text. They
are more important than you might guess at first glance.

---

[16]  Everyone, really – not just us.
[17]  Did you know Montana is Big Sky country?

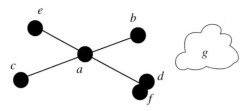

Figure 1.10. Another picture for the matroid on the columns of *C*.

An equally good picture representing the same dependences is given in Figure 1.10. Again, the order of the points on the lines is irrelevant – we only care about depicting the dependences faithfully.

Matroids that come from matrices are very important, but they aren't the entire matroid story. (If that were the case, the subject would be completely contained within linear algebra.) A reasonable question is:

> How do we draw a matroid if there are no vectors?

The answer to this question occupies the next section.

## 1.3 Geometries

One of the most appealing aspects of matroid theory is that there are many different approaches to the subject. In Section 1.2, you saw how dependent (and independent) subsets of vectors may be represented by pictures of points and lines. We turn this around now:

> Given a configuration of points and lines in the plane, how do we figure out the dependent and independent sets of the matroid it's supposed to represent?

To do this, we will need to determine the independent sets of the matroid directly from the geometry. For example, the pictures of Figures 1.9 and 1.10 determine the same matroid, and it's relatively easy to determine the independent sets of the matroid directly from the geometry without referring back to any of the vectors. As we saw in Example 1.7, depicting dependent singletons (*loops*) and dependent doubletons (*multiple points*) is not quite geometric. Technically, if we disallow these two conditions we restrict ourselves to *simple matroids*, also called *combinatorial geometries*.

Simple matroid

Combinatorial geometry

In a simple matroid, each element of the ground set is independent; it is represented geometrically by a point. In a non-simple matroid, some elements of the ground set may be dependent; these are loops and are

Figure 1.11. Matroid for
Example 1.8.

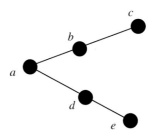

A *point* is an independent set of
size 1

represented geometrically by enclosing them in that silly cloud. When
we say "*e* is a *point* of a matroid," we *always* assume *e* is not a loop.

Pairs of points may behave differently in simple and non-simple
matroids. In a simple matroid, each pair of points is independent and
thus spans a line. In a non-simple matroid, the situation is not quite so
simple;[18] some pairs may be dependent. These dependent pairs correspond to multiple points which are represented geometrically by piling
up two (or more!!) points on top of each other. See Exercise 22 for
further details.

Multiple points can come in
any size

We summarize this discussion with the following procedure to determine the independent and dependent sets in a matroid from the geometric
picture. We first read the independence information in the case that the
matroid is simple and then describe the modifications needed if the
matroid is not simple.

---

**Determining the independent sets of a simple rank 3
matroid from a point–line incidence geometry**

- The empty set is an independent set.
- Every point is independent.
- Every pair of points is an independent set.
- A triple of points is an independent set if and only if the three
  points are *not* collinear.
- No set with more than three points is independent.

**Modifications for a non-simple matroid**

- Every element in the cloud is a loop, a one-element dependent
  set.
- Every pair of elements of a multiple point is a two-element
  dependent set.

---

**Example 1.8.** Let *M* be the matroid defined on the points $\{a, b, c, d, e\}$
shown in Figure 1.11.

---

[18]  Witty footnote is left to the reader.

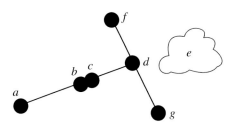

Figure 1.12. Matroid for Example 1.9.

Let's list the independent sets in this matroid directly from our picture:

- Ø is independent.
- Each point is independent: $a, b, c, d, e$.
- Each pair of points is independent: $ab, ac, \dots de$.
- Each subset of size 3 is independent except $abc$ and $ade$.

Sets with more than three elements are dependent since the rank of this matroid is 3.

What if there are loops and multiple points in the geometry?

**Example 1.9.** Now let $M$ be the matroid defined on the points $\{a, b, \dots, g\}$ shown in Figure 1.12.

Again, we use our procedure for finding all the independent sets:

- Ø is independent.
- Each point is independent except for the loop $e$.
- Each pair of points is independent, except for $bc$ (and pairs including $e$).
- Each subset of size 3 not containing a smaller dependent set is independent except $abd$, $acd$ and $dfg$.

Note the matroid in Figure 1.12 matches the matroid we constructed from the matrix $C$ in Example 1.7. We say the two matroids are *isomorphic*. It's possible to relabel the points of $M$ to match each other – for instance, the point $a$ of Figure 1.9 corresponds to the point $d$ in Figure 1.12.

We'll make a bigger deal about this later. Isomorphisms are pervasive throughout mathematics, and the definition of an isomorphism between two matroids is (morally) the same as it is for groups, rings, fields, and so on.

We can *simplify* the matroid $M$ of Figure 1.12 by eliminating the loops and coalescing all the multiple points. In this case, we get the matroid $M'$ in Figure 1.13. Appropriately, $M'$ is called the *simplification* of $M$. Simplification

Since axiom (I2) ensures subsets of independent sets in a matroid remain independent, it is often useful to restrict attention to the *maximal independent sets*. It is an easy (and very important) exercise to show Bases = maximal independent that all these maximal independent subsets have the same number of

Figure 1.13. *M'* is the simplification of *M*.

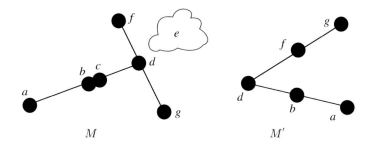

$$M \qquad\qquad M'$$

elements.[19] They are called *bases* and you will see them throughout the text.

It's easy to figure out all the bases of a matroid from the family of all independent sets: in Example 1.8, every subset of size 3 except *abc* and *ade* is a basis of the matroid *M*. Much more about bases appears in Chapter 2.

We introduce another important class of subsets of a matroid and save the rest (and there are plenty of other families of subsets we will care about) for Chapter 2. The subsets that are not independent are dependent – you already knew that. The *minimal dependent sets* are

Circuits = minimal dependent   very important – they are called *circuits*. In Example 1.8, there are only three circuits: *abc, ade* and *bcde*. Every *proper* subset of these three circuits is independent. This gives a quick way to describe a matroid – at least in this example.

Fano plane   **Example 1.10.   Fano plane** Consider the collection of seven points $\{a, b, \ldots, g\}$ in Figure 1.14. This configuration gives a matroid structure in which the independent sets include $\emptyset$, all singletons, all pairs of points and all triples *except* the seven lines *abe, adg, acf, bdf, bcg, cde* and *efg* (note that *efg* is a "line" even though these three points are not collinear as drawn[20]). This leaves $\binom{7}{3} - 7 = 28$ triples[21] which are independent – these maximal independent sets are the bases. Any collection of four or more points is dependent. This matroid is called the *Fano plane*[22] and is denoted $F_7$.

Why is this a matroid? Technically, we need to show that the three independent set axioms of Definition 1.3 are satisfied. In this case, axioms (I1) (non-triviality) and (I2) (closed under subsets) are immediate.

---

[19]  See Exercise 9 and the proof of Theorem 2.6.
[20]  If this seems like a big deal, it is, and much of the history of civilization depends on it.
[21]  If you haven't seen $\binom{n}{k} = \frac{n!}{k!(n-k)!}$ before, see us after class.
[22]  Gino Fano (1871–1952) was a pioneer in projective geometry, publishing numerous textbooks and over 130 papers, including some of the earliest work in finite geometry. Fano held a chair in Turin from 1901 until he was stripped of his position by Fascists in 1938. After WWII, Fano traveled, continuing his teaching and lecturing in Italy and the US.

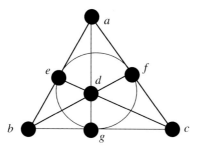

Figure 1.14. The Fano plane $F_7$.

To prove (I3) (augmentation), we need to make sure that whenever we take two independent sets of different sizes, we can augment the smaller set with something from the larger while preserving independence. There are a few cases to check, and most of them are very easy. For instance, if $|A| = 1$ and $|B| = 2$, then the sets must both be independent. At least one of the points in $B$ is not in $A$; call this point $x$. If $A = \{y\}$, then the pair $\{x, y\}$ is independent (since every pair of points is independent). Hence, we can always find something in $B$ to add to $A$ to preserve independence.

In fact, the same argument works when $|A| = 2$ and $|B| = 3$. In this case, since every pair of points is independent, we can choose $A$ to be any such pair. But $B$ must correspond to three non-collinear points, so we can always augment $A$ with some element from $B$. More precisely, if $A = \{x, y\}$, then there is a unique point $z$ so that $x$, $y$ and $z$ are collinear. Since $B$ is independent, $B$ must contain some $w \neq x, y, z$. Then $A \cup \{w\}$ is independent, so $F_7$ is a matroid.

The Fano plane has lots of very nice properties. The lines are simply the three-point dependent sets, so we can describe the geometry (points and lines) in terms of the matroid (independent sets). Two things you might notice are that every point is in exactly three three-point lines and every line has exactly three points. This point–line symmetry is a characteristic of *projective planes*, which you will meet in Chapter 5. Projective planes will also be very important in the study of matroid representation questions in Chapter 6.

It is also possible to represent the Fano plane by vectors, as in Section 1.2:

$$\begin{array}{ccccccc} a & b & c & d & e & f & g \end{array}$$
$$\begin{bmatrix} 1 & 0 & 0 & 1 & 1 & 1 & 0 \\ 0 & 1 & 0 & 1 & 1 & 0 & 1 \\ 0 & 0 & 1 & 1 & 0 & 1 & 1 \end{bmatrix}.$$

There's a funny story about this matrix, although closer inspection will reveal that it isn't really funny, and it's not exactly a story. The problem is that the last three columns, which correspond to the points $e$, $f$ and $g$, need to be linearly dependent to accurately reflect the fact that these

Figure 1.15. The non-Fano
matroid $F_7^-$.

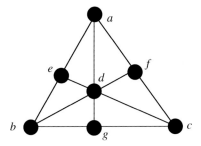

Figure 1.15. The non-Fano matroid $F_7^-$.

The Fano plane is a binary matroid    three points are a dependent set in the matroid. To do this, we note
that $e + f + g = (2, 2, 2)$. So, in order for this matrix to accurately
represent all the lines in the matroid, we require $2 = 0$ for whatever field
we're working in. This suggests that the Fano plane is somehow tied to
binary[23] arithmetic.

You can also see the connection with binary arithmetic by computing
a determinant:

$$\begin{vmatrix} 1 & 1 & 0 \\ 1 & 0 & 1 \\ 0 & 1 & 1 \end{vmatrix} = -2.$$

Since three vectors in $\mathbb{R}^3$ are linearly dependent if and only if the $3 \times 3$
matrix they form has determinant 0, we see $efg$ are collinear if and only
if $2 = 0$.

Here's something to think about: The seven points of the Fano plane
$F_7$ have the property that any line through two of these points contains
a third point. But the line $efg$ was a bit of a cheat, from the Euclidean
point of view. This suggests the following question about Euclidean
geometry:

> Is it possible for a finite set of points in the plane (not all on one
> line) to have no two-point lines?

See Exercise 25 for more on this problem.

There is no *matroid* reason for the triple $efg$ to be a dependent set. This
means we can form another matroid on the seven points $\{a, b, \ldots, g\}$
by simply adding $efg$ to the list of independent sets. Not surprisingly,
Non-Fano matroid    the resulting matroid is called the *non-Fano matroid*, and we denote it
$F_7^-$. See Figure 1.15.

Note that the three-point line $efg$ in $F_7$ has been replaced by three
two-point lines: $ef, eg$ and $fg$. As before, we don't draw these two-point

---

[23] We're using "binary" as a synonym for "modulo 2."

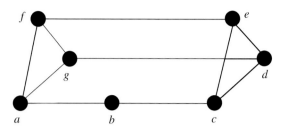

Figure 1.16. A rank 4 matroid.

lines, but they're there. From the matroid viewpoint, a two-point line is an independent set of size 2 with the property that adding any other point of the matroid preserves independence.

A good (if somewhat ill-defined) question might be something like: "When does this work?" More precisely, you might wonder when you can take a single dependent set in a matroid and decree that it's independent, forming a new collection of independent sets that still satisfy (I1), (I2) and (I3). *One such* procedure is called *relaxing a hyperplane*, and the theory is developed in Project P.3.

We can create matroids in higher dimensions, but if the picture involves more than three dimensions, it's pretty hard to visualize. For now, here's an example where the rank is 4.

**Example 1.11.** The picture of Figure 1.16 gives a matroid $M$ on the seven points $\{a, b, \ldots, g\}$ in which every subset of size 1 or 2 is independent, and all subsets of size 3 are independent *except abc*. A subset of size 4 is dependent precisely when the four points are coplanar. So, for example, the subset $abfg$ is independent, but $acdg$ is dependent. Every subset of five or more points is dependent because $r(M) = 4$. We also break our not-drawing-two-point-lines promise[24] to help you see the three-dimensionality of the picture.

A rank 4 matroid

## 1.4 Graphs and matroids

There are close ties between matroid theory and graph theory; indeed, much of the motivation and vocabulary for matroid theory comes directly from graph theory. Before exploring this relationship, we introduce some basic graph terminology.[25] Our approach to graph theory has a matroid bias – this should not be surprising. We begin with a working definition of a graph.

---

[24] So, we lied, but doesn't that lead us to a greater truth? The answer: no.

[25] Graph theory is an extremely active area of research within mathematics, in part because of its numerous applications to network and circuit design. For more information, you might check out one of the standard graph theory texts [7, 38].

Figure 1.17. A graph named *G*.

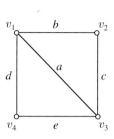

Figure 1.18. Three copies of *G*. The bold edges represent the edge-sets of (i) the path *bc*, (ii) the non-path *be* and (iii) the cycle *abc*.

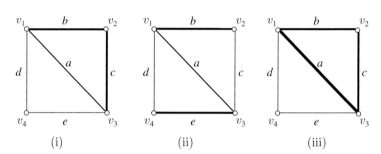

### 1.4.1 Graph basics

Graph A *graph* is a finite set of vertices *V* along with a finite set *E* of edges which join two vertices. Usually, an edge joins two different vertices, but we allow an edge to join a vertex to itself; such an edge is called a *loop*. Two (or more) distinct edges may join the same two vertices – such edges are called *multiple edges*. We say two edges are *incident* if they meet at a common vertex. Two vertices are *adjacent* if they share an edge. (Some authors use the term *multigraph* to indicate the possible presence of loops and multiple edges. We simply use the broad term *graph* to include loops and multiple edges, though.)

**Example 1.12.** The graph *G* in Figure 1.17 has vertex-set $V = \{v_1, v_2, v_3, v_4\}$ and edge-set $E = \{a, b, c, d, e\}$. We note that the dots in this picture are vertices of the graph; **not** points of a matroid. Edges of the graph are **not** lines in a geometry.[26] We'll return to this issue at the end of this section. We also note that the way we draw the edges, straight or curved, is not important for us.

In many applications of graph theory, it is useful to think of the vertices as cities and the edges as roads. In this example, starting at $v_1$, we could first take road *b* to $v_2$, then road *c* to $v_3$. Such a journey will be

Paths referred to as a *path* in *G* and abbreviated $v_1, b, v_2, c, v_3$ or, even more succinctly, simply *b, c*; see Figure 1.18(i). The two edges *b, e* do not form a path since these two edges are not incident (Figure 1.18(ii)). We also exclude sequences where edges are repeated: for instance, we will

---

[26] We apologize for shouting, but this is a very important distinction. Throughout this text, our graphs will have hollow vertices to better highlight this distinction.

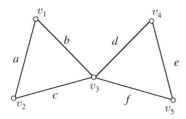

Figure 1.19. The edges $a, b, c, d, e, f$ form two edge-disjoint cycles. These edges do not form a path or a single cycle.

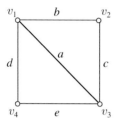 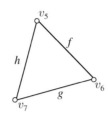

Figure 1.20. A disconnected graph $H$.

never call $\{b, c, e, c\}$ a path.[27] We also exclude situations where *vertices* are repeated (except for cycles – see the next paragraph); so $b, c, a, d$ is not a path, either. For instance, in Figure 1.19, the sequence of edges $a, b, d, e, f, c$ is *not* a path because the vertex $v_3$ is repeated.

If our path $b, c$ continues by adding the edge $a$ to our trip, we get a *cycle*, i.e., a path that starts and ends at the same vertex; see Figure 1.18(iii). Some care has to be taken here: in Figure 1.19, the sequence of edges $a, b, d, e, f, c$ a *not* a cycle – it's two cycles. Cycles

Although paths and cycles are defined as *sequences* of edges, we generally won't care about the order of the edges. Thus, we'll be perfectly happy calling $\{b, c, d, e\}$ a cycle in $G$, even though the sequence $b \rightarrow c \rightarrow d \rightarrow e$ of roads travelled is impossible. For brevity, we'll write $bcde$ instead of $\{b, c, d, e\}$, as in Section 1.2.

A graph is *connected* if there is a path between every pair of vertices; the graph is *disconnected* otherwise. Connected

**Example 1.13.** The graph $G$ in Figure 1.17 is connected, while the graph $H$ in Figure 1.20 is not; for example there is no path from $v_2$ to $v_6$ in $H$.

A common mistake is to say something like:

"But $H$ isn't a graph – it's two graphs."

A graph theorist would say:

"$H$ is a disconnected graph composed of two connected components. Each component is a called a subgraph of $H$."

---

[27] This is usually called a *walk* in graph theory books.

Figure 1.21. Three subgraphs
of $G$ (top): (i) a tree, (ii) a
disconnected forest, (iii) a
spanning tree of $G$.

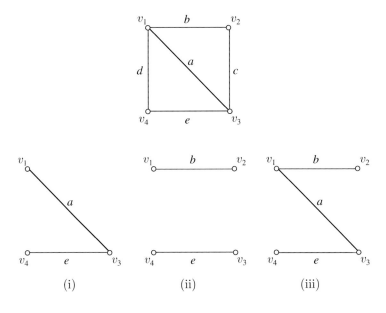

(i)                          (ii)                          (iii)

We agree with the graph theorist in this case – graphs are allowed to be
disconnected.

Subgraph    A graph $H$ is a *subgraph* of a graph $G$ if every vertex and edge of
$H$ also belongs to $G$. The cycles and paths above are both examples of
subgraphs.

From the matroid viewpoint, cycles will play a central role. A con-
Tree, forest, spanning tree    nected graph that contains no cycles is called a *tree*. A (connected or
disconnected) graph with no cycles is called a *forest*.[28] If $G$ is a con-
nected graph, then a tree that contains all the vertices of $G$ is called a
*spanning tree* of $G$.

**Example 1.14.** Consider the graph $G$ in Figure 1.21. It contains lots
of subgraphs that are trees or forests. Figure 1.21 shows a tree (i), a
(disconnected) forest (ii) and a spanning tree (iii) of $G$.

If you like this sort of activity, you might enjoy counting all of the
spanning trees, forests and trees in $G$. For instance, there are $\binom{5}{3} - 2 =$
$10 - 2 = 8$ spanning trees (we choose any three edges of $G$ in $\binom{5}{3}$ ways,
then exclude the two cycles $abc$ and $ade$). The total number of trees
includes the eight spanning trees, but it also includes eight two-edge
paths as well as the five single edges and $\emptyset$. This gives a total of 22 trees.

To count forests, we include the 22 trees we just found,[29] but we add
two disconnected forests: $cd$ and $be$. This gives us 24 forests in $G$.

---

[28]  What else would a bunch of trees be called? The entire subject is replete with colorful
botanical terms.

[29]  A forest is allowed to be composed of a single tree. You may wish to contemplate the
philosophical implications of this fact.

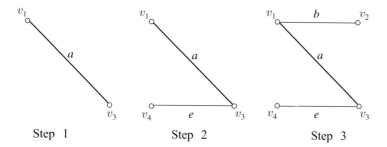

Figure 1.22. Creating a spanning tree of $G$, one edge at a time.

Step 1     Step 2     Step 3

All of our counting focuses on edges. For example, the subgraph consisting of the single edge $e$ is a forest, but so is the subgraph consisting of the edge $e$ together with the vertex $v_1$ of $G$ (see Figure 1.21). We do not view these as different subgraphs. This view is somewhat at odds with some graph theory texts, but it's the right way to think about graphs from the matroid viewpoint.

In connected graphs, spanning trees are maximal *acyclic* subsets of edges, while *cycles* are minimal non-acyclic subsets.[30] The collection of all spanning trees and the collection of all cycles will be important when we connect graphs to matroids later in this section.

Before we give you the matroid viewpoint, we have a few questions about trees:

(1) Does every graph have a spanning tree?
(2) If $G$ has a spanning tree $T$, how many edges does $T$ have?
(3) How many spanning trees does a graph have?

The answers to the first two questions are in the next theorem:

**Theorem 1.15.** *If $G$ is a connected graph with $n$ vertices, then $G$ has a spanning tree. Furthermore, every spanning tree of $G$ has $n - 1$ edges.*

You are asked to prove this in Exercise 17. The existence of a spanning tree in a connected graph is suggested by the sequence of trees in Figure 1.22.

For question (3), the number of spanning trees in a connected graph is given in Kirchhoff's *Matrix-Tree Theorem*[31] by a beautiful formula involving the determinant of a related matrix. You'll get a chance to prove an extension of this formula in Project P.2.

---

[30] This last point is pretty stupid – of course a cycle is the smallest thing you could create that contains a cycle.

[31] Gustav Kirchhoff (1824–1887) published a proof of this theorem in 1847. He made fundamental contributions to electrical network theory, mathematical physics and astronomy. He was also considered an outstanding teacher.

Figure 1.23. Left: graph *G*;
right: cycle matroid *M*(*G*).

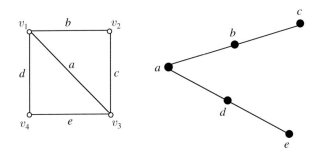

### 1.4.2 Matroids defined from graphs

We now know enough about graphs to connect them to matroids. We define a matroid $M(G)$ associated with the graph $G$ by specifying the ground set and the independent sets. A subset of edges is called *acyclic* if it contains no cycles.

We prove the next theorem in Chapter 4, where we concentrate on the connections between graphs and matroids.

**Theorem 4.1.** *Let E be the set of edges of a graph G and let $\mathcal{I}$ be the collection of all subsets of edges that are acyclic. Then $M = (E, \mathcal{I})$ is a matroid.*

Independent = Acyclic

Equivalently, the independent sets $\mathcal{I}$ are edge-sets of forests of $G$. Proving that $M(G)$ is a matroid involves showing that the acyclic subsets of edges satisfy the independent set axioms (I1), (I2) and (I3).

As was the case with subsets of vectors, it's easy to verify (I1) and (I2). (This is a persistent theme – when proving an object or a class of objects is a matroid, it will usually be easy to prove the object or the class satisfies (I1) and (I2). The real work is proving that (I3) is also satisfied.) Proving (I3) is not too difficult for acyclic subsets of edges, and our proof will use Theorem 1.15. But we postpone[32] the proof to

Cycle matroid

Chapter 4. The matroid $M(G)$ is called the *cycle matroid* of $G$.

**Example 1.16.** Let's describe the matroid associated with the graph $G$ from Example 1.12 (see Figure 1.23). For instance, the set $abe$ doesn't contain any cycle, so it is independent in the matroid $M(G)$, as are all its subsets. The cycle $abc$ is dependent, however. It is pretty easy to see that the minimal dependent sets are simply the three cycles:

Circuits of $M(G)$ = cycles of $G$

$C_1 = abc$, $C_2 = ade$ and $C_3 = bcde$. This is true in general:

> The circuits of the matroid are precisely the cycles of the graph.

[32] In the words of G. B. Stern, "One thing that's good about procrastination is that you always have something planned for tomorrow."

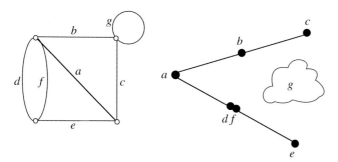

Figure 1.24. A graph with a double edge and a loop, and its cycle matroid.

You can now easily write down the independent sets; they are precisely the subsets of $E = \{a, b, c, d, e\}$ that do *not* contain one of these three cycles.[33] Thus, every subset of size 0, 1 or 2 is independent, and every subset of size 3 except $abc$ and $ade$ is also independent. No set of size 4 or 5 can be independent since any subset of size 4 or 5 (except for the cycle $bcde$) contains one of the two triangles $abc$ or $ade$.

If this matroid looks familiar, it should: the matroid $M(G)$ associated with the graph $G$ is the same matroid as the one from Example 1.8. Figure 1.23 shows a picture of the graph $G$ together with the geometric picture of $M(G)$.

Notice that the maximal independent sets (bases) of $M(G)$ are the spanning trees of $G$. Each spanning tree has three edges, so the rank of the matroid $M(G)$ is 3. We state this observation as a theorem – this follows from Theorem 1.15.

Bases of $M(G)$ = spanning trees of $G$

**Theorem 1.17.** *If $G$ is a connected graph with n vertices, then the rank of the matroid $M(G)$ is $n - 1$, the number of edges in a spanning tree of $G$.*

Let's do another example, shall we?

**Example 1.18.** Consider the graph on the left in Figure 1.24. This graph can be obtained from the graph $G$ of Figure 1.17 by replacing the single edge $d$ with a double edge $df$ and adding a loop $g$. Then the cycle matroid is easy to draw – compare Figures 1.23 and 1.24. (You should recognize the matroid from Example 1.7.)

It's worth noting that loops in a graph $G$ correspond directly to matroid loops in the cycle matroid $M(G)$ (and the term "loop" in matroid theory is taken from graph theory). Note that the location of the loop in the graph is irrelevant from the matroid point of view: if someone covered up the graph in Figure 1.24 so you could only see the picture of the matroid (on the right), you couldn't figure out which vertex of the graph had the loop $g$. Also note the double edge $df$ appears in the

Graph loops are matroid loops

[33] Technically, we should say "edge-sets of cycles" here rather than just "cycles." As we are mainly interested in edges of a graph, we omit the phrase "edge-sets."

Figure 1.25. The uniform
matroid $U_{2,4}$ is not graphic.

drawing of the cycle matroid as a double point. (From the matroid point
of view, $df$ is a two-point circuit.)

The first important question concerning the relation between matroids
and graphs is simply this:

---

Do all matroids come from graphs?

---

More precisely, given a matroid $M$, can we always find a graph $G$ with
$M(G) = M$? (This means the independent sets of the matroid $M$ must
precisely match the cycle-free subsets of edges of $G$.) The satisfying[34]
answer to this question is *No!*

Graphic matroid 　　Matroids that do arise as cycle matroids of graphs are called *graphic*.

**Example 1.19.**　We'll show the matroid $U_{2,4}$ in Figure 1.25 is not

$U_{2,4}$ is not graphic 　graphic by attempting to construct a graph $G$ with $M(G) = U_{2,4}$. We can
assume that if such a graph exists, it is connected (see Proposition 4.3),
so let's try to build such a $G$. The ground set of $M$ is $\{a, b, c, d\}$,
so $G$ has four edges. Since the bases ($=$ spanning trees) all have two
edges, $G$ must have three vertices by Theorem 1.15. Further, all subsets
of size 0, 1 and 2 are independent and every subset of size 3 or 4 is
dependent.

Summarizing, if there is a $G$, then:

- $G$ has three vertices and four edges.
- $G$ has no loops and no multiple edges.

Since any graph with three vertices and no loops or multiple edges
has at most three edges, the matroid $U_{2,4}$ is not graphic.

In the argument above, we assumed that if $M$ is a graphic matroid,
then there is a *connected* graph $G$ with $M = M(G)$. Why should this be
true?

Figure 1.26 offers a clue: the disconnected graph $G'$ appearing on
the top in the figure has the same cycle matroid as the connected graph
$G''$ in the bottom. This is easy to see: any acyclic subset of edges in $G'$
remains acyclic in $G''$, and vice versa. The process of creating $G''$ from
$G'$ is called *vertex identification*, and the reverse procedure is *vertex
splitting*.[35]

---

[34]　This is good news – if every matroid came from a graph, the entire theory would be a
subfield of graph theory.

[35]　A more detailed discussion of vertex splitting and vertex identification appears in
Section 4.4 of Chapter 4.

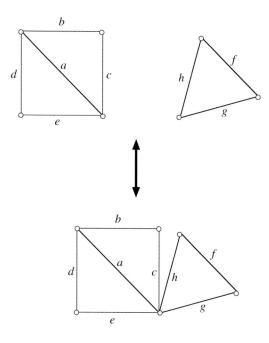

Figure 1.26. Vertex identification/splitting: the disconnected graph $G'$ (top) and the connected graph $G''$ (below) have the same cycle matroid.

We summarize this discussion with a proposition.

**Proposition 4.3.** *Let G be a graph with cycle matroid M(G). Then there is a connected graph $G'$ with $M(G) = M(G')$.*

One more connection is worth pointing out: graphs are representable matroids. This means that there is a matrix whose columns correspond to the edges of the graph. Further, a subset of columns will be linearly independent in that matrix if and only if the corresponding edges are acyclic. Such a matrix is constructed in Example 4.17 of Chapter 4. A more refined discussion appears later, in Theorems 8.13 and 8.28 of Chapter 8.

We end this section with a brief discussion of the difference between two kinds of pictures – ordinary drawings of graphs, and "geometric" pictures of matroids. For example, look back at the two pictures in Figure 1.23 or 1.24. While both pictures have dots and lines joining the dots, they have very different meanings; the pictures on the left in both figures are graphs and the pictures on the right are geometric configurations – matroids.

This is really important!

To help you keep these pictures separate, we will always draw graphs with small, hollow vertices and matroids with large, filled points. Another difference is that in a graph, we are interested in the edges and not the vertices, so pictures of graphs usually have the edges labeled, while our geometric pictures of matroids have the points (dots) labeled. It should also be clear from the context whether we are working with a graph or with a geometry. While these pictures look similar, we want to

Table 1.1. *Job interests for each applicant.*

| Applicants | Job |
|---|---|
| Alice, Darla | Cook |
| Bob, Carla, Darla | Food taster |
| Darla, Ed | Kitchen cleaner |

emphasize this difference as strongly as humanly possible, so there is no confusion (at least not with the pictures).

---

The key differences are:

- In a graph, the matroid is defined on the edges, and dependence comes from cycles.
- In a geometric configuration, the matroid is defined on the points, and dependence is "geometric."

---

## 1.5  Bipartite graphs and transversal matroids

Matroids arise "naturally" in several settings. In the 1940s, Richard Rado[36] discovered that there is a close connection between matroids (which were brand new) and matchings in bipartite graphs.

*Bipartite graph*   Let's begin with a few definitions. A *bipartite graph $B$* is a graph with vertex-set $V = X \cup Y$ partitioned into two non-empty disjoint sets $X$ and $Y$ so that no edges of $B$ join two vertices in the same part of the partition, i.e., you can't have an edge from one vertex of $X$ to another vertex of $X$, and similarly for $Y$. A subset of edges $M$ is called a *matching*

*Matching*   if no two edges of $M$ share any vertex. Each matching uses the same number of vertices in $X$ as it does in $Y$. Write $M = (I, J)$ for $I \subseteq X$ and $J \subseteq Y$ for the vertices $M$ uses in $X$ and $Y$, respectively.

Bipartite graphs arise in many different settings. Our first example is fairly typical – the job assignment problem.

**Example 1.20.**  Suppose you are opening a restaurant. You need to staff three positions: cook, food taster and kitchen cleaner. Five people (we'll imaginatively call them Alice, Bob, Carla, Darla and Ed) are interested in employment. Table 1.1 shows who is interested in each of the jobs.

---

[36]  Richard Rado (1906–1989) completed two doctorates: one in his native Germany with Schur, and a second in Cambridge with Hardy. Rado was elected a fellow of the Royal Society in 1978 for his work in combinatorics. Rado was also a professional-level concert pianist, and with his wife, who was a first rate singer, gave many private and public recitals.

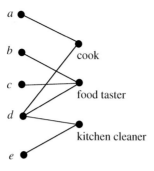

Figure 1.27. A bipartite graph models the job assignment choices.

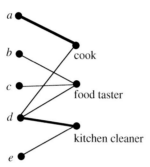

Figure 1.28. A matching: Alice cooks, Darla cleans.

It's easy to draw a bipartite graph $B$ for this problem: let $X = \{a, b, c, d, e\}$ represent the five people in the problem, and let $Y = \{$cook, food taster, kitchen cleaner$\}$ be the three jobs. See Figure 1.27.

What do matchings in the graph mean for your restaurant? They correspond precisely to legal assignments of people to jobs. For example, suppose we hire Alice to cook and Darla gets the coveted kitchen cleaner job. Then we get the matching in Figure 1.28.

You can check the following subsets of people can be hired: Ø, any single person, all pairs except $bc$, and all triples except $abc, bcd, bce$. Obviously, no subset of more than three people can be hired. If your goal is to staff all three positions, then there are lots of different assignments that would work in this case. It's also clear that you can't hire both Bob and Carla.[37]

The collection of all the possible matchings gives us our connection to matroids. Theorem 7.2, which we prove in Chapter 7, shows that matchings in bipartite graphs always give rise to matroids.

**Theorem 7.2.** *Let $G$ be a bipartite graph with vertex partition $X \cup Y$. Let $\mathcal{I}$ be the collection of subsets $I \subseteq X$ such that the vertices $I$ are precisely those vertices of $X$ in some matching $M = (I, J)$ of $G$. Then $\mathcal{I}$ satisfies the independent set axioms (I1), (I2), and (I3).*

Matchings give matroids

---

[37] They would probably fight – food fight!

Figure 1.29. Transversal
matroid associated with job
assignments of Example 1.20.

Figure 1.30. Another bipartite
graph $B$ and the associated
transversal matroid $M(B)$.

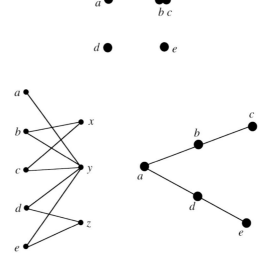

Transversal matroid    The matroid $M(B)$ associated with the bipartite graph $B$ is called
a *transversal matroid*. As usual, proving Theorem 7.2 requires us to
verify (I1), (I2) and (I3). As with ordinary graphs, it's not hard to see
that axioms (I1) and (I2) are true, but proving that (I3) is always satisfied
in this situation will again require more work. See Theorem 7.2, where
we prove the theorem by relating a matrix to a bipartite graph.[38]

For now, let's look back at Example 1.20. The theorem tells us we
have a matroid on the set $\{a, b, c, d, e\}$. The independent sets are all
those subsets that participate in matchings. Thus, the following subsets
are independent: $\emptyset$, all singletons, all pairs except $bc$, and all triples
except $abc, bcd, bce$. No subset with more than three elements can be
independent.

For practice, it's probably worth drawing a picture of the matroid
$M(B)$ associated with the bipartite graph $B$. In this case, we get the
matroid in Figure 1.29, with the double point $bc$.

For no real reason, here's another example.

**Example 1.21.**  This time, let $G$ be the bipartite graph shown on the left
in Figure 1.30. The ground set of our matroid is again $\{a, b, c, d, e\}$, and
a subset is independent if those vertices participate in some matching
in the graph. In this graph, you can check $ac$ is independent since we
can pair $a$ with $y$ and $c$ with $x$ in the graph, but $abc$ is dependent since
this subset is only joined to the set $\{x, y\}$, so there is no matching. The
(familiar) matroid is drawn on the right in Figure 1.30.

---

[38] A *direct* proof that (I3) is satisfied that uses an *augmenting path algorithm* appears in
Exercise 13 of Chapter 7.

**Example 1.22.** Let $M$ be the uniform matroid $U_{r,n}$ with $r > 0$. (Recall – the uniform matroid $U_{r,n}$ has a ground set of $n$ points and every subset of size $r$ is a basis.) We would like to cook up a bipartite graph $B$ so that $M(B) = U_{r,n}$; this means we need every subset of $B$ of size $r$ or smaller to participate in a matching, but no subset of size $r + 1$ to be matchable.

Creating the bipartite graph is easy: let $|X| = n$, $|Y| = r$ and join every vertex of $X$ to every vertex of $Y$. (This is the *complete bipartite graph $K_{n,r}$.*) Then it's clear that every subset of vertices of $X$ of size $r$ or smaller will be independent in the transversal matroid $M(K_{n,r})$, but no subset of $r + 1$ or more points of $X$ is independent.

*Uniform matroids are transversal*

Before leaving transversal matroids for now, here are a few things to ponder.

(1)  Is every matroid transversal?
(2)  Is there a connection between transversal matroids and representable matroids?
(3)  Is there a quick way to go from the bipartite graph to the picture of the matroid (as in Figure 1.30)?

[The answers are: (1) No. (2) Yes. (3) Yes. Did that help?] We'll return to transversal matroids in Chapter 7, where we give some justification for these answers.

# Exercises

## Section 1.2 and 1.3 – Introduction to matroids

(1)  This is a history question. Try to guess when the image in Figure 1.31 was used as the cover of a catalog of matroid extensions.
(2)  Am I a matroid? For each of the following, determine whether or not $(E, \mathcal{I})$ is a matroid. If it is a matroid, draw the geometry. If it is not a matroid, can you easily modify the collection to yield a matroid? If you're up to it, try to prove your answers. (As usual, we write $ab$ instead of $\{a, b\}$, etc.)
   (a)  Let $E = \{a, b, c\}$ with $\mathcal{I} = \{\emptyset, a, c, ab, ac\}$.
   (b)  Let $E = \{a, b, c\}$ with $\mathcal{I} = \{\emptyset, a, b, c, ab\}$.
   (c)  Let $E = \{a, b, c, d\}$ with $\mathcal{I} = \{\emptyset, a, b, c, ab, ac, abc\}$.
   (d)  Let $E = \{a, b, c, d, e, f\}$ with $\mathcal{I}$ consisting of $\emptyset$, all singletons, all doubletons and $abc, bcd, cde, def$.
(3)  Let $E = \{a, b, c, d\}$.
   (a)  Define $\mathcal{I}$ to be $\emptyset$ along with all subsets of $E$ that contain either $a$ or $b$ (or both). Does this define a matroid, i.e., does $\mathcal{I}$ satisfy (I1), (I2) and (I3)? If so, draw a picture of the matroid; if not, explain how the axioms fail.

Figure 1.31. Can you determine
what decade this matroid
catalog was produced?

(b) This time, define $\mathcal{I}$ to be $\emptyset$ along with all subsets of $E$ that contain both $a$ and $b$. Does this define a matroid? If so, draw it; if not, explain what goes wrong.

(4) This exercise is concerned with the uniform matroid $U_{r,n}$. Let $n$ be a positive integer, $E = \{a_1, a_2, \ldots, a_n\}$ and let $r \leq n$. A set $I \subseteq E$ will be independent if and only if it has $r$ or fewer elements.

    (a) Show $U_{r,n}$ is a matroid.

    (b) Draw the geometry of the matroid $U_{2,6}$. Repeat for $U_{3,6}$. For each of these, try to find a matrix whose column dependences match the matroid dependences.

(5) Small matroids.

    (a) Describe all (unlabeled) matroids on one, two, three or four elements, then draw the geometry for each matroid (you can "describe" each matroid by specifying its independent sets, for example). (Useful fact: there are two one-element matroids, four two-element matroids, eight three-element matroids and 17 four-element matroids.)

    (b) How many of the matroids in each class are simple, i.e., have no loops or multiple points?

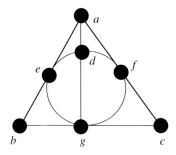 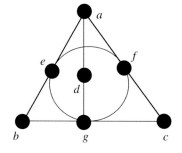

Figure 1.32. Left: the "Hallows matroid" is not a matroid. Right: moving point $d$ produces a matroid. See Exercise 10.

(c) For each of the matroids you just found, find a matrix whose column dependences match the matroid dependences.

(6) Let $A = \begin{bmatrix} 1 & 1 & -1 & 0 & 0 & 0 \\ 0 & 1 & 0 & 1 & 5 & 0 \\ 0 & 1 & 0 & 0 & 0 & 1 \end{bmatrix}$ $\begin{matrix} a & b & c & d & e & f \end{matrix}$. Draw a picture of the matroid $M[A]$ and then draw its simplification. Is $M[A]$ a paving matroid (see Exercise 7 for the definition).

(7) A matroid of rank $r$ is called a *paving matroid* if all of its circuits have size at least $r$.        Paving matroid
   (a) Show that if $C$ is a circuit of any rank $r$ matroid, then $|C| \le r + 1$.
   (b) Which of the following are paving matroids? Explain your answer by finding the rank of the matroid along with the sizes of all circuits.
      (i) The Fano matroid $F_7$; see Figure 1.14.
      (ii) The non-Fano matroid $F_7^-$; see Figure 1.15.
      (iii) The matroid $\mathcal{W}^3$; see Figure 1.35 below.

(8) This exercise continues the investigation of paving matroids – see Exercise 7.
   (a) Show that the uniform matroid $U_{r,n}$ is paving for all $r$, $n$ with $0 \le r \le n$.
   (b) Show that all simple matroids of rank 3 are paving.
   (c) Give an example of a simple rank 4 matroid that is not paving.

(9) Let $B_1$ and $B_2$ be maximal independent sets in a matroid $M$, i.e., no independent set $I$ properly contains either $B_1$ or $B_2$. Use property (I3) to show that $|B_1| = |B_2|$.[39]

(10) (a) Show that the "Hallows" point–line configuration on the left in Figure 1.32 is not a matroid. (Hint: concentrate on the two "lines" $adg$ and $defg$.)
    (b) Show that moving the point $d$ off the "line" $defg$ produces a point–line configuration that is a matroid (as shown on the right in Figure 1.32).

---

[39] This is a fundamental property of bases in matroids. It will reappear as axiom (B2′) throughout this text. See Exercise 17 for an application to graphic matroids.

Figure 1.33. The Escher "matroid" is not a matroid. See Exercise 11.

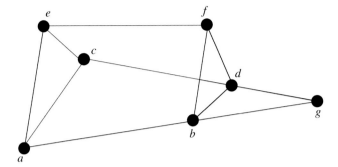

Figure 1.34. The graph *G* for Exercise 12.

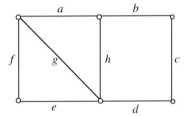

(c) Redraw the matroid on the right in Figure 1.32 with no "curved" lines.

(11) Escher "matroid."[40] The "matroid" of Figure 1.33 shows elements in three-dimensional space. The picture is supposed to suggest that all three-element subsets are independent except *abg* and *cdg*, so that, in particular, the subset *efg* is independent. Show that this family of "independent sets" does not satisfy the independent axiom (I3), so this is *not* a matroid!! (Hint: concentrate on the subsets $\{e, f, g\}$ and $\{a, c, e, f\}$, for example.)[41]

### Section 1.4 – Graphs and matroids

(12) Consider the cycle matroid $M(G)$ of the graph $G$ given in Figure 1.34.

(a) List the circuits of $M(G)$.

(b) Draw the geometry of $M(G)$.

(c) Consider the circuits $C_1 = agh$ and $C_2 = bcdh$. Note $h \in C_1 \cap C_2$. Find a circuit $C_3$ contained in $C_1 \cup C_2 - h$. (This property is the *circuit elimination property*, and we will see it again in Chapter 2 as axiom (C3).)

(13) For all the matroids on $E$ with $|E| \leq 4$, show that the uniform matroid $U_{2,4}$ is the only one that is not graphic. (You were asked to find all matroids on four or fewer points in Exercise 5. Recall, that

---

[40] Would M. C. Escher have wanted his name used for this figure? We don't know.
[41] The Escher configuration will be important in Example 6.30 of Chapter 6, where we study the *Vamos cube*.

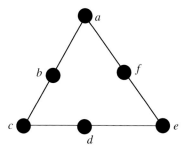

Figure 1.35. Matroid $\mathcal{W}^3$.
See Exercise 14.

in Example 1.19 we showed $U_{2,4}$ is not graphic. Thus, you need to show for each matroid $M$ with four or fewer elements (except $U_{2,4}$), there is a (connected) graph $G$ whose cycle matroid is $M$.)

(14) (a) Consider the matroid $\mathcal{W}^3$ given by the configuration of Figure 1.35. All sets of size 0, 1 and 2 are independent. What other sets are independent?

(b) If we add the dependence $bdf$ to this matroid (i.e., if we remove $bdf$ from the list of independent sets), we get another matroid. Draw the geometry of this new matroid using only straight lines.

(c) Show that the matroid $\mathcal{W}^3$ is not graphic, but the matroid in (b) is graphic.

(15) This exercise asks you to show that neither the Fano plane nor the non-Fano plane is a graphic matroid.

(a) Show that the Fano matroid $F_7$ of Figure 1.14 is not a graphic matroid. (Hint: if $F_7$ were graphic, then it can be represented by a connected graph $G$. Figure out how many edges and vertices $G$ must have.)

(b) Show that the non-Fano matroid $F_7^-$ of Figure 1.15 is not graphic, either.

(16) Find "all" values of $r$ and $n$ so that $U_{r,n}$ is graphic.[42]

(17) Let $G$ be a connected graph with $n$ vertices. The goal of this exercise is to show that every spanning tree for $G$ has $n-1$ edges (this is Theorem 1.15).

(a) Show that $G$ has a spanning tree $T$.

(b) Suppose $T$ is a tree with $n$ vertices. Show that $T$ has $n-1$ edges as follows:

(i) First show that if $T$ is a tree with at least one edge, then $T$ must have a *leaf*, i.e., an edge incident to a vertex of degree 1. (Try a proof by contradiction: if every vertex in the tree had degree 2 or more, then show that $T$ must contain a cycle.)

---

[42] There are infinitely many values that work, but that shouldn't discourage you. It makes us giddy.

Figure 1.36. Graph for
Exercise 18.

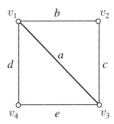

(ii)  Now finish the proof by induction on the number of edges
by removing a leaf.

(c)  Now use part (b) to prove that every spanning tree $T$ of $G$
has $n - 1$ edges. (Note: this result will be important when
we prove the acyclic subsets of edges in a graph satisfy the
independent set axioms – Theorem 4.1.[43])

(18)  (a)  It is difficult to define a matroid structure on the *vertices* of
a graph in a meaningful way. Here is something that *doesn't*
work in general. Define a subset of vertices $A \subseteq V(G)$ in a
graph $G$ to be *independent* if no two vertices of $A$ are joined by
an edge. (This terminology is standard. These subsets are quite
important in many applications – they are also called *stable*
sets.) Let $\mathcal{A}_G$ be the collection of all independent subsets of
vertices. Show that, although $\mathcal{A}_G$ always satisfies (I1) and (I2),
it *is not* the independent sets of a matroid in general. (Hint:
show that (I3) is violated for the graph $G$ given in Figure 1.36.)

(b)  Find $\mathcal{A}_G$ for the graphic matroid $M(G)$ for $G$ in Figure 1.34.

(c)  For *some* graphs $G$, the family $\mathcal{A}_G$ of independent subsets of
vertices may satisfy (I3). For example, the complete graph $K_n$
has no independent subsets of vertices containing more than
one vertex – this family satisfies (I3) trivially. Show this is the
*only* time this works, i.e., show that if $G$ is a connected graph
in which $\mathcal{A}_G$ satisfies (I3), then $G$ is a complete graph. (Hint:
show that if $G$ is connected and not complete, then there are
three vertices $u$, $v$ and $w$ where $uv$ and $uw$ are edges, but $vw$
is not. Use $u$, $v$ and $w$ to show (I3) is violated.)

**Section 1.5 – Transversal matroids and other topics**

(19)  Let $M$ be the transversal matroid on $X = \{a, b, \dots, g\}$ defined on
the bipartite graph from Figure 1.37.

(a)  Draw this matroid. What is the rank of $M$? (Hint: you might
want to look at Figure 1.41 below.)

(b)  Is this matroid graphic? If so, find a graph to represent it – if
not, explain why not.

---

[43]  In fact, Theorem 1.15 is important enough for us to include its proof in Chapter 4.

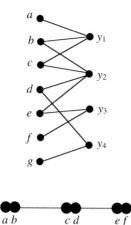

Figure 1.37. Bipartite graph for Exercise 19.

Figure 1.38. The matroid $M$ is not a transversal matroid.

Figure 1.39. $M_1$.

(20) Show that all the matroids on four or fewer points are transversal – see Exercise 5. (In fact, all matroids on five points are transversal – see Exercise 5 in Chapter 7.)

(21) Show that the matroid in Figure 1.38 is not transversal.[44]

(22) Multiple points – done right. Let $E$ be the ground set of a matroid $M$ and define a relation $\sim$ on $E$ as follows: $a \sim b$ if $a = b$ or if $ab$ is a circuit.

(a) Show the relation $\sim$ is an equivalence relation.[45] (Hint: use the independent set property (I3) and a proof by contradiction to show transitivity.) We denote the equivalence class containing the element $a \in E$ by $[a]$.

(b) Define a *multiple point* as an equivalence class of the relation $\sim$ containing more than one element. Find the multiple points in each of the following matroids.

(i) The matroid $M_1$ of Figure 1.39.

(ii) The graphic matroid $M_2 = M(G)$ of Figure 1.40.

(iii) $M_3 = M[B]$ for the matrix $B$

Multiple point

$$B = \begin{bmatrix} a & b & c & d & e & f & g \\ 1 & 2 & -1 & 0 & 0 & 1 & 0 \\ 0 & 0 & 0 & 1 & 5 & 7 & 0 \end{bmatrix}.$$

---

[44] This example is the smallest non-transversal matroid – see Example 7.3 of Chapter 7.
[45] Equivalence relations are defined in Definition 2.28 of Chapter 2.

Figure 1.40. $M_2 = M(G)$.

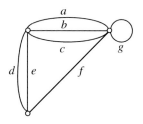

(c) Let $\overline{E}$ be the set of all equivalence classes formed under the relation $\sim$ on $E$ with all loops removed. Let $\overline{I} = \{[a_1], [a_2], \ldots, [a_k]\}$ be a set of distinct equivalence classes. We say $\overline{I}$ is independent if and only if $I = \{a_1, a_2, \ldots, a_k\}$ is independent in $M$. Let $\overline{\mathcal{I}}$ be the collection of all such subsets $\overline{I}$ of $\overline{E}$; show that $\overline{\mathcal{I}}$ satisfies (I1), (I2) and (I3). Conclude that $\overline{M} = (\overline{E}, \overline{\mathcal{I}})$ is a simple matroid.

(d) The definition of $\overline{M}$ given in part (c) can be described easily: form $\overline{M}$ by removing all the loops of $M$ and replacing each multiple point $[a_i]$ with a single point $a_i$. Draw the geometry for $\overline{M}$ for each of the matroids $M_1$, $M_2$ and $M_3$ in (b) above.

(23) This problem asks you to find all the symmetries of the Fano and non-Fano matroids – see Figures 1.14 and 1.15.

(a) Suppose $E$ is the ground set of the Fano plane $F_7$, and $f : E \to E$ is a bijection with the property that $I \subseteq E$ is independent in $F_7$ if and only if $f(I)$ is also independent. The function $f$ is an *automorphism*[46] of the matroid. Of the 7! possible relabelings of the points of the Fano plane, show that there are 168 such automorphisms.

(b) Now suppose $f : E \to E$, but we require $I \subseteq E$ is independent in $F_7^-$ if and only if $f(I)$ is also independent in $F_7^-$. Show that there are precisely 24 such functions.

(c) (*If you know some group theory.*) Show that each of these collections of automorphisms forms a group. Further, show that the automorphisms of $F_7^-$ form an index 7 subgroup of the automorphism group of $F_7$. Finally, show this subgroup is not normal, and interpret the seven cosets of the non-Fano automorphisms in the larger group "geometrically." (If any of this makes sense, then you can do this last question by writing the automorphisms explicitly as permutations of $E$.)

(24) (a) Show that the elements $a$, $b$ and $c$ are interchangeable in the rank 4 matroid of Figure 1.41 by showing the circuits of $M$ remain the same after any relabeling via a permutation of $\{a, b, c\}$.

---

[46] "Automorphism" refers to the shape of your car. Most cars are car-shaped.

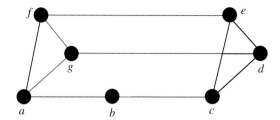

Figure 1.41. Matroid for Exercise 24.

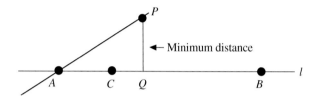

Figure 1.42. Picture for Exercise 25.

(b) Show that there are a total of 48 relabelings of this matroid that preserve the matroid structure (the independent sets, bases, circuits, etc.).

(c) (*If you know some group theory.*) Show that the set of all automorphisms $Aut(M)$ of $M$ form a group, and $Aut(M) \cong S_3 \times (\mathbb{Z}_2)^3$, where $S_3$ is the symmetric group.[47]

(25) Recall that the seven points of the Fano plane $F_7$ of Figure 1.14 have the property that any line through two of these points contains a third point, but to draw this in the plane, we made one "line" a circle. This is unavoidable: Suppose you are given $n > 2$ points in the plane, and suppose the line joining any two of the points in your set always contains another point from your set. In this exercise, you will prove that all of the points are on one line.[48]

(a) Suppose such a configuration exists – we argue by contradiction. There must be some point $P$ and line $l$ with $\text{dist}(P, l) \leq \text{dist}(P', l')$ for any other point $P'$ and line $l'$ in the set. Suppose $A$ and $B$ are two points from your set on $l$. Drop the perpendicular from $P$ to $l$, and call this point $Q$. See Figure 1.42.

(b) Then another point (say $C$) from your set must be on $l$. Show that two of the three points $A, B, C$ must be on the same side of $Q$ on $l$.

---

[47] See Exercise 27 of Chapter 3 for another approach to this problem.

[48] This problem is the famous Sylvester–Gallai theorem, proven by Tibor Gallai in the 1930s. Note that the Fano plane $F_7$ satisfies the Sylvester–Gallai hypothesis, but it requires us to call the three points $\{e, f, g\}$ a "line." One consequence of the Sylvester–Gallai theorem is that it is impossible to draw the Fano plane with all seven lines represented by "straight" lines in the plane, so, a "curved" line is really necessary to draw this matroid.

(c) Show that this gives a point–line pair with distance smaller than the distance from $P$ to $l$, which is a contradiction. This proof of the theorem is due to L. M. Kelly.[49]

---

[49] Paul Erdős, who believed God possesses a Book with the best proofs of all the theorems, thought this proof was "straight from The Book."

# 2

# Cryptomorphisms

Chapter 1 provided an introduction to matroids via a glimpse at several prototypes and their properties. This chapter will give the formal definitions and some of the basic results we will need for the rest of the book.

One of the most attractive features of matroids is the plethora of equivalent definitions. Borrowing from linear algebra, we defined a matroid in terms of *independent* sets in Chapter 1. In this chapter, we continue in this vein and give an equivalent definition in terms of *bases*. Graph theory motivates alternate formulations of matroids in terms of *circuits* and *cocircuits*, while *closed sets*, *closure* and *hyperplanes* come from geometry. We will tie these disparate concepts together using the *rank* function. There are many additional ways to define a matroid; we'll discuss some of them in the chapter and the exercises.

Each formulation (or reformulation) of the definition of a matroid will require a finite, *ground* set $E$ and a family of subsets of $E$ or a function defined on the power set $2^E$. These subsets and functions will satisfy various axioms, and each set of axioms will have one or two easy conditions and one more substantial axiom that characterizes a matroid property.

The central idea of this chapter is that these formulations are logically equivalent, and it is instructive to explicitly convert one axiom system to another. For example, you might wonder how you could determine the independent sets from the bases, or – if you were told the rank of any subset of the matroid, but nothing else – could you still figure out the independent sets? We'll describe a method, called a *cryptomorphism*, for converting from one system to another.

> A **cryptomorphism** is an "equivalent way to define matroids, yet not straightforward"[1]

---

[1] G. Birkhoff coined the term "crypto-isomorphism" in 1967 [4]. The term has now been shortened to "cryptomorphism". Garrett Birkhoff was an American mathematician who was born in Princeton, NJ in 1911. He was educated at Harvard and Cambridge, eventually returning to Harvard in 1933, first as a Fellow and later as a

G.-C. Rota remarked that matroids are "one of the richest and most useful ideas of our day."[2] While this richness stems from the variety of axiom systems (that is to say the number of cryptomorphic definitions of a matroid), and this gives matroid theory much of its power, it can often contribute to some difficulty for students learning the subject. We suggest the reader keep the prototypes from Chapter 1 in mind while learning the formal definitions and refer often to the cryptomorphism charts that appear throughout this chapter.

We have structured this chapter so that the material in Sections 2.1, 2.2 and 2.3 is "essential" in that it will be used throughout the remainder of the text. Sections 2.4, 2.5, 2.6 and 2.7 may be regarded as optional in the sense that studying the material presented there will not be a prerequisite for the material in the remaining chapters. We include this material here for several reasons, but we understand that some readers may not have the time to study these sections on a first pass through this text. The different axiom systems and the various crypotomorphic connections are summarized in the Appendix.

## 2.1  From independent sets to bases and back again

In Chapter 1, we defined the family of independent sets of a matroid – see Definition 1.3. Technically, a *matroid* is a pair of objects $M = (E, \mathcal{I})$, where $E$ is the ground set of $M$ and $\mathcal{I}$ is the collection of independent sets. For completeness, we restate this definition of a matroid formally:

Independent sets

**Definition 2.1.**  A *matroid*[3] $M$ is a pair $(E, \mathcal{I})$ in which $E$ is a finite set and $\mathcal{I}$ is a family of subsets of $E$ satisfying

Non-triviality   (I1) $\mathcal{I} \neq \emptyset$;

Closed under subsets   (I2) if $J \in \mathcal{I}$ and $I \subseteq J$, then $I \in \mathcal{I}$;

Augmentation   (I3) if $I, J \in \mathcal{I}$ with $|I| < |J|$, then there is some element $x \in J - I$ with $I \cup \{x\} \in \mathcal{I}$.

The family $\mathcal{I}$ are the *independent sets of the matroid*.

A family of subsets of a (finite) set is a *set system* if the family satisfies (I1) and (I2). (Topologists call such a set system a *simplicial complex.*)

Set system   **Definition 2.2.**  A pair $(E, \mathcal{I})$ is a *set system* if $\mathcal{I}$ is a non-empty collection of subsets of $E$ that is closed under taking subsets.

---

faculty member. Birkhoff is known for his work in lattice theory and abstract algebra. In his text, *Lattice Theory*, Birkhoff includes a new lattice-theoretic definition of a simple matroid. In the statement of this new definition Birkhoff coined the phrase crypto-isomorphic (known today as cryrptomorphic).

[2]  This remark was made many, many days ago, however.

[3]  Memorize this!

Thus, the independent sets in a matroid are a set system. The three properties (I1), (I2) and (I3) are not the only properties the independent sets of a matroid satisfy. For instance, it's easy to see that properties (I1) and (I2) together imply $\emptyset \in \mathcal{I}$. We can find alternate versions of (I1) and (I3) to *define* the independent sets anew. This is the point of the next proposition, and going through the (easy) proof should be good practice for the (harder) proofs of our cryptomorphisms to come.

**Proposition 2.3.** *Let E be a finite set and let $\mathcal{I}$ be a family of subsets of E. Then the family $\mathcal{I}$ are the independent sets of a matroid if and only if:*

(I1') $\emptyset \in \mathcal{I}$;
(I2) *if $J \in \mathcal{I}$ and $I \subseteq J$, then $I \in \mathcal{I}$;*
(I3') *if $I, J \in \mathcal{I}$ with $|J| = |I| + 1$, then there is some element $x \in J - I$ with $I \cup \{x\} \in \mathcal{I}$.*

*Proof Proposition 2.3.* To prove the proposition, we must show that $\mathcal{I}$ satisfies (I1), (I2) and (I3) if and only if $\mathcal{I}$ satisfies (I1'), (I2), (I3'). We begin with a matroid $M = (E, \mathcal{I})$ satisfying (I1), (I2) and (I3), and we show (I1'), (I2) and (I3') must also be satisfied.

($\Rightarrow$) Since $\mathcal{I} \neq \emptyset$ (this is (I1)), there is a set $X \subseteq E$ with $X \in \mathcal{I}$. Now, since every subset of $X$ is independent (why?), $\emptyset \in \mathcal{I}$. So, (I1') and (I2) are satisfied. Lastly, note (I3') is just a special case of (I3), and so we are (half) done.

($\Leftarrow$) For the converse, let $E$ be a finite set and $\mathcal{I}$ a family of subsets of $E$ that satisfy (I1'), (I2) and (I3'). We will show that $\mathcal{I}$ also satisfies (I1), (I2) and (I3), and hence is the collection of independent sets of a matroid. Since $\emptyset \in \mathcal{I}$, the collection $\mathcal{I}$ is not empty.

Lastly, to prove (I3), let $I, J \in \mathcal{I}$ with $|I| < |J|$. We need to find an element $x \in J - I$ with $I \cup \{x\} \in \mathcal{I}$. Here is the (only) clever bit in this proof: let $I'$ be any subset of $J$ with $|I'| = |I| + 1$. Since $I' \in \mathcal{I}$ (from (I2)), we can use (I3') to find some element $x \in I' - I$ with $I \cup \{x\} \in \mathcal{I}$. Since $I' \subseteq J$, clearly $x \in J - I$, so we are (completely) done. □

Proposition 2.3 gives us a second way to define a matroid via the family of independent sets. The proof of the equivalence between the conditions (I1), (I2), (I3) and the corresponding (I1'), (I2), (I3') would be more straightforward if (I1) $\Leftrightarrow$ (I1') and (I3) $\Leftrightarrow$ (I3'); unfortunately, this is **not true**. See Exercise 11 for an example.

If a friend hands you a family of subsets and asks you for the biggest one, there are at least two ways you might respond. On the one hand, you might give her a *maximal* subset from your family. The maximal subsets in the family are, by definition, those subsets that are not contained in any other sets in the family; they are *maximal with respect to inclusion.*   Maximal vs. maximum
On the other hand, you might look at the number of elements in each

Figure 2.1. This is a picture of all the subsets of $E$, ordered by inclusion. The bases are the maximal independent sets.

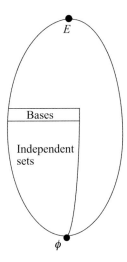

set in your family (the cardinalities of all of your sets) and give her one with the most elements – a *maximum-size subset*.

A set of maximum size must be maximal, but the converse is not true for arbitrary families of subsets. For instance, the family of subsets $\{\emptyset, \{a\}, \{c\}, \{a, b\}\}$ has two *maximal* subsets $\{a, b\}$ and $\{c\}$, but only one *maximum* subset $\{a, b\}$.

The key point for us is the following: these two concepts – maximal and maximum – coincide for the independent sets in a matroid. (This follows immediately from axiom (I3).) Another way to say this is that all the maximal independent sets have the same size. These maximal independent sets in a matroid are the *bases* of the matroid. See Figure 2.1.

**Definition 2.4.** If $M$ is a matroid with independent sets $\mathcal{I}$, then $B$ is a
Bases    basis of the matroid $M$ if $B$ is a maximal independent set.

Let's see how we can *define* a matroid directly in terms of bases. The proof will be similar in style to the proof of Proposition 2.3, but there are important differences because now we have two collections of subsets of $E$, namely $\mathcal{I}$ and $\mathcal{B}$. Here are three things that (we'll eventually prove) are always true for the bases $\mathcal{B}$ of a matroid.

Non-triviality    (B1) $\mathcal{B} \neq \emptyset$.

Clutter    (B2) If $B_1, B_2 \in \mathcal{B}$ and $B_1 \subseteq B_2$, then $B_1 = B_2$.

Weak basis exchange    (B3) If $B_1, B_2 \in \mathcal{B}$ and $x \in B_1 - B_2$, then there is an element $y \in B_2 - B_1$ so that $B_1 - x \cup \{y\} \in \mathcal{B}$.

(A *clutter* is a family of subsets of a set, no one of which is a subset of another.)

These *properties* that the collection $\mathcal{B}$ of bases satisfy will be used to define a matroid in terms of bases. That is, these properties can be taken as an equivalent *axiom system* for matroids. Just as the axiom

system for independent sets mirrors the properties of linear independence in vectors, the axiom system for bases is motivated by properties satisfied by the collection of bases of a (finite-dimensional) vector
space.

Here's an alternate version of (B2):

(B2′) If $B_1$, $B_2 \in \mathcal{B}$, then $|B_1| = |B_2|$.                                    Equicardinal

   (Proving that (B2′) holds for maximal independent sets is the content
of Exercise 9 in Chapter 1.) The next proposition, whose proof is left for
you to complete[4] in Exercise 12, will be useful when we work through
the details of the cryptomorphism between independent sets and bases.

**Proposition 2.5.**  *Let E be a finite set and let $\mathcal{B}$ be a family of subsets of
E. The family $\mathcal{B}$ satisfies (B1), (B2), (B3) if and only if $\mathcal{B}$ satisfies (B1),
(B2′), (B3).*

   We are finally ready to discuss our first cryptomorphism. This
involves the following steps:

(1) Given the independent sets $\mathcal{I}$ in a matroid, *define* a collection of
    subsets $\mathcal{B}$ to be the maximal independent sets:

$$\mathcal{B} = \{B \in \mathcal{I} \mid B \subseteq A \in \mathcal{I} \text{ implies } B = A\}.$$

(2) Next, *prove* that the collection $\mathcal{B}$ satisfies the properties (B1), (B2′)
    and (B3), i.e., (B1), (B2′) and (B3) are *theorems* when we assume
    (I1), (I2) and (I3).
(3) Reverse the procedure: given the collection of bases $\mathcal{B}$ of a matroid,
    *define* a collection of subsets $\mathcal{I}$ to be all subsets of the bases in $\mathcal{B}$:

$$\mathcal{I} = \{I \mid I \subseteq B \text{ for some } B \in \mathcal{B}\}.$$

(4) Now, *prove* the collection $\mathcal{I}$ you just defined satisfies the properties
    (I1), (I2) and (I3), so (I1), (I2) and (I3) are *theorems* when we
    assume (B1), (B2′) and (B3).

   We're still not done.[5] When we define $\mathcal{I}$ in terms of $\mathcal{B}$ and also define
$\mathcal{B}$ in terms of $\mathcal{I}$, we are actually dealing with *functions* on the set of all
families of subsets of $E$. Let $f$ and $g$ be these cryptomorphic functions
with $f(\mathcal{I}) = \mathcal{B}$ and $g(\mathcal{B}) = \mathcal{I}$ (so $f$, $g : 2^{2^E} \to 2^{2^E}$). We have to show
these two cryptomorphisms compose correctly: if we start with $\mathcal{I}$, then
apply the map $f$ followed by the map $g$, we had better end up with the
family we started with, i.e., $g(f(\mathcal{I})) = \mathcal{I}$. Conversely, we also need to
show the reverse composition works correctly, i.e., $f(g(\mathcal{B})) = \mathcal{B}$. See
Figure 2.2.

---

[4] "I hear and I forget. I see and I remember. I do and I understand." Chinese proverb.
[5] Actually, we haven't even started.

Figure 2.2. Cryptomorphism
between independent sets and
bases.

You can think about this entire enterprise as a kind of elaborate game
in logic[6] – you'll end up with a whole bunch of things that are true, but
you'll also have the satisfaction of understanding precisely why we can
use either $\mathcal{I}$ or $\mathcal{B}$ to define a matroid.

**Theorem 2.6.** *Let $E$ be a finite set and let $\mathcal{B}$ be a family of subsets
of $E$ satisfying (B1), (B2), (B3). Then, $(E, \mathcal{B})$ is cryptomorphic to the
matroid $M = (E, \mathcal{I})$ and $\mathcal{B}$ is the collection of bases of a matroid.*

*Proof Theorem 2.6.* In light of Proposition 2.5, we can substitute (B2′)
for (B2) throughout the proof.

Part 1: $\mathcal{I} \to \mathcal{B}$    *Part 1*. From $\mathcal{I}$ to $\mathcal{B}$. Given the matroid $M = (E, \mathcal{I})$; that is $\mathcal{I}$ is a family
of subsets of $E$ satisfying (I1), (I2), (I3); let $\mathcal{B}$ be the maximal subsets
in $\mathcal{I}$, i.e., $\mathcal{B} = \{B \in \mathcal{I} \mid B \subseteq B' \in \mathcal{I}$ implies $B = B'\}$. We need to show
$\mathcal{B}$ satisfies (B1), (B2′), (B3).

(B1) is easy: since $\emptyset \in \mathcal{I}$ and since $E$ is finite, we can find an element
$B \in \mathcal{I}$ which is not properly contained in any other independent set.
Clearly, $B \in \mathcal{B}$, so $B \neq \emptyset$ and (B1) holds.

(B2′) is also easy to check: if there are sets $B_1, B_2 \in \mathcal{B}$ with $|B_1| <
|B_2|$, then, since $B_1, B_2 \in \mathcal{I}$, there is an element $x \in B_2 - B_1$ so that
$B_1 \cup \{x\} \in \mathcal{I}$ (by (I3)). But, $B_1$ is a proper subset of $B_1 \cup \{x\}$, which
contradicts our definition of $\mathcal{B}$, so $|B_1| = |B_2|$. Thus, (B2′) holds for $\mathcal{B}$.

Next, we need to show (B3) holds: let $B_1, B_2 \in \mathcal{B}$ with $x \in B_1 - B_2$.
(We can always find such an $x$ if $B_1 \neq B_2$ – this follows from the
way we defined $\mathcal{B}$ from $\mathcal{I}$.) Again, since $B_1, B_2 \in \mathcal{I}$, we can use the
independent set axioms. By (I2), $B_1 - x \in \mathcal{I}$. Now we apply (I3) to the
sets $B_1 - x$ and $B_2$. Since $|B_1 - x| < |B_2|$ (why?), there is an element
$y \in B_2 - (B_1 - x)$ with $B_1 - x \cup \{y\} \in \mathcal{I}$.

If we can show $B_1 - x \cup \{y\} \in \mathcal{B}$, we'll have verified (B3), which is
what we're shooting for right now. But $|B_1 - x \cup \{y\}| = |B_1| = |B_2|$, so
if $B_1 - x \cup \{y\}$ is in $\mathcal{I}$ but not in $\mathcal{B}$, then there is a $B_3 \in \mathcal{B}$, with $B_1 - x \cup
\{y\} \subset B_3$ and $|B_1| = |B_1 - x \cup \{y\}| < |B_3|$, which contradicts (B2′).
Thus, for all $B_1, B_2 \in \mathcal{B}$ with $x \in B_1 - B_2$, there is a $y \in B_2 - B_1$ so
that $B_1 - x \cup \{y\} \in \mathcal{B}$, and (B3) holds. Thus, we have shown axioms
(B1), (B2′), (B3) follow from (I1), (I2), (I3).

Part 2: $\mathcal{B} \to \mathcal{I}$    *Part 2*. From $\mathcal{B}$ to $\mathcal{I}$. Let $\mathcal{B}$ be a family of subsets of $E$ satisfying (B1),
(B2′), (B3) and let $\mathcal{I} = \{I \mid I \subseteq B$ for some $B \in \mathcal{B}\}$. We need to show
$\mathcal{I}$ satisfies (I1), (I2), (I3) and hence $(E, \mathcal{I})$ is a matroid.

---

[6] You could argue this is true for all of mathematics.

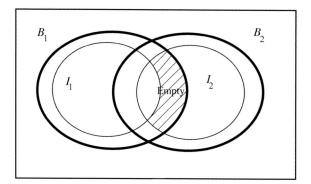

Figure 2.3.
$I_1 \cap I_2 = B_1 \cap I_2$.

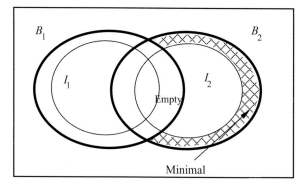

Figure 2.4. Choose $B_2$ so that $|B_2 - (I_2 \cup B_1)|$ is minimal.

Since $\mathcal{B} \neq \emptyset$ (by (B1)) and $\mathcal{B} \subseteq \mathcal{I}$, it follows that $\mathcal{I} \neq \emptyset$ and (I1) holds. To verify (I2), we need to show that if $I' \subseteq I$ for some $I \in \mathcal{I}$, then $I' \in \mathcal{I}$. Now by the way $\mathcal{I}$ is constructed, we know $I \subseteq B$ for some $B \in \mathcal{B}$. But then $I' \subseteq I \subseteq B$, and so $I' \in \mathcal{I}$, too, and (I2) holds. Thus, $\mathcal{I}$ is a set system.

Proving (I3) holds involves a bit more work; this is the hardest part of the proof, and it involves several steps. We give a proof by contradiction. Let $I_1, I_2 \in \mathcal{I}$ with $|I_1| < |I_2|$ and suppose that (I3) does not hold. We will show this implies $|I_1| \geq |I_2|$ – a contradiction.

Since $I_1, I_2 \in \mathcal{I}$, there are $B_1, B_2 \in \mathcal{B}$ with $I_1 \subseteq B_1$ and $I_2 \subseteq B_2$. We may assume that $I_1 \cap I_2 = B_1 \cap I_2$ (or, equivalently, $I_2 - B_1 = I_2 - I_1$) since if $x \in (I_2 \cap B_1) - I_1$, then $I_1 \cup \{x\} \subseteq B_1$ and axiom (I3) holds. See Figure 2.3.

Now there may be many choices for $B_2$ with $I_2 \subseteq B_2$; among all of these possible choices, choose one so that $|B_2 - (I_2 \cup B_1)|$ is minimal. See Figure 2.4.[7]

We now claim $B_2 - (I_2 \cup B_1) = \emptyset$. If $x \in B_2 - (I_2 \cup B_1)$, then by reversing the roles of $B_1$ and $B_2$ in (B3), there is some element $y \in B_1 - B_2$ so that $B_3 = B_2 - x \cup \{y\} \in \mathcal{B}$. (See Figure 2.5.) But, $I_2 \subseteq B_3$

---

[7] The idea of choosing a set so that some cardinality is minimized (or maximized) is quite useful, and we will use it again.

Figure 2.5. $B_3$ contradicts
minimality:
$B_2 - (I_2 \cup B_1) = \emptyset.$

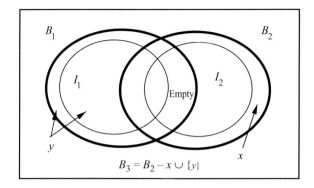

Figure 2.6. $B_2 - B_1 = I_2 - I_1$
and $B_1 - B_2 \subseteq I_1 - I_2.$

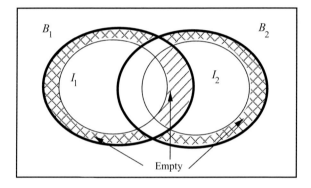

and $|B_3 - (I_2 \cup B_1)| < |B_2 - (I_2 \cup B_1)|$, which contradicts our choice
of $B_2$. Thus $B_2 - (I_2 \cup B_1) = \emptyset$.

By a similar argument,[8] we can choose $B_1$ so that $B_1 - (I_1 \cup B_2) = \emptyset$.
In particular, this follows by selecting $B_1 \in \mathcal{B}$ with $I_1 \subseteq B_1$ so that
$B_2 - (I_2 \cup B_1) = \emptyset$ and $|B_1 - (I_1 \cup B_2)|$ is minimal. Then, as before,
if $x \in B_1 - (I_1 \cup B_2)$, then by (B3) there is $y \in B_2 - B_1$ so that
$B_4 = B_1 - x \cup \{y\} \in \mathcal{B}$. But, $I_1 \subseteq B_4$ and $|B_4 - (I_1 \cup B_2)| < |B_2 -
(I_1 \cup B_2)|$, contradicting the minimality of $|B_1 - (I_1 \cup B_2)|$.

Let's examine the sets $B_2 - B_1$ and $B_1 - B_2$ (see Figure 2.6):

$$B_2 - B_1 = I_2 - B_1 = I_2 - I_1$$

and

$$B_1 - B_2 = I_1 - B_2 \subseteq I_1 - I_2.$$

Now (B2′) gives $|B_1| = |B_2|$, so $|B_2 - B_1| = |B_1 - B_2|$. But this imme-
diately gives $|I_2 - I_1| \leq |I_1 - I_2|$ and hence $|I_2| \leq |I_1|$. This contradic-
tion finally establishes (I3), so we are done with this part.

---

[8] The phrase "similar argument" appears frequently in math books. We typically won't
include an argument after invoking this phrase – that's the point of invoking it – but we
do so here so you know exactly what to expect later on.

*Compositions.* Finally we must show that our two cryptomorphisms compose correctly, i.e., $g \circ f = i_{\mathcal{I}}$ and $f \circ g = i_{\mathcal{B}}$. Part 3: The compositions

$g \circ f$: We are given a matroid $M = (E, \mathcal{I})$, and, as above, we let $f(\mathcal{I}) = \mathcal{B}$ be all the maximal members of $\mathcal{I}$ and $g(\mathcal{B}) = \mathcal{I}'$ be all subsets of all elements of $\mathcal{B}$. That is, $\mathcal{I} = \{I \mid I \subseteq B \text{ for some } B \in \mathcal{B}\}$. We must show $\mathcal{I} = \mathcal{I}'$.

The main idea here is direct: given any collection of subsets of a set that is closed under taking subsets (i.e., the property that all the subsets of every member of the collection are also in the collection), the composition $\mathcal{I} \rightarrow f(\mathcal{I}) \rightarrow g(f(\mathcal{I}))$ will always return $\mathcal{I}$. Now if $I \in \mathcal{I}$, then $I$ is in some maximal $B$, which, in turn, contains $I$. Thus, $I \in \mathcal{I}' = g \circ f(\mathcal{I})$. For the reverse inclusion, we need to show no new sets creep in during this composition. But if $I' \in \mathcal{I}'$, then $I' \subseteq B$ for some maximal $B \in \mathcal{I}$. By (I2), $I' \in \mathcal{I}$, so $\mathcal{I} = \mathcal{I}'$ and $(g \circ f)(\mathcal{I}) = \mathcal{I}$.

$f \circ g$: We are given a matroid $(E, \mathcal{B})$ with $\mathcal{B}$ satisfying (B1), (B2'), (B3). Let $\mathcal{B}' = (f \circ g)(\mathcal{B})$ be the maximal members of $\mathcal{I} = g(\mathcal{B})$; that is $\mathcal{B}' = \{B \in \mathcal{I} \mid B \subseteq B' \in \mathcal{I} \text{ implies } B = B'\}$. We must show $\mathcal{B} = \mathcal{B}'$. The main idea here is again rather simple: given any collection of subsets of a set, as long as no set in the collection contains another set in the collection, the procedure $\mathcal{B} \rightarrow g(\mathcal{B}) \rightarrow f(g(\mathcal{B}))$ will always return $\mathcal{B}$.

First suppose $B \in \mathcal{B}$. Then we need to show $B \in \mathcal{B}'$, i.e., $B$ is a maximal member of $g(\mathcal{B}) = \mathcal{I}$. But, if this were not the case, $B$ would be a non-maximal member of $\mathcal{I}$, so $B \subsetneq B'$ for some $B'$, which contradicts (B2'). Hence, $\mathcal{B} \subseteq \mathcal{B}'$. Now, to finish, we need to show the reverse inclusion, i.e., $\mathcal{B}' \subseteq \mathcal{B}$. If $B' \in \mathcal{B}'$, then $B' \in \mathcal{I}$, so $B' \subseteq B$ for some $B \in \mathcal{B}$. Since $B'$ is a maximal member of $\mathcal{I}$ and $B \in \mathcal{I}$ as well, we get $B = B'$, and so $B' \in \mathcal{B}$. This finishes the proof. $\square$

The last part of the proof of Theorem 2.6 is really a "matroid-free" fact about families of subsets of a set – see Exercise 27. Although the proof of Theorem 2.6 is long, the three bite-sized pieces[9] are easy enough to digest.

There are lots of other ways to define a matroid – via *circuits, flats, closure operator, hyperplanes, rank function* and more – and each equivalent definition requires a complete cryptomorphism. It would be overkill to reproduce careful proofs of all the equivalent definitions in this text[10] – but it's worth going through some of the steps for some of the cryptomorphisms. In subsequent sections we give other cryptomorphisms, some of which we'll prove and some we'll leave as activities for you in the exercises. As a sample of the relationships between bases, independent sets, dependent sets and circuits, see Figure 2.7.

---

[9] If these pieces are truly bite-sized, then, as our Spanish teacher told one of us, "¡que boca tan grande!"

[10] Trust us on this – we're not kidding.

Figure 2.7. Relations for
cryptomorphisms between
independent sets, bases,
dependent sets and circuits.

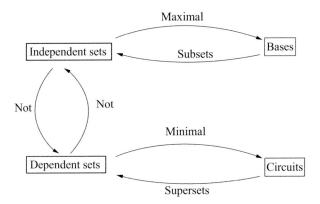

### Brief philosophical diversion

Given that someone has proven that all of the various definitions of a
matroid are equivalent by producing appropriate cryptomorphisms, you
might well ask:

> "Why bother? Why can't I simply learn what circuits, flats, hyperplanes,
> etc. are and why they are important?"

This question applies not just to cryptomorphisms, but mathematics in
general. You might as well ask: How does one learn mathematics? Why
does one write and study mathematical proofs? or What does it mean
to understand a mathematical concept? These are deep philosophical
questions. Our short answer is that learning a topic is like an onion; there
are many layers: definitions and examples, applications and properties,
relevant theorems and proofs, generalizations and special cases. All of
these are useful in understanding a mathematical idea. For more on the
philosophy of mathematics see [13], for example.

We believe that you will gain a deeper understanding of matroids if
you read and work through as many of the steps of the proofs as you
can. On the other hand, there is (obviously) a limit to how much time
you can spend wrestling with cryptomorphisms, so we need to strike a
balance.

For practice, here's a matroid to play with.

**Example 2.7.** Let $M$ be the matroid on the six-element set $E =
\{a, b, c, d, e, f\}$ with bases $\mathcal{B} = \{ace, ade, bce, bde, cde\}$ (see Fig-
ure 2.8).

Figure 2.8. The matroid from
Example 2.7.

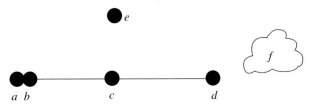

Some curious features of this matroid are the behavior of the elements
$e$ and $f$. Note that $e$ is in every basis of the matroid. Equivalently, it can
be added to any independent set it's not in to create another independent
set. Such an element is called an *isthmus* or *coloop* of the matroid. On the
other hand,[11] the element $f$ is in *no* basis – in fact, $f$ is in no independent
set at all. This element is called a *loop* – you met loops in Chapter 1.
Both *isthmus* and *loop* are terms borrowed from graph theory.[12]

For the record, here are the definitions.

**Definition 2.8.** Let $M$ be a matroid on the ground set $E$. An *isthmus*
is an element $x \in E$ that is in every basis. A *loop* is an elements $x \in E$     Isthmus, loop
that is in no basis.

Isthmuses are also called *coloops* by many matroid authors. This terminology anticipates matroid duality, which is introduced in Section 3.3.

## 2.2 Circuits and independent sets

The minimal dependent sets in a matroid are called *circuits*.

**Definition 2.9.** Let $M$ be a matroid. If $C$ is dependent, but every *proper*
subset of $C$ is independent, we call $C$ a *circuit* in the matroid.     Circuits

For the matroid in Example 2.7, there are precisely four circuits:

$$\mathcal{C} = \{f, ab, acd, bcd\}.$$

Listing the circuits is often an efficient way to describe a matroid. It's
also pretty easy to reconstruct the independent sets from the list of
circuits.

Note that "minimal" and "minimum" are quite different for dependent     "Minimal" $\neq$ "minimum" for
sets: the circuits can have very different sizes (unlike bases). The loop     dependent sets
$f$ is a circuit by itself, but the isthmus $e$ is not contained in any circuit
of $M$. This is one way (of many) to characterize loops and isthmuses;
see Exercises 6 and 7.

We can play the same game with circuits as we did with independent
sets and bases. In particular, we can list three properties the family of
circuits of a matroid satisfy, then use these properties to *define* a matroid.
Here is the outline for this approach.

---

[11] This phrase only works for "either–or" arguments. Assignment: find a nice, homey
phrase that describes one of three alternatives.

[12] The term "loop" is familiar enough, but some authors use "bridge" instead of
"isthmus" in graph theory. An *isthmus* is a narrow land bridge that connects two
larger land masses. You will almost always find the word "isthmus" followed by the
words "of Panama" in standard (non-mathematical) literature. This remains true,
despite the fact that the Panama Canal effectively destroys the isthmus property.

First, we need to identify the three key properties the family $\mathcal{C}$ of all circuits satisfies:

Non-triviality  (C1) $\emptyset \notin \mathcal{C}$;

Clutter  (C2) if $C_1, C_2 \in \mathcal{C}$ and $C_1 \subseteq C_2$, then $C_1 = C_2$;

Circuit elimination  (C3) if $C_1, C_2 \in \mathcal{C}$ with $C_1 \neq C_2$, and $x \in C_1 \cap C_2$, then $C_3 \subseteq C_1 \cup C_2 - x$ for some $C_3 \in \mathcal{C}$.

Next, we need to obtain $\mathcal{C}$ from $\mathcal{I}$, and vice versa:

> - $\mathcal{I} \to \mathcal{C}$: The circuits $\mathcal{C}$ are the minimal subsets of $E$ that are not independent.
> - $\mathcal{C} \to \mathcal{I}$: The independent sets $\mathcal{I}$ are all the subsets of $E$ that *don't* contain a circuit $C \in \mathcal{C}$.

If you like lots of mathematical symbols, you could write the circuits in terms of the independent sets:

$$\mathcal{C} = \{C \subseteq E \mid C \notin \mathcal{I} \text{ and if } I \subsetneqq C \text{ then } I \in \mathcal{I}\}.$$

Reversing this to get $\mathcal{I}$ from $\mathcal{C}$:

$$\mathcal{I} = \{I \subseteq E \mid \forall \, C \in \mathcal{C}, C \not\subseteq I\}.$$

Then the cryptomorphism between circuits and independent sets is expressed as a theorem:

**Theorem 2.10.** *Let $E$ be a finite set and let $\mathcal{C}$ be a family of subsets of $E$ satisfying (C1), (C2), (C3). Then $(E, \mathcal{C})$ is cryptomorphic to the matroid $M = (E, \mathcal{I})$ and $\mathcal{C}$ is the collection of circuits of $M$.*

The proof of this theorem is left for your problem-solving pleasure (see Exercise 28). The structure of the proof is identical to the structure of the proof of Theorem 2.6. In particular, you assume you have a matroid defined in terms of its independent sets, then you define the circuits to be the minimal dependent sets, proving they satisfy (C1), (C2) and (C3). You then need to reverse this procedure, defining the independent sets from the family of circuits, showing these independent sets indeed satisfy (I1), (I2) and (I3). Finally, you need to show the cryptomorphisms compose correctly.

There are several important variations on the axioms for bases and circuits. The most interesting (and useful) are both stronger versions of the last axioms, (B3) and (C3):

Strong basis exchange  (B3′) If $B_1$ and $B_2$ are bases with $x \in B_1 - B_2$, then there is an element $y \in B_2 - B_1$ so that both $B_1 - x \cup \{y\}$ and $B_2 - y \cup \{x\}$ are bases.

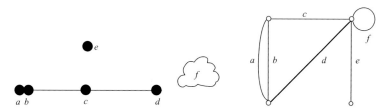

Figure 2.9. The matroid $M$ (left) and a graph $G$ with $M = M(G)$ (right).

(C3′) If $C_1$ and $C_2$ are circuits with $y \in C_1 - C_2$ and $x \in C_1 \cap C_2$, then there is a circuit $C_3 \subseteq C_1 \cup C_2$ with $y \in C_3$ and $x \notin C_3$.

Strong circuit elimination

It should be clear that (B3′) $\Rightarrow$ (B3) and (C3′) $\Rightarrow$ (C3). Strong basis exchange (B3′) and strong circuit elimination (C3′) are both true for all matroids. Exercise 14 guides you to a proof of (B3′), and Exercise 46 of Chapter 3 is devoted to proving (C3′). Strong basis exchange is used in proving Proposition 3.17 on duality in Chapter 3, and strong circuit elimination is used in proving *circuit transitivity* in Theorem 3.36, also in Chapter 3.

## 2.3 Rank, flats, hyperplanes and closure

In this section, we define several additional matroid concepts having a geometric flavor. Two of these definitions will be the familiar collection of a family of subsets of $E$ with a certain property (flats and hyperplanes), while another two will be defined in terms of a function defined on $2^E$ (rank and closure). As before, we can *define* a matroid directly in terms of these new concepts, but we'll look at a few examples before jumping into the deep end of the pool.[13] For each new term we define, we'll relate it to previously defined matroid concepts and describe the new term for geometries and graphs.

The example we will use throughout this section is the matroid $M$ on the six-element set $E = \{a, b, c, d, e, f\}$ with bases $\mathcal{B} = \{ace, ade, bce, bde, cde\}$ – this is the same matroid we met in Example 2.7. In Figure 2.9, we also give a graph $G$ with $M = M(G)$.

**Example 2.11.** Let $M$ be the matroid on the left of Figure 2.9. Then $B$ is a basis of $M$ if $B$ consists of three non-collinear points. Note that the isthmus $e$ is in every such set. For the graph $G$ on the right of Figure 2.9, the bases correspond to the spanning trees of $G$, and this time, the edge corresponding to the isthmus $e$ is an edge whose removal separates one vertex of $G$ from the rest of the graph.

Let's also investigate the circuit exchange property (C3). Let $C_1 = \{a, b\}$ and $C_2 = \{a, c, d\}$. Then $b \in C_1 \cap C_2$, so (C3) tells us there is

---

[13] This seems to be an appropriate metaphor for lots of things. If you jump in, make sure you come up for air on occasion.

another circuit $C_3$ with $C_3 \subseteq (C_1 \cup C_2) - b = \{a, c, d\}$. But $\{a, c, d\}$ is a circuit, so $C_3 = \{a, c, d\}$ satisfies property (C3).

It's worth looking at what this means in the graph $G$. In this case, we have two cycles; the double edge $ab$ and the triangle $bcd$. Then these two circuits intersect in the edge $b$, and it's clear $C_1 \cup C_2 - b$ is a circuit. In fact, for graphs, $(C_1 \cup C_2) - (C_1 \cap C_2)$ will be a *disjoint union of circuits* – see Theorem 4.5.

There is some freedom in how the graph $G$ was constructed in Example 2.11; for instance, we could have placed the loop at any vertex, and we could have moved the isthmus to other vertices, too (as long as it didn't form a cycle with any of the other edges). The general question of "When do two different graphs give the same cycle matroid?" was answered by Whitney; we treat this topic in Section 4.4.

### 2.3.1 Rank

Given any subset $A$ of the ground set $E$ of the matroid, we can look at the size of all the independent sets that are contained in $A$. The largest such independent subset of $A$ is its *rank*.

Rank function

**Definition 2.12.** Let $M = (E, \mathcal{I})$ be a matroid and let $A \subseteq E$. The rank of $A$, written $r(A)$, is the size of the largest independent subset of $A$:

$$r(A) := \max_{I \subseteq A} \{|I| : I \in \mathcal{I}\}.$$

The *rank of the matroid* $r(M)$ is just $r(E)$. The rank of the matroid was introduced in Chapter 1; it is just the size of any basis of $M$ (which are all the same size by our "maximal = maximum" property for independent sets). Technically, the rank function $r$ is a map from the set of all subsets of the ground set $E$ of the matroid to the non-negative integers:

$$r : 2^E \rightarrow \mathbb{N} \cup \{0\}.$$

We investigate some important properties of the rank function in Section 2.5.

**Example 2.13.** We'll continue examining the matroid in Figure 2.9. This time we'll find the rank of various subsets. A complete list would involve listing the rank of each of the $2^6 = 64$ subsets, which would be rather laborious. But we can organize the data intelligently to make life a bit more pleasant.

- $r(A) = 0$: $r(\emptyset) = 0$ is always true, but any subset of loops will also have rank 0. In this case, there are two sets with rank 0: $\emptyset$ and $f$.
- $r(A) = 1$: All singletons *except* the loop $f$ have rank 1, but adding the loop $f$ to a subset will never increase its rank, so $af, bf, cf, df$ and $ef$ also have rank 1. Finally, the sets $ab$ and $abf$ also have rank 1.

Table 2.1. *The number of*
*subsets of each rank.*

| Rank | 0 | 1 | 2 | 3 |
|---|---|---|---|---|
| # of subsets | 2 | 12 | 30 | 20 |

- $r(A) = 2$: Let's ignore the loop for a moment. A set has rank 2 if it spans a line in the geometry. Among the points $a, b, c$ and $d$, choosing any two (or more) *except ab* will span the line containing these four points. There are 10 such subsets of $a, b, c, d$. Rank 2 sets that also include the point $e$ are more restrictive: $r(abe) = r(ae) = r(be) = r(ce) = r(de) = 2$. These correspond to the two-point lines we generally don't draw in our matroid pictures. Finally, as before, we can add the loop $f$ to any of these sets without changing the rank.
- $r(A) = 3$: $r(A) = 3$ precisely when $A$ contains a basis. This means we must include the point $e$ and we must span the line containing $a, b, c$ and $d$.

We list the number of subsets of each rank in Table 2.1.

Note that $2 + 12 + 30 + 20 = 64$, so we have listed all the subsets.

Maximal sets of a given rank will be important later in this section. Foreshadowing a bit, we list the maximal rank 2 subsets: $\{abef, cef, def, abcdf\}$. You can check that each of these subsets spans a line in the geometry, but adding any point to one of these subsets increases its rank.

In the example above, we saw that since $f$ is a loop, if we add $f$ to any set, the rank does not increase. For isthmuses, adding the isthmus $e$ to a subset $A$ causes the rank to increase (provided the isthmus was not in $A$). We state these two observations as a proposition.

**Proposition 2.14.** *Let M be a matroid on the ground set E with rank function r.*

(1) *An element $x \in E$ is a loop if and only if for all $A \subseteq E$ with $x \notin A$, we have $r(A \cup \{x\}) = r(A)$.* Loop characterization via rank

(2) *An element $x \in E$ is an isthmus if and only if for all $A \subseteq E$ with $x \notin A$, we have $r(A \cup \{x\}) = r(A) + 1$.* Isthmus characterization via rank

See Exercises 6 and 7 for these and several other characterizations of loops and isthmuses.

As with independent sets and bases, we can define a matroid in terms of the rank function. The *definition* is given in the following *theorem*.

Figure 2.10. Cryptomorphism between independent sets and rank.

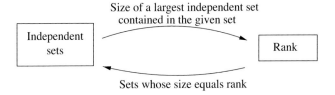

**Theorem 2.15.**   *Let E be a finite set with an integer-valued function r defined on subsets of E. Then r is the rank function of a matroid if and only if for $A, B \subseteq E$:*

Normalization   (r1)  $0 \le r(A) \le |A|$;

Increasing   (r2)  *if $A \subseteq B$, then $r(A) \le r(B)$;*

Semimodular   (r3)  $r(A \cup B) + r(A \cap B) \le r(A) + r(B)$.

We prove Theorem 2.15 in Section 2.5. It should be clear that the rank function of a matroid satisfies (r1) and (r2). For the *semimodular* property (r3), let's look at the matroid in Example 2.13, pictured in Figure 2.9. Let $A = \{a, b, c, e\}$ and $B = \{a, b, d\}$. Then you can check that $r(A) = 3, r(B) = 2, r(A \cup B) = 3$ and $r(A \cap B) = 1$. Then $r(A) + r(B) = 5$, and $r(A \cup B) + r(A \cap B) = 4$, so the property (r3) is satisfied.

Given the collection of independent sets, it's easy to find (but tedious to write out) the rank of every subset of $E$. What about the reverse procedure? As a mini-example, suppose we are given the ground set $E = \{a, b, c\}$ of a matroid $M$ and the rank of each subset of $E$:

$$r(\emptyset) = 0; \quad r(a) = r(b) = r(c) = r(ab) = 1;$$
$$r(ac) = r(bc) = r(abc) = 2.$$

How could we find the independent sets of $M$? (Pause to think.)

Independent set characterization via rank

Here's how: we made the observation earlier that the rank of an independent set equals its size; we note here that these are the *only* sets with this property. Thus, the independent sets are simply those sets $I$ with $r(I) = |I|$ (see Figure 2.10). In our mini-example, the independent sets are $\{\emptyset, a, b, c, ac, bc\}$.

You might try the following recipe for making a matroid. Let $E$ be a finite set and define a "rank" function $r : 2^E \to \mathbb{N} \cup \{0\}$ however you like. Then define $\mathcal{I}$ as above: $I$ is "independent" if $r(I) = |I|$. In general, this procedure will not produce a matroid unless the rank function satisfies properties (r1), (r2) and (r3) of Theorem 2.15. For an example where this fails, see Exercise 15.

## 2.3.2  Flat or closed set

Our next concept comes from geometry. A *flat* in a matroid is a subset that is *rank-maximal*: if you add anything new to a flat, its rank increases.

**Definition 2.16.** Let $E$ be the ground set of the matroid $M$. A subset $F \subseteq E$ is a *flat* if $r(F \cup \{x\}) > r(F)$ for any $x \notin F$.

Flats

Flats are also called *closed sets*, a term borrowed[14] from topology. We'll usually use the term "flat" in this text. Intuitively, flats are the points, lines, planes and higher-dimensional (hyper)planes of a geometry.

**Example 2.17.** We return to the matroid in Example 2.13, pictured in Figure 2.9. Let's list all the flats:

- rank 0 flats: There's only one, and it's $f$. Note: there is always exactly one rank 0 flat; if $M$ has no loops, then the rank 0 flat is $\emptyset$, while if $M$ has loops then it is the set of all loops.
- rank 1 flats: $abf, cf, df, ef$.
- rank 2 flats: In the discussion following Example 2.13, we determined all the maximal sets of rank 2: $abef, cef, def, abcdf$. These four lines are precisely the rank 2 flats.
- rank 3 flats: The entire matroid is the only rank 3 flat: $abcdef$.

**Important things to notice**

(1) The loop $f$ is in every flat (which should make sense – it won't ever increase the rank of a set). (See Exercise 6.)
(2) Removing the isthmus $e$ from a flat $F$ that contains it yields another flat, and adding $e$ to any flat that $e$ is not in also yields another flat. More concisely, if $e$ is an isthmus and $F$ is a flat, then $e \notin F \Rightarrow F \cup \{e\}$ is a flat, and $e \in F \Rightarrow F - \{e\}$ is a flat. (See Exercise 7.)
(3) $\emptyset$ will be a flat precisely when $M$ has no loops.
(4) $E$ is always a flat in any matroid; it is the unique flat of rank $r(M)$.

While it is pretty easy to pick out the flats from the picture of the geometry, it is a bit more difficult to find the flats of a graphic matroid from its graphic representation. To describe the flats of a graphic matroid, we consider a graph $G = (V, E)$ and a subset $F$ of the edges $E$. Note that the subset of edges $F \subseteq E$ gives a subgraph of $G$, and that subgraph may have several connected components. Then, loosely speaking, $F$ is a flat in a graphic matroid if adding any edge to $F$ reduces this number of connected components.

Flats in a graphic matroid

More precisely, we let $\Pi$ be a partition of the vertices of $G$, and then let $F_{\Pi}$ be those edges of the graph both of whose endpoints are contained in the same part, called a *block*, of the partition. Then $F_{\Pi}$ is a flat of $M(G)$.[15]

Block of a partition

**Example 2.18.** The vertex partition $\Pi = \{\{v_1, v_2\}, \{v_3\}, \{v_4\}\}$ of the graph $G$ in Figure 2.9 gives the flat $abf$ while the partition $\Pi = \{\{v_1, v_2, v_3\}, \{v_4\}\}$ gives $abef$. Conversely, the flat $f$ is the collection of edges associated with the partition $\Pi = \{\{v_1\}, \{v_2\}, \{v_3\}, \{v_4\}\}$.

---

[14] stolen.
[15] Much more on graphic matroids appears in Chapter 4.

Figure 2.11. Top: a graph $G$
with vertex partition
$\Pi = \{16, 234, 57\}$. Bottom:
the associated flat
$F_\Pi = \{b, c, f, h, k\}$.

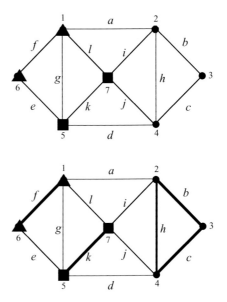

For another example, consider the graph in Figure 2.11. Then
the vertex partition $\Pi = \{16, 234, 57\}$ corresponds to the flat $F_\Pi = \{b, c, f, h, k\}$. Note that adding any edges not in $F_\Pi$ to $F_\Pi$ will increase
the rank (by reducing the number of connected components).

We'll examine the fundamental properties/axioms that flats satisfy
in Section 2.5, but, as a quick preview, we'll state one very important
property for flats:

**Proposition 2.19.** *If $F_1$ and $F_2$ are flats in a matroid, then so is $F_1 \cap F_2$.*

The proof is left for you in Exercise 24(b).

As we noted above, flats have the property that they are *rank-maximal*,
i.e., adding an element to a flat it is not already in increases its rank by
one. We state this as a proposition and we include the proof, but you
might try to cover up the proof to see if you can construct your own
argument.

**Proposition 2.20.** *Let $F$ be a flat of a matroid $M$ with $x \notin F$. Then
$r(F \cup \{x\}) = r(F) + 1$.*

*Proof Proposition 2.20.* First, by our definition of flats, we know
$r(F \cup \{x\}) > r(F)$. If $r(F \cup \{x\}) = k$, then there is an independent set
$I$ of size $k$ contained in $F \cup \{x\}$. If $x \notin I$, then $I \subseteq F$ and $r(F) \geq k = r(F \cup \{x\})$, a contradiction.

Thus, $x \in I$ and $I - x$ is an independent set (by (I2)) of size
$k - 1$ contained in $F$. Hence, $r(F) \geq k - 1 = r(F \cup \{x\}) - 1$, i.e.,
$r(F \cup \{x\}) \leq r(F) + 1$. Then, since $r(F) < r(F \cup \{x\}) \in \mathbb{Z}$, we have
$r(F \cup \{x\}) = r(F) + 1$.   □

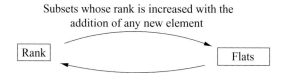

Subsets whose rank is increased with the
addition of any new element

Rank

Flats

The length of a maximal chain of flats contained in the set

Figure 2.12.
Cryptomorphism between
flats and the rank function.

We've defined flats in terms of the rank function – how can we reverse this procedure to get the rank function of the matroid from its list of flats?

**Example 2.21.** As an example, suppose we are given the ground set $E = \{a, b, c\}$ of a matroid and the collection of flats $\mathcal{F} = \{\emptyset, ab, c, abc\}$. Can we figure out the rank of any subset?

The idea is straightforward: to compute the rank of a subset $A \subseteq E$, follow these simple steps:

*From flats to rank*

(1) Find the smallest flat that contains $A$ – call this flat $F_A$.
(2) Find a *maximal chain*[16] of flats $F_0 \subsetneq F_1 \subsetneq \cdots \subsetneq F_r = F_A$, where $r(F_0) = 0, r(F_1) = 1$, and so on.
(3) Then $r(A)$ is just the length of this chain. $r(A)$ will then be one less than the number of flats appearing in the maximal chain.

To continue with our example, we note that since $\emptyset$ is a flat, there are no loops. This gives us $r(a) = r(b) = r(c) = 1$. Now, since $a$ and $b$ are not flats, we know there is no flat between $\emptyset$ and $ab$, so $r(ab) = 1$ (so $ab$ is a multiple point). The maximal chain of flats

$$\emptyset \subsetneq ab \subsetneq abc$$

tells us $r(M) = 2$. You can also check $r(ac) = r(bc) = 2$.

Figure 2.12 gives the relation between flats and the rank function. The appropriate setting for the study of flats is *lattice theory*, which we defer until Section 2.4.

### 2.3.3 Hyperplanes

Next, we consider special types of flats – *hyperplanes*; those maximal flats that are *not* the entire matroid (which is always a flat).

**Definition 2.22.** Let $E$ be the ground set of a matroid $M$. A subset $H \subseteq E$ is a *hyperplane* if $H$ is a flat of $M$ and if $r(H) = r(M) - 1$.

*Hyperplanes*

**Example 2.23.** Like flats, hyperplanes are also easy to pick out from the (geometric) picture of the matroid. In Figure 2.9, the entire matroid has rank 3, so the hyperplanes are just the rank 2 flats: the

---

[16] See Section 2.4 for more on chains of flats.

Figure 2.13. Cryptomorphism
between flats and hyperplanes.

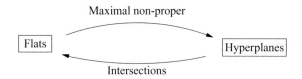

Maximal non-proper

Flats        Hyperplanes

Intersections

lines (with the loop included); $abcdf, abef, cef, def$. In the graph $G$, the hyperplanes correspond to the sets of edges corresponding to the partitions: $\{\{v_1, v_2, v_3\}, \{v_4\}\}$; $\{\{v_1, v_2, v_4\}, \{v_3\}\}$; $\{\{v_1, v_3, v_4\}, \{v_2\}\}$; $\{\{v_1, v_3\}, \{v_2, v_4\}\}$.

Hyperplanes are flats of rank $r(M) - 1$; note that different hyperplanes can have different sizes. Thus, the *maximal = maximum* property that holds for bases (maximal independent sets) does not hold for hyperplanes. It's obvious that we can determine all the hyperplanes of a matroid from its collection of flats – the hyperplanes are those flats $H$ so that no flat $H'$ satisfies $H \subsetneq H' \subsetneq E$, or, said another way, the maximal flats of $\mathcal{F} - E$.

Can this procedure be reversed? Can we reconstruct the flats of the matroid if we are given the ground set $E$ and the collection of hyperplanes?[17]

Flats from hyperplanes

It turns out that every flat is the intersection of some collection of hyperplanes of the matroid. We'll prove this later (see Proposition 2.56), but the procedure should be familiar: it mimics what we know to be true in Euclidean space. (Every line is the intersection of (two) planes, every point is the intersection of (three) planes. More general statements hold in higher dimensions.)

Thus, by forming all possible intersections of various collections of hyperplanes, we can construct all the flats of the matroid, which is what we wanted to do.[18] See Figure 2.13.

**Example 2.24.** Let's verify that the flats in the example above can be constructed as the intersection of hyperplanes. First, the intersection of a hyperplane with itself gives that hyperplane – so we get all the hyperplanes in this way. Secondly, we get the flat $f$ (as the intersection of all hyperplanes) and $E$ (as an empty intersection).

This leaves the rank 1 flats. Here is how to get them all as hyperplane intersections:

$$abf = abcdf \cap abef \qquad cf = abcdf \cap cef$$
$$df = abcdf \cap def \qquad ef = cef \cap def.$$

---

[17] These questions (perhaps like these footnotes) can be predictable. We wouldn't have asked this question if the answer were "no," would we?

[18] We can regard the flat $E$ as arising from an *empty intersection* of the hyperplanes, but this is a technicality that you should not worry about at all.

You can check that these intersections correspond to the intersection of two lines in Figure 2.9.

Defining a matroid in terms of its hyperplanes is another of our cryptomorphisms. See Section 2.6.2 for information about hyperplanes from the axiomatic viewpoint.

Hyperplanes will also be important when we study *duality* in matroids – see Section 3.3. For now, we define the complement of a hyperplane to be a *cocircuit*. Cocircuits are also called *bonds*, a term borrowed from graph theory. For a one-sentence preview (keeping in mind that a circuit is minimal with respect to not being contained in a basis), consider the following:

Cocircuit

Bond

> $C^*$ is a cocircuit if and only if $C^*$ is minimal with respect to not being contained in a basis complement.

### 2.3.4 Closure operator

Given a subset $A \subseteq E$, the *closure* of $A$, written $\overline{A}$, will be another subset of $E$. Thus, the map $A \mapsto \overline{A}$ is a function from the power set $2^E$ to itself.

Closure operators are important in topology and real analysis. In that context, the closure $\overline{A}$ can be defined as the intersection of all the closed sets containing $A$, or, alternatively, as the *smallest* closed set containing $A$.

For matroids, the closure of $A \subseteq E$, also written $\overline{A}$, is just the unique smallest flat containing $A$. Alternatively, we can define $\overline{A}$ to be the intersection of all flats that contain $A$. The fact that there is a unique flat containing $A$ requires proof, as does the equivalent formulation in terms of intersections of flats.

**Lemma 2.25.** *Let $M$ be a matroid on the ground set $E$ with flats $\mathcal{F}$. Let $A \subseteq E$. Then*

(1) *There is a unique flat $F \in \mathcal{F}$ such that*
  (a) $A \subseteq F$, *and*
  (b) *If $A \subseteq F'$ for some flat $F' \in \mathcal{F}$, then $F \subseteq F'$.*
(2) *The flat $F$ from part (1) satisfies $F = \bigcap\limits_{F' \in \mathcal{F}} \{F' : A \subseteq F'\}$.*

We now define the matroid closure operator.

**Definition 2.26.** Let $M$ be a matroid with flats $\mathcal{F}$ and let $A \subseteq E$. Then the *closure of $A$*, written $\overline{A}$, is defined by

Closure

$$\overline{A} = \bigcap_{F \in \mathcal{F}} \{F : A \subseteq F\}.$$

Figure 2.14. Cryptomorphism
between closure and flats.

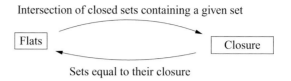

Intersection of closed sets containing a given set

Flats                                              Closure

Sets equal to their closure

Although we leave the proof of Lemma 2.25 for you in Exercise 23,
we remark that its proof depends only on the fact that the intersection of
flats is a flat. Restating this, if $\mathcal{F}$ is *any* family of subsets closed under
intersection, then the associated closure operator will satisfy these two
properties. This tells us the existence of a closure operator does not
depend on the matroid structure. Further, it also tells us that we will
need to require more from the closure operator to define a matroid.

In our running example (the matroid in Example 2.7), let's verify
$\overline{ad} = abcdf$ using the definition. The only flats $F$ that contain $ad$ are
$abcdf$ and $E$, and so we're done. What about $\overline{a}$? The flats containing $a$
are the lines $abef$ and $abcdf$ (and, of course, $E$ itself). Thus $\overline{a} = abf$.

From the definition, it is clear that the flats in a matroid are precisely
*Flats from closure*    those sets $F$ with $\overline{F} = F$. The connection between closure operators
and flats (closed sets) is shown in Figure 2.14.

If $A \subseteq E$, you can also check $A \subseteq \overline{A}$ and $\overline{\overline{A}} = \overline{A}$. These two properties are important in defining a matroid in terms of its closure operator,
but they are not the whole story. We give the full cryptomorphic description of a matroid in terms of its closure operator by connecting the
closure operator to the rank function in Section 2.5.2.

## 2.3.5 Cryptomorphism summary

There are so many different ways to describe matroids that G.-C. Rota
once said, "It is as if one were to condense all trends of present-day
mathematics onto a single finite structure, a feat that anyone would *a
priori* deem impossible, were it not for the mere fact that matroids do
exist."

There are dozens of cryptomorphic ways to describe a matroid. We've
concentrated our attention on the most important ones here. For a matroid
$M$, we have defined seven key concepts: independent sets, bases, circuits,
the rank function (all of which are motivated by linear algebra or graph
theory), flats, hyperplanes, and the closure operator (all of which have
geometric or topological motivation). We showed that a matroid can be
defined in terms of its independent sets via properties (I1), (I2) and (I3),
then gave a cryptomorphism to show how to define a matroid in terms
of its bases (Theorem 2.6).

But this works for other pairs of concepts; for instance, in Theorem 2.10, we give the cryptomorphism between independent sets and

*circuits*. This shows how to define a matroid via its circuits.[19]

$$\mathcal{C} \xleftrightarrow{\text{Thm 2.6}} \mathcal{I} \xleftrightarrow{\text{Thm 2.10}} \mathcal{C}$$

$$\mathcal{I} \xleftrightarrow{\text{Thm 2.15}} \text{Rank } r \xleftrightarrow{\text{Thm 2.50}} \text{Closure} \xleftrightarrow{\text{Thm 2.52}} \mathcal{F} \xleftrightarrow{\text{Thm 2.58}} \mathcal{H}$$

The rest of this chapter is devoted to cryptomorphisms, proofs, properties and connections among our seven key concepts. We will need one more structure: *geometric lattices*. These structures occupy Section 2.4, and they give us a nice way to display all the flats of a matroid. They also give us yet another way to define matroids.

We connect several concepts through their cryptomorphic descriptions in Sections 2.5 and 2.6: the rank function and independent sets in Section 2.5.1, closure and the rank function in Section 2.5.2, flats and the closure operator in Section 2.6.1, and hyperplanes and flats in Section 2.6.2.

We can also define the independent sets of a matroid via the *greedy algorithm*. This connection to optimization is presented in Section 2.7, and provides us with another way to understand why matroids are special.

We feel these connections are important, and working through the details of the remaining cryptomorphisms is a valuable exercise.[20] But, *you can safely skip the rest of this chapter*, using it as a reference as needed. This should not impede your progress through the rest of the text.

As another reference, we have provided a summary of the various axioms characterizing matroids and their connections in the Appendix.

## 2.4 Lattice of flats

The flats of a matroid, ordered by inclusion, form a very pretty structure, called a *geometric lattice*.[21] Before defining lattices, we define partially ordered sets, a more general object.

### 2.4.1 Partially ordered sets

A *binary relation* $R$ on a set $E$ is a subset of $E \times E$, where $E \times E = \{(a, b) \mid a, b \in E\}$ is the collection of all ordered pairs of elements of $E$. If we call our relation $R$, we write $(a, b) \in R$ and we say "*a* is *related to b* under $R$."

---

[19] Well, it will once you provide a proof of Theorem 2.10. See Exercise 28.
[20] You could infer this from the fact they are in this text.
[21] The concept of a geometric lattice was introduced by G. Birkhoff [4] who was "stimulated by the 'matroid' concept." In fact Birkhoff also called these lattices, "matroid lattices."

A binary relation $R$ on a set $E$ may enjoy some of the following properties:

- reflexive: for all $a \in E$, $(a, a) \in R$.
- symmetric: for $a, b \in E$, if $(a, b) \in R$ then $(b, a) \in R$.
- antisymmetric: for $a, b \in E$, if $(a, b) \in R$ and $(b, a) \in R$ then $a = b$.
- transitive: for $a, b, c \in E$, if $(a, b) \in R$ and $(b, c) \in R$ then $(a, c) \in R$.

**Example 2.27.** Let $E = \{a, b, c, d\}$ and $R = \{(a, a), (b, b), (c, c), (d, d), (a, b), (a, c), (a, d)\}$. What properties does R have? You can check $R$ is reflexive, antisymmetric and transitive, but not symmetric.

**Definition 2.28.** A binary relation $R$ on a set $E$ is an *equivalence relation* if $R$ is reflexive, symmetric and transitive.

*Equivalence relation*

Equivalence relations are extremely important throughout mathematics. They also arise in matroids (see Exercise 22 from Chapter 1). The most important equivalence relation for matroids involves circuits. Let $R$ be the relation on the ground set $E$ of a matroid $M$ defined as follows: $a$ is related to $b$ if and only if there is a circuit $C$ containing both $a$ and $b$. Then $R$ is an equivalence relation – see Theorem 3.36 of Chapter 3.

**Definition 2.29.** A *partial order* is a binary relation that is reflexive, antisymmetric and transitive. A *poset*[22] $(E, \preceq)$ is a set equipped with a partial order.

*Poset*

We note that the relation in Example 2.27 is a partial order. The next example presents an important class of posets.

**Example 2.30.** For a positive integer $n$, let $D_n$ be the set of all positive divisors of $n$. For instance, $D_{12} = \{1, 2, 3, 4, 6, 12\}$. Define a partial order on $D_n$ by divisibility, denoted "$|$". For example, 2 is related to 4 since $2|4$ but 4 is not related to 6, since $4 \nmid 6$. You should verify that $(D_n, |)$ is a poset.

The notion of a *cover* will be very important for us when we get back to matroids. For distinct elements $x$ and $y$ in a poset, we say $y$ *covers* $x$ if $x \preceq y$, but there are no elements of the poset between $x$ and $y$.

**Definition 2.31.** Let $(E, \preceq)$ be a poset with $x, y \in E$. For $x \neq y$, we say $y$ *covers* $x$ if $x \preceq y$, and, for all $z \in E$, if $x \preceq z \preceq y$, then $z = x$ or $z = y$. If $y$ covers $x$ we also say $x$ *is covered by* $y$ and denote this $x \lessdot y$.

*Covers*

It's not hard to draw a picture that describes a partial order. These pictures, called *Hasse diagrams*,[23] are constructed as follows: each element of $E$ appears as a point (represented as a dot in the picture), and, if

*Hasse diagram*

---

[22] "Poset" is an acronym for "Partially Ordered SET."
[23] Hasse diagrams are named after the German mathematician Helmut Hasse (1889–1979). Hasse popularized the use of these diagrams in his writing, but as is often the case in mathematics, he was not the first to use these diagrams.

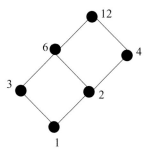

Figure 2.15. The Hasse diagram for the poset $(D_{12}, |)$.

$x$ is covered by $y$, then we draw a line segment directed upwards from $x$ to $y$. It's an elementary exercise using transitivity to re-create the entire partial order from the covering relations. In particular, this means that if $x \preceq y$, then there is an upward path from $x$ to $y$ in the Hasse diagram. Figure 2.15 shows a Hasse diagram for $D_{12}$. You might try to find a Hasse diagram for $D_n$ for your favorite $n$.

We need a few more definitions before we turn our attention back to matroids.

An element $x$ is *maximal* if there is no element above it in the Hasse diagram, i.e., if $x \preceq y$, then $x = y$. Minimal elements are defined in a similar way. You can prove (see Exercise 35) that every finite poset must have at least one maximal element and at least one minimal element.

*Maximal and minimal elements*

If $P$ has exactly one maximal element $x$, then $y \prec x$ for all $y \neq x$. Such an element is usually denoted $\hat{1}$ and is called the *greatest* element of the poset. You can figure out the analogous statement about minimal elements – if $P$ has a *least* element, it's denoted $\hat{0}$.

*Greatest and least elements*

**Definition 2.32.** A *chain* in the poset $P$ is a collection $\{x_1, x_2, \ldots, x_k\}$ of distinct elements of $P$ with $x_1 \preceq x_2 \preceq \cdots \preceq x_k$. A chain is *maximal* if it is not contained in any longer chain. A chain $x_1 \preceq x_2 \preceq \cdots \preceq x_k$ is *saturated* if $x_{i+1}$ covers $x_i$ for $i = 1, 2, \ldots, k-1$. The *length* of a saturated chain $x_1 \preceq x_2 \preceq \cdots \preceq x_k$ is $k$. A poset is *graded* if, for every pair of elements $x$ and $y$, all the saturated chains beginning at $x$ and ending at $y$ have the same length.

*Chains*

*Maximal chains*

*Graded posets*

Maximal chains of flats were introduced in Section 2.3.2. If the poset is graded, then we can define a rank function $\rho$ on the elements of the poset. This will turn out to be the key connection between posets and matroids.

The next example should help illustrate these ideas.

**Example 2.33.** Let $[n] = \{1, 2, \ldots, n\}$ and define a poset $B_n$ on all subsets of $S$, where $A \preceq B$ if and only if $A \subseteq B$. $B_n$ is called a *Boolean algebra* or *Boolean lattice*. You can verify the following elementary facts:

- $\hat{0} = \emptyset$ and $\hat{1} = S$.
- $B_n$ is graded with rank function given by $\rho(A) = |A|$. (Note $\rho(\emptyset) = 0$.)
- $B$ covers $A$ if and only if $B = A \cup \{x\}$ for some $x \notin A$.
- The number of maximal chains in $B_n$ is $n!$.
- If $|A| = k$, then $A$ covers $k$ elements of $B_n$ and $A$ is covered by $n - k$ elements.

One semi-amusing consequence of the facts about chains is the following. Let $A_1, \ldots, A_m$ be the collection of subsets of $S$ of cardinality $k$. Then the number of maximal chains passing through a given $A_i$ is $k!(n-k)!$ since there are $k!$ chains from $\emptyset$ to $A_i$ and $(n-k)!$ chains from $A_i$ to $S$. Since every one of the $n!$ maximal chains in the poset passes through exactly one of these $A_i$, we have $mk!(n-k)! = n!$, or, in a slightly more familiar form:

$$ m = \frac{n!}{k!(n-k)!}. $$

This gives a proof that the number of subsets of size $k$ is $\binom{n}{k} = \frac{n!}{k!(n-k)!}$.

## 2.4.2 Geometric lattices

The goal of the rest of this section is to understand the following statement:

> The flats of a matroid form a geometric lattice under inclusion.

We will first define a *lattice* and then describe what it means for a lattice to be *geometric*. We'll need the following operations on pairs of elements of a poset.

Join **Definition 2.34.** Let $(E, \preceq)$ be a poset with $x, y \in E$. The *join* of $x$ and $y$, denoted $x \vee y$, is the least element of $\{z : x \preceq z \text{ and } y \preceq z\}$, if it Meet exists. The *meet* of $x$ and $y$, denoted $x \wedge y$, is the greatest element of $\{w : w \preceq x \text{ and } w \preceq y\}$, if it exists.

**Example 2.35.**

(1) Let $P$ be the poset with Hasse diagram given in Figure 2.16. Then $a \vee b = d$ and $d \wedge e = b$, but $a \vee c$ and $a \wedge c$ are not defined. This poset is also not graded (the two maximal chains $a \lessdot d$ and $b \lessdot c \lessdot e$ have different lengths).

(2) For the poset $B_n$ on all subsets of $[n]$ from Example 2.33, it's immediate that $A \vee B = A \cup B$ and $A \wedge B = A \cap B$.

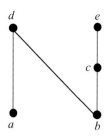

Figure 2.16. Poset *P* for
Example 2.35(1): $a \vee c$ and
$a \wedge c$ are not defined.

(3) For the poset $D_n$ of Example 2.30, you can (and you probably should – see Exercise 36) show $a \vee b = lcm(a, b)$ and $a \wedge b = gcd(a, b)$.

The join is often called the *least upper bound* (l.u.b.) or *supremum* (sup), and the meet is often called the *greatest lower bound* (g.l.b.) or *infimum* (inf). The operations $\wedge$ and $\vee$ are associative, so writing $a \vee b \vee c$ or $a \wedge b \wedge c$ is unambiguous. As mentioned above, often these operations are not defined for all pairs of elements. We are interested in posets in which these operations are defined for all pairs.

l.u.b., sup, g.l.b., inf

**Definition 2.36.** If $L = (E, \preceq)$ is a finite poset such that every pair of elements has a join and a meet, then $L$ is a *lattice*.

Lattice

The next result is a nice property of lattices – see Exercise 40.

**Proposition 2.37.** *A finite, non-empty lattice $L$ has a least element $\hat{0}$ and a greatest element $\hat{1}$.*

Since a lattice has a least and a greatest element, we can talk about the elements that cover $\hat{0}$, and those that are covered by $\hat{1}$.

**Definition 2.38.** The elements of a lattice $L$ that cover $\hat{0}$ are called the *atoms* of $L$, while the elements that are covered by $\hat{1}$ are called the *coatoms* of $L$.

Atoms

Coatoms

In Example 2.30, the atoms are the primes dividing $n$. For the lattice $B_n$ from Example 2.33, the atoms are the subsets of $[n]$ of size 1.

**Definition 2.39.** A lattice $L$ in which every element can be written as the join of atoms is said to be *atomic*.

Atomic lattice

In the Boolean lattice, every subset $A$ is a join of all the singletons it contains, so $B_n$ is atomic. The lattice $(D_{12}, |)$ is not atomic: for example, 4 is not a join of atoms. Question to (briefly) ponder: for which values of $n$ is the divisor lattice $(D_n, |)$ atomic? See Exercise 38.

To finish the connection to matroids, we need a rank function for our lattice.

Figure 2.17. The Hasse
diagram for the lattice *L* in
Example 2.42.

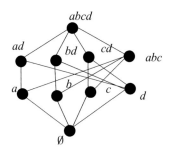

## Definition 2.40.

(1) Let $(E, \preceq)$ be a graded poset with a least element $\hat{0}$, and define the *rank* $\rho(x)$ of an element $x \in E$ to be the length of a saturated chain from $\hat{0}$ to $x$.

Lattice rank function

Semimodular law

(2) Let $L$ be a lattice with rank function $\rho$. Then $\rho$ is *semimodular* if, for all $x, y \in L$,

$$\rho(x \vee y) + \rho(x \wedge y) \le \rho(x) + \rho(y).$$

Modular law

(3) Let $L$ be a lattice with rank function $\rho$. Then $\rho$ is *modular* if, for all $x, y \in L$,

$$\rho(x \vee y) + \rho(x \wedge y) = \rho(x) + \rho(y).$$

A lattice is *semimodular* or *modular* if its rank function $\rho$ is. Modular lattices are semimodular, but the converse is not generally true. Boolean lattices are modular: $|A \cup B| + |A \cap B| = |A| + |B|$. See Exercise 47 for more on modular lattices.

Now that we have introduced various properties of posets and lattices, we are finally able to give the definition of a geometric lattice.

**Definition 2.41.** A lattice $L$ which is semimodular and atomic is a
Geometric lattice *geometric* lattice.

**Example 2.42.** Consider the lattice $L = (\mathcal{F}, \subseteq)$ where

$$\mathcal{F} = \{\emptyset, a, b, c, d, ad, bd, cd, abc, abcd\}.$$

The Hasse diagram for $L$ is given in Figure 2.17. To show $L$ is a geometric lattice, you need to check that each set appearing in $\mathcal{F}$ is a join of atoms, and that the rank function satisfies the semimodular law. We leave out all the details, but they aren't hard. (In fact, $L$ is a modular lattice.)

**Example 2.43.** For the matroid in our running example of Figure 2.18, the lattice of flats $L$ is shown.

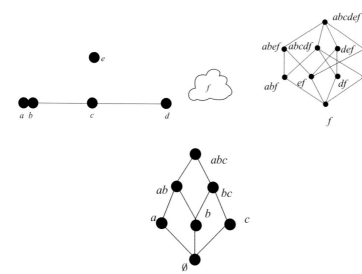

Figure 2.18. The matroid $M$ and its lattice of flats $L$.

Figure 2.19. The Hasse diagram for the lattice in Example 2.44.

**Example 2.44.** Consider the lattice $L = (\mathcal{F}, \subseteq)$ where $\mathcal{F} = \{\emptyset, a, b, c, ab, bc, abc\}$. The Hasse diagram for $L$ is given in Figure 2.19. Then $L$ is *not* a geometric lattice. To see this, note that $a \vee c = abc$. Then the semimodular property fails for $x = a$ and $y = c$:

$$3 = \rho(a \vee c) + \rho(a \wedge c) > \rho(a) + \rho(c) = 2.$$

Lattices are an interesting topic of study in their own right, but their connection with matroids is our main interest now.

**Theorem 2.45.** *Let* $M = (E, \mathcal{F})$ *be a matroid with flats* $\mathcal{F}$. *Then* $(\mathcal{F}, \subseteq)$ *is a geometric lattice.*

The collection $\mathcal{F}$ given in Example 2.42 are the flats of a matroid on $E = \{a, b, c, d\}$. For fun, you might like to draw a picture of this matroid and verify that your flats match the elements of the lattice. (Hint: since $r(abc) = 2$, the points $a$, $b$ and $c$ will be collinear. Now add $d$.)

## 2.5 Tying it together with the rank function

In Section 2.3 we defined several new matroid concepts and we gave a few hints about the cryptomorphisms that connect them. In this section, we prove Theorem 2.15, giving the full cryptomorphism between rank and independent sets, thereby showing that a rank function that satisfies (r1), (r2) and (r3) defines a matroid.

We use the rank function as a bridge between the "linear algebraic" concepts – independent sets, bases and circuits – and the "geometric" concepts – flats, hyperplanes and closure. In addition to the proof of the cryptomorphism between the rank function and independent set axioms,

Figure 2.20. A rank 4 matroid.

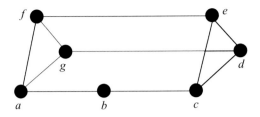

we also outline the cryptomorphism between the rank function and the closure operator axioms.

In Section 2.6, we finish our study of cryptomorphisms by showing flats, hyperplanes and the closure operator are cryptomorphic. In doing so, we will show each of these concepts can be used as the starting point of the definition of a matroid and the other concepts can be determined from this starting point.

### 2.5.1  Rank and independence

We restate Theorem 2.15 for quick reference:

**Theorem 2.15.** *Let $E$ be a finite set with an integer-valued function $r$ defined on subsets of $E$. Then $r$ is the rank function of a matroid if and only if for $A, B \subseteq E$:*

Normalization  (r1)  $0 \leq r(A) \leq |A|$;

Increasing  (r2)  *if $A \subseteq B$, then $r(A) \leq r(B)$;*

Semimodular  (r3)  $r(A \cup B) + r(A \cap B) \leq r(A) + r(B)$.

**Example 2.46.** Consider the rank 4 matroid in Figure 2.20. Then (r1) and (r2) are trivial to check, so we concentrate on (r3), the semimodular property. For many choices of $A$ and $B$, (r3) will hold with equality (for instance, if $A$ is any point and $B$ is any line). But if $A = ce$ and $B = af$, then the inequality is strict: $r(A) + r(B) = 2 + 2 > 3 + 0 = r(A \cup B) + r(A \cap B)$.

Before we prove Theorem 2.15, we investigate various equivalent reformulations of (r1), (r2) and (r3). For instance, it is easy to determine the rank of the empty set using (r1). Since $0 \leq r(\emptyset) \leq 0$, we immediately get the following axiom/property:

$$(\text{r1}')\quad r(\emptyset) = 0.$$

Thus, (r1) implies (r1').

We list three "local" properties of the rank function:

Local normalization  (r1')  $r(\emptyset) = 0$.

Unit rank increase  (r2')  For all $A \subseteq E$ and $x \in E$, we have $r(A) \leq r(A \cup x) \leq r(A) + 1$.

Local semimodularity  (r3')  For $x, y \notin A$, if $r(A) = r(A \cup x) = r(A \cup y)$, then $r(A) = r(A \cup x \cup y)$.

You (see Exercise 18) can also show that the axiom set (r1), (r2), (r3) is equivalent to the axiom set (r1′), (r2′), (r3′).[24] These equivalent axiom systems for the rank function will allow us more flexibility and make it a bit easier to construct our proofs. The next lemma follows easily from axiom (r3), but it can also be deduced directly from the primed axiom set (r1′), (r2′), (r3′). We leave this task to you in Exercise 17.

**Lemma 2.47.** *If $A \subseteq B$, then $r(A \cup x) - r(A) \geq r(B \cup x) - r(B)$.*

We'll also need the following generalization of (r3′) in our cryptomorphism proof below. You are politely requested to prove this in Exercise 16.

**Proposition 2.48.** *Let $r$ be the rank function for a matroid. For any $A \subseteq E$ and $x_1, x_2, \ldots, x_n \in E$, we have:*

(r3″) *If $r(A) = r(A \cup x_1) = r(A \cup x_2) = \cdots = r(A \cup x_n)$, then*

$$r(A \cup \{x_1, x_2, \ldots, x_n\}) = r(A).$$

The proof of Theorem 2.15 is quite similar in form to the proof of Theorem 2.6 in Section 2.1:

(1) $\mathcal{I} \to r$: Given the independent sets $\mathcal{I}$ in a matroid, *define* the rank $r$ of a set $A$ as the size of the largest independent set contained in $A$ and *prove* that the rank function $r$ satisfies the properties (r1), (r2) and (r3).

(2) $r \to \mathcal{I}$: Now reverse the procedure: given the rank function $r$ of a matroid, *define* a collection of subsets $\mathcal{I}$ to be those subsets $A$ satisfying $|A| = r(A)$. For convenience, we'll use the axiom system (I1′), (I2) and (I3) to characterize the independent sets. Then we *prove* the collection $\mathcal{I}$ just defined satisfies the properties (I1′), (I2) and (I3).

(3) Lastly, prove the cryptomorphisms compose correctly. This means that $\mathcal{I} \to r \to \mathcal{I}'$ gives $\mathcal{I} = \mathcal{I}'$ and $r \to \mathcal{I} \to r'$ satisfies $r = r'$.

*Proof Theorem 2.15.*

*Part 1.* Let $M = (E, \mathcal{I})$ be a matroid and define $r(A)$ to be the size of      Part 1: $\mathcal{I} \to r$
the largest independent set contained in $A$. We show $r$ satisfies (r1), (r2) and (r3).

As usual the first two axioms are straightforward. By our definition of the rank function $r$ from the family $\mathcal{I}$, we immediately have $0 \leq r(A) \leq |A|$, so (r1) holds. If $A \subseteq B \subseteq E$, then every independent subset of $A$ will be an independent subset of $B$, so

$$r(A) = \max\{|I| : I \subseteq A; I \in \mathcal{I}\} \leq \max\{|J| : J \subseteq B; J \in \mathcal{I}\} = r(B)$$

and (r2) is satisfied.

---

[24] This was first proved by Whitney in his paper [42] that introduced matroids.

Figure 2.21. $r(A \cap B) = |I_C|$ and $r(A \cup B) = |I_a| + |I_B| + |I_C|$, but $r(A) \geq |I_A| + |I_C|$ and $r(B) \geq |I_B| + |I_C|$.

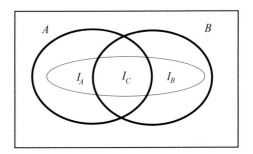

Next we show semimodularity. Let $A$ and $B$ be subsets of $E$ and let $I_C$ be a basis of $A \cap B$, i.e., a maximal independent subset of $A \cap B$. Use axiom (I3) repeatedly to extend $I_C$ to a basis $I$ of $A \cup B$. Let $I_A = I \cap (A - B)$ and $I_B = I \cap (B - A)$. By construction, $r(A \cap B) = |I_C|$ and $r(A \cup B) = |I| = |I_A| + |I_B| + |I_C|$. Now, $I_A \cup I_C$ is an independent set (since it is a subset of $I$) contained in $A$. Thus, $r(A) \geq |I_A| + |I_C|$. Similarly, $r(B) \geq |I_B| + |I_C|$. Combining these results yields the desired result; $r(A) + r(B) \geq r(A \cup B) + r(A \cap B)$. See Figure 2.21 for a helpful Venn diagram.

*Part 2: $r \to \mathcal{I}$*  **Part 2.** Now we are given a rank function $r$ defined on subsets of a finite set $E$ satisfying (r1), (r2) and (r3). Define the family $\mathcal{I}$ to be those subsets $I$ of $E$ for which $r(I) = |I|$. We will show $\mathcal{I}$ is the family of independent sets of a matroid by showing $\mathcal{I}$ satisfies (I1'), (I2) and (I3).

Now by (r1), we know $r(\emptyset) = 0$, so $r(\emptyset) = |\emptyset|$, which means $\emptyset \in \mathcal{I}$. This proves (I1').

To prove (I2) is also satisfied, let $J \in \mathcal{I}$ and let $I \subseteq J$. We'll use (r3) with $A = J - I$ and $B = I$ to show $r(I) = |I|$. Since $J \in \mathcal{I}$, we know $r(J) = |J|$, and so

$$r(A \cup B) + r(A \cap B) = r(J) + r(\emptyset) = |J|.$$

Now by (r1), we know $r(S) \leq |S|$ for all $S \subseteq E$, so

$$r(A) + r(B) = r(J - I) + r(I) \leq |J - I| + |I| = |J|.$$

Putting the two pieces together we get:

$$|J| = r(J) + r(\emptyset) \leq r(J - I) + r(I) \leq |J - I| + |I| = |J|.$$

Thus, we must have equality throughout and so $r(J - I) = |J - I|$ and $r(I) = |I|$; in particular, $I \in \mathcal{I}$.

Finally, we prove the family $\mathcal{I}$ satisfies (I3). As in our proof of Theorem 2.6, we give a proof by contradiction. Let $I, J \in \mathcal{I}$ with $|I| < |J|$ and write $J - I = \{x_1, x_2, \ldots, x_k\}$ for some $k \geq 1$. By assumption, we know $r(I) = |I|$ and $r(J) = |J|$.

Now suppose for all $x_i \in J - I$, property (I3) fails: $I \cup x_i \notin \mathcal{I}$. Then, by definition of $\mathcal{I}$ from the rank function, we must have

$$|I| = r(I) = r(I \cup x_1) = r(I \cup x_2) = \cdots = r(I \cup x_k).$$

Now if $|J - I| = 1$, we get $I \cup x_1 = J$, so $r(J) < |J|$, a contradiction.

Thus $|J - I| > 1$, so $J - I = \{x_1, x_2, \ldots, x_k\}$ for some $k > 1$. Then by (r3″) in Proposition 2.48, $r(I \cup \{x_1, x_2, \ldots, x_k\}) = |I|$. But $I \cup \{x_1, x_2, \ldots, x_k\} = J$, so $|J| = r(J) \le |I| < |J|$, a contradiction.

*Part 3. $r \to \mathcal{I} \to r$*: Given a rank function satisfying (r1), (r2) and (r3), Part 3: The compositions we first create the family $\mathcal{I}$ as above: $I \in \mathcal{I}$ if $r(I) = |I|$. Then we use the family $\mathcal{I}$ to define a (potentially new) rank function, which we'll (temporarily) call $s$: $s(A) = \max_{I \subseteq A}\{|I| : I \in \mathcal{I}\}$. Our immediate goal is to show that this procedure returns the same rank function we started with, i.e., that $r$ and $s$ are identical as functions: $r(A) = s(A)$ for all $A \subseteq E$.

So let $A \subseteq E$. We show $r(A) = s(A)$ for all $A$ by showing $s(A) \le r(A)$ and $r(A) \le s(A)$. For the first inequality, note that $s(A) = |I| = r(I)$ for some $I \subseteq A$. By (r2), we immediately have $r(I) \le r(A)$, or, rewriting this in terms of the function $s$, we have

$$s(A) = |I| = r(I) \le r(A).$$

For the reverse inequality, we need to show $s(A) \ge r(A)$. Suppose this is false, i.e., $s(A) < r(A)$ for some $A \subseteq E$. Then, for all $I \in \mathcal{I}$ with $I \subseteq A$, we must have $r(A) > |I|$. Let $I$ be a maximum-size such set. Then, for all $x \in A - I$, we have $I \cup x \notin \mathcal{I}$. Thus $r(I \cup x) = r(I)$ for all $x \in A - I$, so, by (r3″) in Proposition 2.48, we have $r(A) = r(I) = |I|$, which is a contradiction.

$\mathcal{I} \to r \to \mathcal{I}$: This is the last thing we need to show, but your patience is rewarded at this point with a very easy argument. Here goes: first, we are given the independent sets $\mathcal{I}$, then we define the rank function as above, where $r(A)$ is the size of the largest independent subset of $A$. Finally, we let the family $\mathcal{I}'$ represent those subsets with $r(I) = |I|$.

To finish this off, we need to show $\mathcal{I}' = \mathcal{I}$. But if $I \in \mathcal{I}$, then $r(I) = |I|$ by definition of $r$, so $I \in \mathcal{I}'$. Conversely, if $I \in \mathcal{I}'$, then $r(I) = |I|$, so $I$ is the largest independent subset of $I$ (by definition of $r$, again). Thus $I \in \mathcal{I}'$. □

So what's the story with part 3 of this proof; why do we even need it? Let's look at an example where things screw up.

**Example 2.49.** One step in proving that the cryptomorphisms of Theorem 2.15 compose correctly involved showing the two rank functions $r(A)$ and $s(A)$ were identical when moving from $r$ to $\mathcal{I}$ to $s$. We used axiom (r2) when we proved $s(A) \le r(A)$ for all subsets $A$. Was this really necessary?

To answer this question, define a rank function on $E = \{a, b\}$ as follows: $r(\emptyset) = r(ab) = 0, r(a) = r(b) = 1$. Then axiom (r2) is violated, so this is not the rank function of a matroid. Forming $\mathcal{I}$ as in the cryptomorphism gives $\mathcal{I} = \{\emptyset, a, b\}$. Then using this family to define a function $s$ gives: $s(\emptyset) = 0, s(a) = s(b) = s(ab) = 1$. In particular, $r(ab) < s(ab)$.

Moral: we needed (r2).

More counterexamples that depend on the rank axioms appear in the exercises (see Exercise 30), but we also point out that the last part of the proof, the composition $\mathcal{I}$ to $r$ to $\mathcal{I}'$, is better behaved: given *any* collection of subsets $\mathcal{I}$, applying $\mathcal{I} \to r \to \mathcal{I}'$ as in the proof will always satisfy $\mathcal{I} = \mathcal{I}'$; see Exercise 29.

## 2.5.2 Rank and closure

We next define a matroid in terms of its closure operator.

**Theorem 2.50.** *Let $E$ be a finite set with closure operator $A \mapsto \overline{A}$ defined on subsets of $E$. Then, the closure operator is the closure operator of a matroid if and only if for $A, B \subseteq E$:*

Increasing        (cl1) $A \subseteq \overline{A}$.

Monotonic        (cl2) *If $A \subseteq B$, then $\overline{A} \subseteq \overline{B}$.*

Idempotent        (cl3) $\overline{\overline{A}} = \overline{A}$.

Maclane–Steinitz exchange        (cl4) *If $p \in \overline{A \cup q} - \overline{A}$, then $q \in \overline{A \cup p}$.*

Property (cl4) is the Maclane–Steinitz exchange axiom. As usual, the last axiom/property is what distinguishes the class of matroids from a more general mathematical object.

**Example 2.51.** For practice, let's look at the Maclane–Steinitz exchange axiom (cl4) in a few examples. In the matroid in Figure 2.18, choose $A = ab$, $p = c$ and $q = d$. You can check $\overline{A} = abf$, $\overline{A \cup c} = abcdf$ and $\overline{A \cup d} = abcdf$, so $c \in \overline{A \cup d} - \overline{A}$. Then the exchange axiom (cl4) requires $d \in \overline{A \cup c}$, which is true.

Now consider the rank 4 matroid given in Figure 2.20. This time, let's pick $A = bc$ and let $q$ be any point $q \notin \overline{A}$, say $q = f$. What are the choices for the point $p$? We know $\overline{A} = abc$ and $\overline{A \cup f} = abcef$, so $e$ is the only point (besides $f$) in $\overline{A \cup f} - \overline{A}$. But $f \in \overline{A \cup e} = \overline{A \cup f} = abcef$, so (cl4) is satisfied here.

We outline the proof of Theorem 2.50 below, but leave all of the details to you to work through in Exercise 31. The structure of this proof constructs cryptomorphisms between the rank function and the closure operator. As before, the structure of the proof will be the same as that of Theorems 2.6 or 2.15. See Figure 2.22.

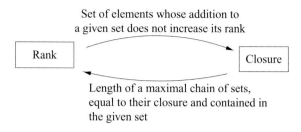

Figure 2.22.
Cryptomorphism between
rank and closure.

Set of elements whose addition to
a given set does not increase its rank

Rank

Closure

Length of a maximal chain of sets,
equal to their closure and contained in
the given set

Outline of proof for Theorem 2.50

(1) Given the rank function $r$ of a matroid, *define* the closure $\overline{A}$ of a set $A$ as the set of elements that can be added to $A$ without increasing its rank:

$$\overline{A} = \{x \in E : r(A \cup x) = r(A)\}.$$

Prove that this closure operator satisfies the properties (cl1), (cl2), (cl3) and (cl4).

(2) Now reverse the procedure: given the closure operator of a matroid, *define* $r(A)$ to be the size of the smallest subset $I$ of $A$ satisfying $\overline{I} = \overline{A}$:

$$r(A) = \min_{I \subseteq A}\{|I| \mid \overline{I} = \overline{A}\}.$$

Prove the function $r$ just defined satisfies the properties (r1), (r2) and (r3).

(3) Lastly, prove the cryptomorphisms compose correctly.

Defining closure in terms of the rank function is "natural" in a certain sense, but defining the rank from the closure takes a detour through independent sets: we need to find the largest independent subset of $A$, which is equivalent to the smallest subset of $A$ with the same closure as $A$. This is not especially satisfying, but, as a thought experiment, we could also define rank in terms of closure via the length of maximal chains of flats (where the flats are just those sets $A$ satisfying $\overline{A} = A$.) It's debatable whether this approach is more satisfying.

## 2.6 Cryptomorphisms between flats, hyperplanes and closure

In this section we define a matroid in terms of flats and hyperplanes. In each case, we give a cryptomorphic description with the closure operator.

Brief advertisement: this veritable menagerie of cryptomorphic definitions actually gives us quite a bit of flexibility in studying and working with matroids. For example, if you are given a geometric representation of a matroid, it might be more natural to consider flats or hyperplanes

Figure 2.23. The geometric
motivation for axiom (F3).

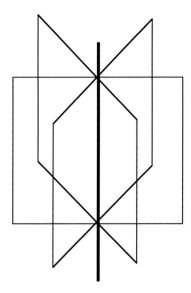

than circuits, say. Having several different descriptions at our disposal
allows us to choose the definition that best fits our situation.

### 2.6.1 Flats

As usual, we begin with a ground set $E$ and a collection $\mathcal{F}$ of subsets of
$E$, called *flats*, and define a matroid directly in terms of axioms for the
family $\mathcal{F}$.

**Theorem 2.52.** *Let $E$ be a finite set and let $\mathcal{F}$ be a family of subsets of
$E$. Then the family $\mathcal{F}$ are the flats of a matroid if and only if:*

(F1) $E \in \mathcal{F}$.
(F2) *If* $F_1, F_2 \in \mathcal{F}$, *then* $F_1 \cap F_2 \in \mathcal{F}$.
(F3) *If* $F \in \mathcal{F}$ *and* $\{F_1, F_2, \ldots, F_k\}$ *is the set of flats that cover $F$, then*
$\{F_1 - F, F_2 - F, \ldots, F_k - F\}$ *partition* $E - F$.

The geometric motivation behind the partitioning axiom (F3) comes
from the following idea. Given a line in $\mathbb{R}^3$, the planes that contain this
line partition the rest of $\mathbb{R}^3$. Figure 2.23 shows this interpretation of
(F3) in $\mathbb{R}^3$. (We only included three of the infinite number of planes that
contain the given line.)
Let's examine (F3) in a few examples.

**Example 2.53.** Consider the rank 4 matroid given geometrically in
Figure 2.20. The flats of the matroid include the points, lines and planes
in the picture. To show how (F3) works, we'll let $F = ae$ and consider
those flats that *cover $F$* in the geometric lattice.

Table 2.2. *Points not on line ae and the unique planes that contain them.*

| Point | b | c | d | f | g |
|---|---|---|---|---|---|
| Plane | abcef | abcef | ade | abcdf | aeg |

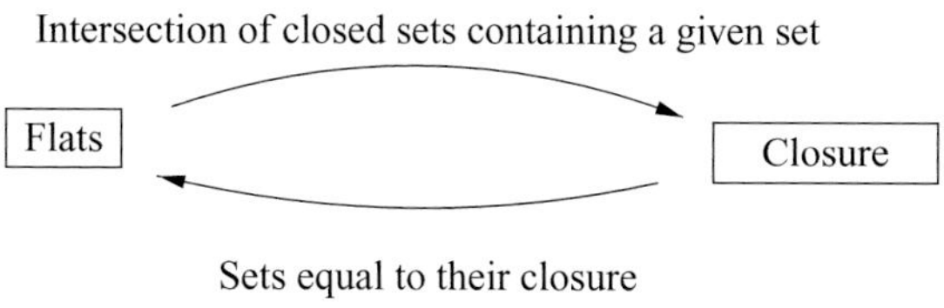

Intersection of closed sets containing a given set

Flats ⟶ Closure

Sets equal to their closure

Figure 2.24. Cryptomorphism between flats and closure.

In this example, since $F$ is a line, the flats that cover $F$ are just the planes that contain it. There are three such planes: $F_1 = ade$, $F_2 = abcef$ and $F_3 = aeg$. (Note that planes $ade$ and $aeg$ are not explicitly drawn in the picture, but they are rank 3 flats.) Then, we can see the collection $\{F_1 - F, F_2 - F, F_3 - F\}$ partitions $\{E - F\}$. Geometrically, this means each point $p$ not on the line $F$ is in a unique plane containing the point $p$ and the line $F$. See Table 2.2 for this partition.

**Example 2.54.** Consider the lattice $L = (\mathcal{F}, \subseteq)$ in Figure 2.19. This is not a geometric lattice, so the elements of the lattice are not the flats of a matroid. We can check that axiom (F3) is violated here: if $F = a$, then the only element of the lattice that covers $F$ is $F_1 = ab$. Then $F_1 - F$ does not partition $E - F$.

Now we prove Theorem 2.52 (see Figure 2.24).

*Proof Theorem 2.52.*

*Part 1.* We begin with a finite set $E$ and a matroid closure operator satisfying (cl1)–(cl4) and define

Part 1: Closure → $\mathcal{F}$

$$\mathcal{F} = \{F \subseteq E \mid F = \overline{F}\}.$$

We need to show this collection satisfies (F1), (F2) and (F3).

For (F1), we know $E = \overline{E}$ since $E \subseteq \overline{E}$ by (cl2). To show (F2), we let $F_1, F_2 \in \mathcal{F}$ and show $F_1 \cap F_2 \in \mathcal{F}$. By our cryptomorphic description of $\mathcal{F}$, this means we need to show $F_1 \cap F_2 = \overline{F_1 \cap F_2}$.

First, by (cl3), we have $F_1 \cap F_2 \subseteq \overline{F_1 \cap F_2}$. Now we need to show the reverse inclusion, i.e., $\overline{F_1 \cap F_2} \subseteq F_1 \cap F_2$. Since $F_1 \cap F_2 \subseteq F_1$ and $F_1 \cap F_2 \subseteq F_2$, (cl2) gives

$$\overline{F_1 \cap F_2} \subseteq \overline{F_1} \quad \text{and} \quad \overline{F_1 \cap F_2} \subseteq \overline{F_2}.$$

But, since $F_1, F_2 \in \mathcal{F}$, we know $\overline{F_1} = F_1$ and $\overline{F_2} = F_2$. Making this substitution yields $\overline{F_1 \cap F_2} \subseteq F_1$ and $\overline{F_1 \cap F_2} \subseteq F_2$, so $\overline{F_1 \cap F_2} \subseteq F_1 \cap$

$F_2$, as required. (Note that, so far, we have not used the Maclane–Steinitz exchange property (cl4).)

To complete this part of the proof, we show (F3) holds. Let $F \in \mathcal{F}$ and suppose $F_1, F_2, \ldots, F_k$ are the flats that cover $F$. We must show $\{E - F_1, E - F_2, \ldots, E - F_k\}$ partition $E - F$. We do this by showing that, for every $x \in E - F$, there is a unique $F_i$ that contains $x$. This will follow immediately from the following two claims, the first showing the existence of $F_i$, and the second showing uniqueness.

*Claim 1.* Existence: if $x \notin F$, then $\overline{F \cup x}$ covers $F$. Suppose not. Then $F \subsetneq F' \subsetneq \overline{F \cup x}$ for some $F' \in \mathcal{F}$, and we let $y \in F' - F$. Clearly,[25] $x \notin F'$. Now we are set up to use (cl4) (for the first time): we have $x, y \notin F$ and $y \in \overline{F \cup x}$. Thus, by (cl4), we must have $x \in \overline{F \cup y}$. But $\overline{F \cup y} \subseteq F'$ and $x \notin F'$, which is a contradiction.

*Claim 2.* Uniqueness: if $\overline{F \cup x} \neq \overline{F \cup y}$, then $(\overline{F \cup x}) \cap (\overline{F \cup y}) = F$. To see this, suppose $z \in (\overline{F \cup x}) \cap (\overline{F \cup y})$. If $z \notin F$, then, as in our proof of the first claim, we use (cl4) to get $\overline{F \cup z} = \overline{F \cup x}$ and $\overline{F \cup z} = \overline{F \cup y}$, a contradiction. So $z \in F$ and the claim is proven.

Part 2: $\mathcal{F} \to$ closure *Part 2.* For this part, we are given the pair $(E, \mathcal{F})$ where $\mathcal{F}$ satisfies (F1), (F2) and (F3). For $A \subseteq E$, define a closure operator by

$$\overline{A} = \bigcap_{F \in \mathcal{F}} \{F \mid A \subseteq F\}.$$

We need to show this closure operator satisfies (cl1), (cl2), (cl3) and (cl4).

The first few axioms are immediate. For any $A$, let $\mathcal{F}_A$ be the family of flats whose intersection is $\overline{A}$, that is, $\mathcal{F}_A := \{F \in \mathcal{F} \mid A \subseteq F\}$ and $\overline{A} = \bigcap \{F \mid F \in \mathcal{F}_A\}$. Then since every $F \in \mathcal{F}_A$ contains $A$, we have $A \subseteq \overline{A}$ and (cl1) holds.

For (cl2), assume $A \subseteq B$. If $F \in \mathcal{F}_B$, then $F \in \mathcal{F}_A$, so $\mathcal{F}_B \subseteq \mathcal{F}_A$. Thus, $\bigcap \{F \mid F \in \mathcal{F}_B\} \supseteq \bigcap \{F \mid F \in \mathcal{F}_A\}$ or equivalently, $\overline{B} \supseteq \overline{A}$ and (cl2) holds.

To show (cl3), we first note that for all $A \subseteq E$, $\overline{A} \in \mathcal{F}$, since $\overline{A}$ is a finite intersection of flats. (This follows from (F2) by induction.) But, if $F \in \mathcal{F}$, we know that the flats that cover $F$ partition $E - F$, and so the intersection of these sets is $F$. Thus, the intersection of all flats that contain $F$ equals $F$ and $\overline{F} = F$. Hence, $\overline{\overline{A}} = \overline{A}$.

Finally, we show the exchange axiom (cl4) is satisfied. Let $p \in \overline{A \cup q} - \overline{A}$ and suppose $q \notin \overline{A \cup p}$. Since $A \subseteq A \cup p$, we know from (cl2)[26] that $\overline{A} \subseteq \overline{A \cup p}$ and similarly $\overline{A} \subseteq \overline{A \cup q}$. But, if $p \in \overline{A \cup q} - \overline{A}$

---

[25] Here "clearly" means, think about it a bit and you'll see why.
[26] You may wonder if it is legal to use (cl2) at this point in the proof. But we've already shown (cl1), (cl2) and (cl3) are true (assuming (F1), (F2) and (F3)), so we are free to use them to show (cl4) holds.

and $q \notin \overline{A \cup p}$ (and hence $q \notin \overline{A}$ and $q \in \overline{A \cup q}$ ), it follows that both these containments are strict, that is:

$$\overline{A} \subsetneqq \overline{A \cup p} \text{ and } \overline{A} \subsetneqq \overline{A \cup q}.$$

Next, we consider the flat $\overline{A}$. By (F3),[27] there is a unique flat $F$ covering $\overline{A}$ with $p \in F$. Since $\overline{A} \subsetneqq \overline{A \cup p} \subseteq F$, it follows that $\overline{A \cup p} = F$. Similarly, $\overline{A \cup q}$ is the unique flat covering $\overline{A}$ and containing $q$. But, $p \in \overline{A \cup q} - \overline{A}$ and $p \in \overline{A \cup p} - \overline{A}$. Since the flats that cover $\overline{A}$ partition $E - \overline{A}$, it must be the case that $\overline{A \cup p} = \overline{A \cup q}$, contradicting the assumption that $q \notin \overline{A \cup p}$.

*Part 3.* The final step of the proof is to show the cryptomorphisms compose correctly. For this we need to show

Part 3: Compositions

$$\overline{A} = \bigcap_{A \subseteq F} \{F \mid F = \overline{F}\}$$

and

$$F \in \mathcal{F} \text{ if and only if } F = \bigcap_{K \in \mathcal{F}} \{K \mid F \subseteq K\}.$$

For the first composition, clearly $\overline{A} \supseteq \bigcap\{F \mid A \subseteq F = \overline{F}\}$ since $\overline{A}$ is one of the sets $F$ in the intersection ($A \subseteq \overline{A} = \overline{\overline{A}}$). Also, for all $F$ with $A \subseteq F = \overline{F}, \overline{A} \subseteq \overline{F} = F$ and $\overline{A} \subseteq \cap\{F \mid A \subseteq F = \overline{F}\}$.

For the second composition, the forward implication is obvious and the reverse implication follows from Proposition 2.19 that a finite intersection of flats is a flat. This finally completes the proof. □

The proof of Theorem 2.52 follows a typical cryptomorphic pattern. In part 1, we proved the flat properties (F1) and (F2) are always satisfied in a matroid by using the closure axioms (cl1), (cl2) and (cl3). We didn't need (cl4) until the end of part 1 – when we proved the covering property (F3). The same pattern appears in part 2 – again, we didn't need to assume (F3) until we got around to proving (cl4), the Maclane–Steinitz exchange axiom. In fact, the equivalence between the three closure axioms (cl1), (cl2), (cl3) and the flat axioms (F1) and (F2) is a *topological* fact. As usual, matroids don't really enter into the game until we get to the final axiom (exchange for closure, and covering for flats).

## 2.6.2 Hyperplanes

Hyperplanes are maximal *proper* flats. (A flat is *proper* if it isn't the entire ground set $E$.) Thus, $H$ is a hyperplane if and only if $E$ covers $H$ in the lattice of flats of the matroid $M$.

[27] This is the first appearance in this part of the proof of (F3).

There is an analogy we like here: we can describe a matroid in terms of its independent sets, or we can restrict to the *maximal* independent sets, i.e., the bases. Our first cryptomorphism (Theorem 2.6) showed that these two families work equally well in defining a matroid.

Playing the same game with flats and hyperplanes, we can define a matroid in terms of flats, or, equally well, in terms of the maximal proper flats, i.e., the hyperplanes. In analogy form:

Independent set *is to* Basis *as* Flat *is to* Hyperplane.

Connecting hyperplanes to the other matroid concepts we've covered so far is worthwhile, even if only to review all of the other concepts (quickly). Proofs are deferred to Exercise 22.

**Proposition 2.55.** *Let M be a matroid on the ground set E. Then the following are equivalent for a subset $H \subseteq E$:*

(1) *H is a hyperplane.*
(2) *H is a maximal proper flat.*
(3) $r(H) = r(E) - 1$ *and* $r(H \cup x) = r(E)$ *for all* $x \notin H$.
(4) *H is maximal with respect to not containing a basis.*
(5) $\overline{H} = H$, *and* $\overline{H \cup x} = E$ *for all* $x \notin H$.
(6) *H is covered by E in the lattice of flats.*
(7) *H has a unique cover in the lattice of flats.*

What's missing from Proposition 2.55?[28] Most notably, we've omitted any mention of circuits. Although there is an intimate relation between these two concepts, the connecting idea is matroid duality, the focus of Section 3.3 in Chapter 3.

When we presented the lattice of flats, one of the key facts was the following:

The lattice of flats is atomic, i.e., every flat is a join of atoms.

An upside-down version of this fact is also true:

**Proposition 2.56.** *Let M be a matroid on the ground set E. Then the lattice of flats is coatomic, i.e., every flat is the intersection of some collection of hyperplanes.*

*Proof Proposition 2.56.* The key to the proof is the flat property (F3) – the flats that cover a given flat partition the complement of that flat. The proof proceeds by induction on the *corank* of the flat, i.e., $r(M) - r(F)$.

So let $F$ be a flat. If $r(M) - r(F) = 0$, then $F = E$ and we are done, since $E$ is the intersection of (the empty collection of) hyperplanes. The proposition is also obviously true for flats of corank 1 – these are just the hyperplanes themselves.

---

[28] A proof?

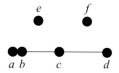

Figure 2.25. Matroid $M$ for Example 2.57 and 2.59.

So assume the proposition is true for any flat of corank $k$ for $k \geq 1$. We need to prove that any flat of corank $k + 1$ is also an intersection of hyperplanes.

Now let $F$ be a flat with $r(F) = r(M) - (k + 1)$. Let $\{H_1, H_2, \ldots, H_p\}$ be the collection of hyperplanes containing $F$, i.e., $F \subseteq H_i$ for hyperplanes $H_i$, where $1 \leq i \leq p$. Then it's clear that $F \subseteq \bigcap_{i=1}^{p}\{H_i\}$. We need to show $F \supseteq \bigcap_{i=1}^{p}\{H_i\}$ to get $F$ as an intersection of hyperplanes and finish the proof.

Suppose this is false. Then there is some $x \notin F$ with $x \in H_i$ for all $i$, $1 \leq i \leq p$. We know $F$ is not a hyperplane, so, by Proposition 2.55(7), $F$ is covered by at least two distinct flats $F_1$ and $F_2$. Then,

- Since $x \in E - F$, by (F3), there is a unique cover of $F$ that contains $x$. Thus, $x$ is not in both $F_1$ and $F_2$. Assume $x \notin F_1$.
- Since $F \subseteq F_1$, every hyperplane that contains $F_1$ also contains $F$. Assume $F_1$ is contained in the hyperplanes $H_1, H_2, \ldots, H_s$ for $s < p$.
- $F_1$ has corank $k$, so, by induction, $F_1$ is an intersection of hyperplanes, i.e., $F_1 = \bigcap_{i=1}^{s}\{H_i\}$. Thus, $x \notin H_j$ for some $j$, $1 \leq j \leq s$.
- But $x \in H_i$ for all $1 \leq i \leq p$.

Connecting the dots completes the proof.                                □

One consequence of this fact is the following:

A point $x$ is in every hyperplane if and only if $x$ is a loop, i.e., $x \in \overline{\emptyset}$.

Given that geometric lattices are coatomic, one might be tempted to turn a geometric lattice upside-down. The resulting lattice will be atomic (since the original was coatomic), but is it semimodular?

Let's look at an example.

**Example 2.57.** Let $M$ be the matroid in Figure 2.25.

The lattice of flats for $M$ is given in Figure 2.26. Now let $H_1 = ef$ and $H_2 = abcd$ be two hyperplanes of $M$. Then, in the upside-down lattice, the "rank" function $r'$ would satisfy $r'(H_1) = r'(H_2) = 1$, but $r'(H_1 \cap H_2) = r'(\emptyset) = 3$. (This "rank" function is just the corank function in the original matroid $M$.) Thus, the upside-down lattice is *not* the lattice of flats of a matroid, and we've more-or-less reached a dead end. In Exercise 47, we (= you) investigate when the inverted lattice is geometric.

**Theorem 2.58.** *Let $E$ be a finite set and let $\mathcal{H}$ be a family of subsets of $E$. Then the family $\mathcal{H}$ are the hyperplanes of a matroid if and only if:*

Figure 2.26. The lattice of flats
for *M*.

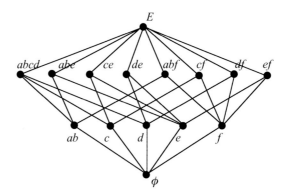

(H1)  $E \notin \mathcal{H}$.

Clutter    (H2)  *If $H_1, H_2 \in \mathcal{H}$ and $H_1 \subseteq H_2$, then $H_1 = H_2$.*

(H3)  *For all distinct $H_1, H_2 \in \mathcal{H}$ and for all $x \in E$, there exists $H \in \mathcal{H}$
with $(H_1 \cap H_2) \cup x \subseteq H$.*

The proof is left for you to ponder (Exercise 32). Note, the structure
of the proof should mimic that of the other cryptomorphisms, such as
the proof of Theorem 2.10.

As a welcome alternative to the proof, we check the axioms for the
matroid given in Example 2.57. Axioms (H1) and (H2) should have a
familiar feel – (H1) is a non-triviality condition and (H2) is the clutter
axiom. (H3) is the "matroid" axiom.

**Example 2.59.** Consider the hyperplanes $H_1 = abcd$, $H_2 = abe$ and
$x = f$ of the matroid in Figure 2.25. Axiom (H3) asserts the existence of
a hyperplane $H$ that contains $abf$. In this case, $H = abf$ is a hyperplane,
so (H3) is satisfied (and the required hyperplane $H$ was unique).

For practice, let's try this one more time: let $H_1 = ce$ and $H_2 =
df$, with $x = a$. Then we have $H_1 \cap H_2 = \emptyset$, so we only need find a
hyperplane that contains the point $a$. This time, there are lots of choices –
there are three hyperplanes that satisfy (H3): $abcd$, $abe$, and $abf$.

### 2.6.3 Comments about complements

Here is a truly transparent observation: if you can define a matroid in
terms of a family of subsets $\mathcal{F}$, then you can also define a matroid in
terms of the *complements* of the members of $\mathcal{F}$. Table 2.3 shows, in
convenient tablular form, three popular family complements.[29]

Recall that flats are also called closed sets. Then the term "open
set" makes sense if you've studied topology.[30] The family of basis

---

[29]  The homophone "compliment" offers some especially tempting jokes here, but we
(for once) resist this urge.

[30]  Actually, it makes sense without regard to your mathematical background.

Table 2.3. *Complementary families.*

| Family | Complementary family |
|---|---|
| Flats $\mathcal{F}$ | Open sets $\mathcal{O}$ |
| Bases $\mathcal{B}$ | Basis complements $\mathcal{B}^C$ |
| Hyperplanes $\mathcal{H}$ | Cocircuits $\mathcal{C}^*$ |

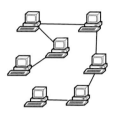

Figure 2.27. Networking computers.

complements is extremely important: it turns out that the complements of the bases also satisfy the basis axioms (B1), (B2) and (B3). That will form the foundation of the theory of duality, which we study in Section 3.3 in Chapter 3.

## 2.7 Application to optimization: the greedy algorithm

The are many instances in which we want to connect a collection of objects (computers, houses, cities) in some optimal way. For example, suppose we want to connect a group of computers with cables. If we think of the computers as vertices and the cables between them as edges, then we are searching for an optimal connected graph.

If we add the requirement that we want to use the smallest number of cables to connect the computers, then it is easy to see we won't have any cycles. Thus, what we seek is a spanning tree; see Figure 2.27.

Perhaps the cable we are using can be cut to any length. If this is the case then we might ask to connect the computers in such a way that the total length of cable used is minimized. Note that this requires us to know something about the distances between computers. We keep track of these distances in our graph by adding a label to each edge called the *weight* of that edge. Such a graph is called a *weighted graph*. Thus, our solution is a spanning tree with minimum total weight; see Figure 2.28.

Weight

**Example 2.60.** The city planners of the city AlmostUtopia want to put bike paths along all the roads within the city. Unfortunately, unlike the neighboring city of Utopia, they don't have enough money to add bike paths to all the roads. The City Council members must decide where to put the bike paths. They decide that they want bikers to be able to travel

Figure 2.28. (a) A computer network, with edge weights. (b) A minimum-weight spanning tree of the network.

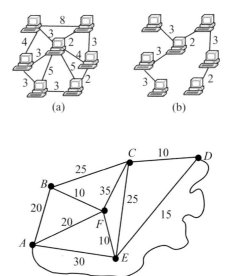

(a)                    (b)

Figure 2.29. AlmostUtopia with bike traffic data.

to all the major intersections of their city (labeled *A*, *B*, *C*, *D*, *E*, *F* in Figure 2.29). Additionally, they would like to add the bike paths to the roads most traveled[31] (by bikes that is).

After an extensive study, data about the daily bike traffic per road was obtained. In Figure 2.29, the edge weights represent an average of the number of bikers per hour traveling on a given road. Help the city planners solve their problem by finding a maximum-weight spanning tree of AlmostUtopia.[32]

The examples above are special cases of a more general matroid optimization problem, stated below. We notice that one example asked for the weight to be minimized while in the other it is maximized. We solve the maximization problem, and note that the "greedy" procedure given also solves the minimization problem.

This problem is usually called the *Minimum Spanning Tree* (MST) problem. The MST problem has a long and rich history, see [15] for details. The first record of the problem in the literature seems to be in 1926, when it was mentioned by Otakar Borůvka in the context of constructing an optimal power-line network in Southern Moravia.

The matroid version goes like this: given a matroid $M = (E, \mathcal{I})$ with weight function $w : E \to \mathbb{R}$ we define the *weight of a subset* $w(A) = \sum_{x \in A} w(x)$. Then our networking problems above motivate the following general matroid problem.

---

[31] The City Council thinks this will make all the difference.

[32] Answer: the maximum-weight spanning tree has weight equal to 130.

> ### Matroid optimization problem
> Given a matroid $M = (E, \mathcal{I})$ with weight function $w : E \to \mathbb{R}$, find a basis $B \in \mathcal{B}$ of maximum weight.

The amazing thing about this problem is that it can be solved quite easily. We present an algorithm, the *greedy* algorithm,[33] that solves the matroid optimization problem. The proof that this really works is given in Theorem 2.62.

> ### Greedy algorithm
> Input: A finite set $E$, a weight function $w : E \to \mathbb{R}$ and a family $\mathcal{I}$ of subsets of $E$.
>
> Order the elements of $E$: $e_1, e_2, \ldots, e_n$ so that $w(e_i) \geq w(e_j)$ for $i \leq j$.
>
> Set $B := \emptyset$.
>
> For $i = 1$ to $n$,
>
> $$\text{if } B \cup e_i \in \mathcal{I}, \text{ then set } B := B \cup e_i.$$
>
> Output: $B$, a maximal member of $\mathcal{I}$ of maximum weight.

Informally, we can think of the greedy algorithm as repeated use of the rule:

*Choose the best you can at every step, without doing anything stupid.*

Mathematically, the rule is "choose the element of largest weight that doesn't create a circuit with the elements you've already chosen." Saying the greedy algorithm solves the optimization problem means the basis $B_G$ produced by that algorithm satisfies $w(B_G) \geq w(B)$ for all bases $B$. Even more striking is the following: the greedy algorithm *characterizes* matroids.

Before stating the theorem, we give another application to the job assignment problem that we saw in Chapter 1.

**Example 2.61.** Before hiring the staff for your new restaurant, you decide to interview the applicants. You rate each applicant on a scale from 1 to 5, with 5 being the highest rating. You are interested in hiring the three highest-rated people for three positions: cook, food taster and

---

[33] This algorithm is also called Kruskal's algorithm, after Joseph Kruskal who published it in 1956 [20]. Kruskal is an American mathematician who worked for many years as a researcher with Bell Labs. This research group was started at the Company's inception in 1925 and has contributed greatly to advances in both mathematics and communications. (Bell Labs is now Alcatal-Lucent.)

Figure 2.30. A bipartite graph models the job assignment choices.

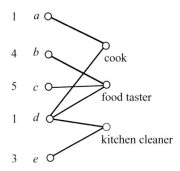

kitchen cleaner. The qualifications and ratings of the five applicants, Alice, Bob, Carla, Darla and Ed, are given in Figure 2.30. (*Transversal matroids* defined from bipartite graphs were introduced in Section 1.5. Much more information about these matroids appears in Chapter 7.)

The three top-rated potential employees are Bob, Carla and Ed. Unfortunately, hiring these three people will only allow us to staff two positions.[34] In matroid language, the set $\{a, c, e\}$ is not an independent set in the associated transversal matroid. Thus, the problem is the same as our bike path problem in that we are looking for a basis in the transversal matroid of maximum weight.

You can achieve your goal of staffing all three positions with the highest-rated employees by using the greedy algorithm:

- First hire Carla, since she has your highest rating. Since she only applied for the food tester job, that is what you must hire her for.
- This means your next highest-rated applicant, Bob, can't be hired at all ($\{b, c\}$ is a circuit in the matroid). Your next best choice is Ed, and you can hire him to clean the kitchen.
- Lastly, consider Alice and Darla, whom you perceive as equally valuable employees, and pick one randomly to be your cook.[35]

Congratulations! You have now hired the highest-rated staff for your three positions.

Recall that a family of subsets $\mathcal{I}$ is a set system if it satisfies (I1) and (I2). Then the greedy algorithm *characterizes* the independent sets of a matroid in the following, cryptomorphic way.

**Theorem 2.62.** *Let $(E, \mathcal{I})$ be a set system. For all weight functions $w : E \rightarrow \mathbb{R}$ the greedy algorithm produces a maximal member of $\mathcal{I}$ of maximum weight if and only if the family $\mathcal{I}$ also satisfies (I3), i.e., $\mathcal{I}$ is the collection of independent sets of a matroid.*

---

[34] Could we train either Bob or Carla to cook? Probably, but that is not the solution we are pursuing.
[35] This never happens in real life.

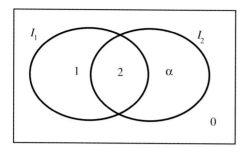

Figure 2.31. The weights of elements of $E$.

*Proof Theorem 2.62.* ($\Leftarrow$) Given a matroid $M = (E, \mathcal{I})$, let $w$ be any weight function on $E$ and let $B_G$ be the output of the greedy algorithm for this matroid and weight function. We first note that by construction, $B_G \in \mathcal{I}$. Since the algorithm only terminates when $B_G \cup e \notin \mathcal{I}$ for all $e \in E - B_G$, we have $B_G$ is a maximal member of $\mathcal{I}$, i.e., a basis of $M$.

Next we must show $w(B_G) \geq w(B)$ for all bases $B$. Order the elements of $B$ and $B_G$ by non-increasing weight, and suppose $B_G = \{e_1, e_2, \ldots, e_r\}$ and $B = \{f_1, f_2, \ldots, f_r\}$. If $w(e_i) \geq w(f_i)$ for all $i$, then we're done, so suppose otherwise. Let $k$ be the smallest index with $w(e_k) < w(f_k)$, so $w(e_i) \geq w(f_i)$ for all $i < k$.

Let $I = \{e_1, e_2, \ldots, e_{k-1}\} \subseteq B_G$ and $J = \{f_1, f_2, \ldots, f_k\} \subseteq B$. We note $I, J \in \mathcal{I}$ with $|I| < |J|$. Thus, by (I3), there is an element $f_j \in J$ for some $1 \leq j \leq k$ with $I \cup f_j \in \mathcal{I}$.

Now $w(f_j) \geq w(f_k) > w(e_k)$. Since both $I \cup f_j$ and $I \cup e_k$ are independent, both $e_k$ and $f_j$ are elements that could be added to $I$ to create a larger independent set. But then the greedy algorithm would have chosen $f_j$ instead of $e_k$ at this stage, which is a contradiction. Thus, for all bases $B$, $w(B) \leq w(B_G)$.

($\Rightarrow$) We will prove the contrapositive: given a set system $(E, \mathcal{I})$, if $\mathcal{I}$ does not satisfy (I3), then there is a weight function $w : E \to \mathbb{R}$ so that the greedy algorithm does not produce a maximal member of $\mathcal{I}$ of maximum weight.

Since $(E, \mathcal{I})$ does not satisfy (I3), there are sets $I_1$ and $I_2$ with $|I_1| < |I_2|$ so that $I_1 \cup x \notin \mathcal{I}$ for all $x \in I_2 - I_1$. We'll use these sets to define our weight function $w$ as follows:

$$
w(e) = \begin{cases} 2 & e \in I_1 \cap I_2 \\ 1 & e \in I_1 - I_2 \\ \alpha & e \in I_2 - I_1 \\ 0 & \text{otherwise} \end{cases}
$$

where $\alpha$ is chosen so that $0 < |I_1 - I_2|/|I_2 - I_1| < \alpha < 1$. See Figure 2.31.

What does the greedy algorithm do to this set system and weight function? First, it chooses all elements of weight 2, and then it adds all the elements of weight 1. So all of $I_1$ has been chosen at this stage of the greedy algorithm.

Now we attempt to add elements of weight $\alpha$. But $I_1 \cup x \notin \mathcal{I}$ for all $x$ with $w(x) = \alpha$, so none of these elements can be added to $I_1$. Thus, the greedy algorithm is forced to add elements of weight 0 to extend $I_1$ to a maximal independent set. So, the greedy algorithm produces a maximal member of $\mathcal{I}$, say $B_1$, with

$$w(B_1) = 2|I_1 \cap I_2| + 1|I_1 - I_2| + 0.$$

On the other hand, $I_2$ is contained in a maximal independent set, say $B_2$ with

$$w(B_2) \geq w(I_2) = 2|I_1 \cap I_2| + \alpha|I_2 - I_1|.$$

By the choice of $\alpha$, $w(B_2) \geq 2|I_1 \cap I_2| + \alpha|I_2 - I_1| > 2|I_1 \cap I_2| + |I_1 - I_2| = w(B_1)$, which is a contradiction. $\qquad\square$

**Example 2.63.** Let $E = \{a, b, c, d\}$ and $\mathcal{I} = \{\emptyset, a, b, c, d, ab, ac, ad, acd\}$ be a set system with weight function: $w(a) = 2$, $w(b) = 1$, $w(c) = w(d) = \frac{3}{4}$. (This weight function is based on the one used in the proof of Theorem 2.62.) We verify that the greedy algorithm does not work in this case: the output from the greedy algorithm is the maximal set $ab$ of weight 3. By inspection, we can see that the maximal set $acd$ has weight $3\frac{1}{2}$.

The greedy algorithm fails, so, by Theorem 2.62, the set system $\mathcal{I}$ is not the independent sets of a matroid. You can see this by observing that (I3) fails for the sets $acd$ and $ab$.

The greedy algorithm is fast and easy to implement. Unfortunately, it doesn't work on many important optimization problems. We mention the *Traveling Salesperson Problem* (TSP) as a famous example:

Traveling salesperson problem

Given a network with weighted edges, can you find a cycle that contains all the vertices of the network of minimum total weight? A cycle that uses all the vertices of a graph is called a *Hamiltonian* cycle.[36] A traveling salesperson who must visit a certain number of cities (and then return home) is traversing a Hamiltonian cycle in the network determined by those cities.

---

[36] William Rowan Hamilton (1805–1865) was one of the most influential mathematicians and physicists of the nineteenth century. He is best known as the discoverer of the *quaternions* in 1843 while walking along the Royal Canal in Dublin with his wife. Hamilton committed the most famous act of mathematical vandalism in history by carving their defining equations into the Brougham Bridge. He liked to play the parlor game "Around the world." This game involved finding a cycle among 20 "cities" placed at the vertices of a dodecahedron Hamilton kept, presumably in his parlor. Such a cycle is now called a *Hamiltonian cycle*.

Note that this is quite similar to the MST problem; just replace "spanning tree" with "Hamiltonian cycle." Unfortunately, there is no known efficient algorithm for solving TSP.[37]

## Exercises

### Basic matroid concepts

(1) Let $M$ be the matroid on $E = \{a, b, c, d, e\}$ with bases $ac, ad, ae,$ $bc, bd, be, ce, de.$
  (a) Find all of the circuits of $M$.
  (b) Find all of the flats of $M$.
  (c) Draw a picture of this matroid.
  (d) Does $M$ have any isthmuses? Loops? Well? Does it?
  (e) Show that $M$ is graphic by finding a graph $G$ with $M = M(G)$.
  (f) Show that $M$ is representable by finding a $2 \times 5$ matrix whose columns correspond to the points of the matroid. You should be able to do this using only 0's and 1's in your matrix.

(2) Let $M$ be a matroid with ground set $E = \{a, b, c, d, e, f\}$ having exactly three circuits: $abcd, abef, cdef.$
  (a) Draw a picture of this matroid. What's the rank?
  (b) Show $M$ really is a matroid by verifying the three circuit axioms (C1), (C2) and (C3).
  (c) Show that $M$ is graphic by finding a graph $G$ whose cycle matroid $M(G)$ matches $M$.
  (d) Show that this matroid has the property that every circuit is a flat. Find another matroid $M'$ with more than three circuits, with this property. (Hint: play with the graph you drew in part (c).)

(3) Let $M$ be the rank 3 matroid pictured in Figure 2.32. This configuration is called the *Pappus* configuration,[38] and its discovery dates back to Pappus of Alexandria,[39] who is now dead. Pappus knew[40] the following:

---

[37] TSP is a famous example of an *NP-complete* problem. These are a class of problems for which no polynomial-time algorithm (that solves the problem) is known to exist, but the discovery of such an algorithm for any NP-complete problem would automatically give polynomial-time solutions for all the problems in the class. The Clay Mathematics Institute offers $1,000,000 for the resolution of this issue (proving either $P = NP$ by finding a polynomial-time algorithm for an NP-complete problem, or, alternatively, showing no such algorithm exists for any NP problem).
[38] This is an important matroid in the study of representation. See Example 6.20 in Chapter 6.
[39] Pappus lived in the late third century in Egypt. The majority of his most famous work entitled "Collection" has survived. In this work, Pappus investigated a wide range of topics, including arithmetic, geometry and mechanics. One of his research interests was the following question: "Why are hexagons used in a honeycomb?"
[40] How on earth could we have any idea of what Pappus actually "knew"?

Figure 2.32. The Pappus configuration.

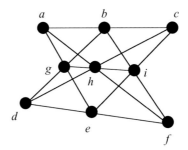

Figure 2.33. The Escher "matroid" is not a matroid.

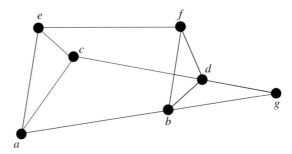

- Start with two three-point lines $abc$ and $def$.
- Now form the point $g$ as the intersection of the two lines $ae$ and $bd$, i.e., $g = ae \cap bd$.
- Continue in this way to form the points $h = af \cap cd$ and $i = bf \cap ce$.
- Then the three points $g, h$ and $i$ must be collinear (in the Euclidean plane).
  - (a) Find all of the hyperplanes of $M$ (these are just the lines of the matroid – there are 18).
  - (b) How many bases does $M$ have?
  - (c) Show that $M$ is not graphic. (Hint: how many vertices and edges would your connected graph have?)
  - (d) Show that every point is in the same number of flats and bases, and determine these two numbers.
  - (e) True/False: every *pair* of points is in the same number of bases.
- (4) The Escher "matroid" was introduced in Exercise 11 in Chapter 1 – see Figure 2.33. Show this configuration is not a matroid using the axiom systems for each of the following.
  - (a) flats;
  - (b) hyperplanes;
  - (c) the closure operator;
  - (d) the rank function.
- (5) Let $M$ be a matroid on the ground set $E$. Call a property of a subset $A \subseteq E$ *internally defined* if you can determine whether $A$ has the

property when looking solely at $A$, i.e., not examining the rest of the matroid. For example, the property "$A$ is independent in $M$" is internally defined since you can determine whether or not $A$ is independent directly from $A$, while the property "$A$ spans $M$" is not internally defined. Determine whether each of the following properties is internally defined.
- (a) $A$ is a basis for $M$.
- (b) $A$ is a circuit.
- (c) $A$ is a cocircuit.
- (d) $A$ is a flat.
- (e) $A$ is a hyperplane.
- (f) Can you determine $r(A)$ internally?

(6) Let $M$ be a matroid on the ground set $E$ with $e \in E$. How many ways can you say "I'm a loop?" Show that the following statements are all equivalent to each other.[41]
- (a) $e$ is in no basis. (This is our definition of a loop – see Definition 2.8.)
- (b) $e$ is a circuit.
- (c) $e$ is in no independent set.
- (d) $r(A \cup e) = r(A)$ for all $A \subseteq E$.
- (e) $e$ is in every flat.
- (f) $e$ is in every hyperplane.
- (g) $e \in \overline{\emptyset}$.

(7) Let $M$ be a matroid on the ground set $E$ with $e \in E$. How many ways can you say "I'm an isthmus?" (Isthmuses are also called *coloops*.) Show that the following statements are all equivalent.
- (a) $e$ is in every basis. (This is our definition of an isthmus – see Definition 2.8.)
- (b) $e$ is in no circuit.
- (c) For any independent set $I$, $I \cup e$ is independent.
- (d) For all $A \subseteq E$ with $e \notin A$, we have $r(A \cup e) = r(A) + 1$.
- (e) For any flat $F$, $F - e$ is a flat.
- (f) $E - e$ is a hyperplane.
- (g) For all $A \subseteq E$ with $e \notin A$, we have $e \notin \overline{A}$.

(8) (a) Let $M$ be a matroid in which every element is either an isthmus or a loop. Prove that $M$ has only one basis.
- (b) Show the converse of part (a) is also true: if $M$ is any matroid with a unique basis, show that every element of $M$ is either an isthmus or a loop.

---

[41] Note: to show a bunch of statements are equivalent, you might show (a) $\Rightarrow$ (b) and (b) $\Rightarrow$ (c) and ... (e) $\Rightarrow$ (a). There are lots of other ways to do this, though. Just make sure your path of implications allows you to get from any of the statements to any other statement. A graph theorist would say "The directed graph of proven implications is strongly connected."

Spanning set

(c) A subset $S \subseteq E$ is *spanning* if $S$ contains a basis. Fill in the blank, then provide a proof: "A matroid $M$ has only one spanning set if and only if every element of the matroid is a(n) _____."

(d) Repeat part (c), replacing "spanning" with "independent."

(9) This exercise is important, and we'll need it later. For instance, see Exercise 13 below.

(a) If $C$ is a circuit and $C^*$ is a cocircuit, prove that $|C \cap C^*| \neq 1$.

Circuit–cocircuit intersection

Recall, a cocircuit is the complement of a hyperplane. (Hint: first show that, if $C$ is a circuit, then $\overline{C - x} = C$ for any $x \in C$. See Proposition 4.8.)

(b) Suppose $M$ is a matroid, $C$ is a circuit and $F$ is a flat. Show that it is impossible for $C$ and the complement of $F$ to meet in exactly one point: i.e., show that $|F^c \cap C| \neq 1$.

(10) Consider the rank $r$ uniform matroid on $n$ points $U_{r,n}$, with $r \leq n$.

(a) Describe all of the bases, flats, hyperplanes and cocircuits (hyperplane complements) of $U_{r,n}$. Give a concise description of the rank function while you're at it.

(b) Find the spanning sets of $U_{r,n}$ (see Exercise 8).

(c) Pick one of the families from part (a) or (b) and prove that your description characterizes uniform matroids. For example, you might try something like: "Let $M$ be a rank $r$ matroid on $n$ points. Then $M$ is a uniform matroid if and only if every set of size $r + 1$ is a circuit."

### Axiom systems – Independent sets and bases

(11) Let $E = \{a, b, c\}$ and $\mathcal{I} = \{a, b, abc\}$. Show (I3$'$) is satisfied, but (I3) is not.

(12) Given a finite set $E$ and a collection of subsets $\mathcal{B}$ of $E$.

(a) Prove Proposition 2.5 by showing the axiom systems (B1), (B2), (B3) and (B1), (B2$'$), (B3) are equivalent.

(b) Give an example to show that the axiom system (B1), (B2) is not equivalent to the system (B1), (B2$'$). (For this problem, produce a family $\mathcal{B}$ satisfying (B1) and (B2), but not (B2$'$). Then find a family satisfying (B1) and (B2$'$), but not (B2). Of course, neither example will be the set of bases for a matroid.)

(13) Let $B$ be a basis of a matroid $M$ and suppose $x \notin B$.

(a) Use the circuit elimination axiom (C3) to show that $B \cup x$

Basic circuit

contains a *unique* circuit. This is called the *basic circuit* determined by $B$ and $x$. (This will be important later – see Proposition 4.7(1).)

(b) Suppose $C$ is a circuit in $M$. Show that there is a basis $B$ and an element $x \notin B$ so that $C$ is the basic circuit determined by $B$ and $x$. Thus, *every* circuit is a basic circuit.

Strong basis exchange

(14) Show that the bases of a matroid satisfy *strong basis exchange*:

(a) (B3') If $B_1, B_2 \in \mathcal{B}$ and $x \in B_1 - B_2$, then there is an element $y \in B_2 - B_1$ so that both $B_1 - x \cup \{y\} \in \mathcal{B}$ and $B_2 - y \cup \{x\} \in \mathcal{B}$.

(Hint: let $C$ be the basic circuit contained in $B_2 \cup \{x\}$ from Exercise 13 and let $C^* = \overline{B_1 - \{x\}}$ be the hyperplane containing $B_1 - \{x\}$. Now apply Exercise 9.)

(b) Show that the three axioms (B1), (B2'), (B3') can be used to define the bases of a matroid.

Strong basis exchange will be important when we study *duality* in Section 3.3 of Chapter 3.

## Axiom systems – Rank function

(15) Let $E = \{a, b, c\}$. Define a "rank" function stupidly, as follows:

$$r(\emptyset) = 0, r(a) = r(b) = r(c) = r(ac) = r(bc) = 1,$$
$$r(ab) = r(abc) = 2.$$

(a) Show that this function satisfies (r1) and (r2), but not (r3). Conclude that this is not a matroid.

(b) Use this function and the definition of independent sets in terms of the rank function to define "independent" sets for this non-matroid. Which of the properties (I1), (I2) and (I3) are violated?

(c) Repeat part (b) for circuits, rank, flats and closure. For each one, use the rank function to define the family of subsets (or the operator in the case of closure). Then figure out which of the defining properties for the family (or closure) are violated.

(d) When we repeat part (b) for hyperplanes, we get two maximal subsets of rank 1: $ac$ and $bc$. Find a matroid on $\{a, b, c\}$ with these two subsets as its hyperplanes. How does this rank function for your matroid differ from $r$?

(e) Now repeat part (d) for bases.

(16) Use rank axiom (r3') to prove Proposition 2.48: the rank function of a matroid satisfies the *generalized* local semimodularity axiom:

(r3'')　If $r(A) = r(A \cup x_1) = r(A \cup x_2) = \cdots = r(A \cup x_n)$,

then

$$r(A \cup \{x_1, x_2, \ldots, x_n\}) = r(A).$$

(17) Prove Lemma 2.47 from the primed rank axioms (r1'), (r2'), (r3'):

If $A \subseteq B$, then $r(A \cup x) - r(A) \geq r(B \cup x) - r(B)$.

[Note: this is easy if you use rank axiom (r3). The restriction on only using the primed axiom set is important for *proving* rank axiom (r3) from (r1'), (r2'), (r3') (see Exercise 18(b)).]

(Extended hint: suppose not. First use (r3″) (see Exercise 16) to show that you may assume $B$ is rank closed, i.e., $r(B \cup x) = r(B) + 1$ for all $x \notin B$. Now, assuming $r(B) - r(A) = k \geq 0$, show there must be a set $\{y_1, y_2, \ldots, y_k\} \subseteq B - A$ satisfying

$$r(A) < r(A \cup \{y_1\}) < r(A \cup \{y_1, y_2\}) < \cdots$$
$$< r(A \cup \{y_1, y_2, \ldots, y_k\}) = r(B).$$

Letting $A' := A \cup \{y_1, y_2, \ldots, y_k\}$, use (r2′) and (r3″) again to show that adding $(B - A') \cup x$ to $A'$ does not increase the rank, yielding a contradiction.)

(18) The two rank axiom systems (r1), r(2), (r3) and (r1′), (r2′), (r3′) are equivalent. The goal of this (long) problem is to prove this.
  (a) Assume the axioms (r1), (r2) and (r3).
    (i) Prove (r1′): $r(\emptyset) = 0$. (Note: this follows directly from (r1).)
    (ii) Prove (r2′): (Unit rank increase) For all $A \subseteq E$ and $x \in E$,

$$r(A) \leq r(A \cup x) \leq r(A) + 1.$$

    We point out that (r2′) does *not* follow from (r1) and (r2) – you will need (r3) for your proof. See Exercise 19.
    (iii) Prove (r3′): If $r(A) = r(A \cup x) = r(A \cup y)$, then $r(A \cup \{x, y\}) = r(A)$.
  (b) Now assume the primed versions (r1′), (r2′), (r3′), and prove (r1), (r2) and (r3).
    (i) Prove (r1): $0 \leq r(A) \leq |A|$. (Note: this follows from (r1′) and (r2′).)
    (ii) Prove (r2): If $A \subseteq B$, then $r(A) \leq r(B)$.
    (iii) Prove (r3): $r(A) + r(B) \geq r(A \cup B) + r(A \cap B)$. (Suggestion: use Lemma 2.47 and induction on $|B - A|$.)

(19) More rank axiom games. The two sets of rank axioms (r1), r(2), (r3) and (r1′), (r2′), (r3′) are equivalent – see Exercise 18. But we cannot freely exchange an axiom with its primed version. Show that the three axioms (r1), (r2) and (r3′) do *not* imply unit rank increase (r2′) by constructing a counterexample with $E = \{a, b\}$. Conclude that the three axioms (r1), (r2) and (r3′) do not define a matroid.[42]

## Axiom systems – Closure and flats

(20) Gaining some closure. Let $M$ be a matroid on $E$.
  (a) Here are two false statements about the closure operator:

$$\overline{A \cup B} = \overline{A} \cup \overline{B} \qquad \overline{A \cap B} = \overline{A} \cap \overline{B}.$$

---

[42] A rank function satisfying (r1), (r2) and (r3′) defines a *greedoid*. Really.

Recall: to prove $X = Y$, you must show $X \subseteq Y$ and $X \supseteq Y$. Hence, there are four containments to investigate in this problem. Of those four statements, two are true and two are false. Find proofs for the ones that are true and give specific counterexamples for the false containments.

(b) Use (a) to show that if $\rho$ is an integer-valued function defined on subsets of $E$ satisfying (r1), (r2) and, for all flats $F_1$ and $F_2$,

$$\rho(F_1) + \rho(F_2) \geq \rho(F_1 \cup F_2) + \rho(F_1 \cap F_2),$$

then $\rho$ is the rank function of a matroid. (That is, show that if $\rho$ is semimodular on flats, then $\rho$ is semimodular on all sets.)

(21) Here are two statements concerning rank and closure: for all $A, B \subseteq E$,

$$r(\overline{A \cup B}) = r(\overline{A} \cup \overline{B}) \qquad r(\overline{A \cap B}) = r(\overline{A} \cap \overline{B}).$$

This time, one of these statements is true and the other is false. Figure out which is which, prove the true one, and give a counterexample for the other. Then fix the false one by replacing "=" with "≤" or "≥," and prove your inequality.

(22) Prove the characterizations of hyperplanes given in Proposition 2.55.

(23) The goal of this exercise is a proof of Lemma 2.25, i.e., the closure $\overline{A}$ is both the unique smallest flat containing $A$ *and* the intersection of all flats containing $A$. Let $M$ be a matroid on the ground set $E$, and let $A \subseteq E$.

(a) Show there is a unique *smallest* flat containing $A$: show that there is a unique flat $F$ such that $A \subseteq F$ and, for any other flat $F'$, if $A \subseteq F'$, then $F \subseteq F'$. (Hint: use the fact that the intersection of flats is a flat: Proposition 2.19, which is proven in Exercise 24(b).)

(b) Let $\overline{A}$ be the flat from part (a). Show that $\overline{A} = \bigcap_{F \in \mathcal{F}} \{F : A \subseteq F\}$. (Hint: use Proposition 2.19 and mathematical induction to show $\bigcap_{F \in \mathcal{F}} \{F : A \subseteq F\}$ is a flat that contains $A$.)

(24) The goal of this problem is to prove the intersection property (F2) for flats from the rank axioms.

(a) Suppose $A \subseteq B$ and $r(B \cup x) = r(B) + 1$. Use the rank axiom (r3) to show $r(A \cup x) \geq r(A) + 1$. Conclude from axiom (r2′) that $r(A \cup x) = r(A) + 1$.

(b) Use part (a) and Definition 2.16 of a flat in terms of the rank function to prove property (F2): Proposition 2.19: If $F_1$ and $F_2$ are flats, then so is $F_1 \cap F_2$.

**Cryptomorphisms**

(25) This problem deals with the connections between circuits and independent sets.

    (a) Suppose $C_1$ and $C_2$ are two different circuits in some matroid $M$. Show $C_1 \cap C_2$ must be independent.

    (b) Suppose $C_1$ and $C_2$ are both circuits in some matroid $M$, and let $e \in C_1 \cap C_2$. Show that $(C_1 \cup C_2) - e$ cannot be an independent set. (Hint: suppose $(C_1 \cup C_2) - e$ is independent. Let $I_1 = C_1 \cap C_2$ and $I_2 = (C_1 \cup C_2) - e$. Use the augmentation axiom (I3) repeatedly to (eventually) get a contradiction.)

    (c) Use part (b) to prove circuit elimination (C3): If $C_1$ and $C_2$ are both circuits in some matroid $M$ and $e \in C_1 \cap C_2$, show that $(C_1 \cup C_2) - e$ contains a circuit $C_3$.

(26) Let $M$ be a matroid with circuits $C$ and bases $B$. Show that $C \cup B$ satisfies the circuit axioms (C1), (C2) and (C3). (The matroid whose circuits are obtained from $M$ in this way is called the *truncation* of $M$, denoted $T(M)$. This operation appears in Project P.7.)

(27) Define the functions $f$ and $g$ as in the proof of Theorem 2.6, namely $f(\mathcal{I})$ is the collection of maximal elements of $\mathcal{I}$ and $g(\mathcal{B})$ is the collection of all subsets of elements of $\mathcal{B}$.

    (a) Let $\mathcal{I} = \{a, b, abc\}$ and $\mathcal{B} = \{abc\}$. First notice that $\mathcal{I}$ is not the collection of independent sets of any matroid. Show $g \circ f(\mathcal{I}) \neq \mathcal{I}$.

    (b) Show $f \circ g(\mathcal{B}) = \mathcal{B}$ for any collection of sets $\mathcal{B}$ satisfying (B2).

(28) Prove the cryptomorphism between circuits and independent sets; Theorem 2.10. This is a long exercise:

    (a) First, assume you are given the family $\mathcal{I}$ of independent sets of a matroid, satisfying the axioms (I1), (I2) and (I3). Then define the circuits as the minimal dependent sets, i.e., $C$ is a circuit if $C \notin \mathcal{I}$, but every proper subset of $C$ is in $\mathcal{I}$. Then prove the family of circuits $C$ satisfies the circuit axioms (C1), (C2) and (C3). (Hint: see Exercise 25 for a proof of (C3).)

    (b) Next, assume you are given a family $C$ of subsets of $E$ satisfying (C1), (C2) and (C3). Define the family $\mathcal{I}$ to be all subsets of $E$ not containing any member of $C$. Then show the family $\mathcal{I}$ satisfies (I1), (I2) and (I3).

    (c) Finally, show that these two cryptomorphisms compose correctly: $\mathcal{I} \to C \to \mathcal{I}'$ has $\mathcal{I} = \mathcal{I}'$, and $C \to \mathcal{I} \to C'$ has $C = C'$.

(29) Let $\mathcal{I}$ be *any* family of subsets of a finite set $E$ with $\emptyset \in \mathcal{I}$, and define a rank function $s$ as in our cryptomorphism proof (Theorem 2.15) by:

$$s(A) = \max_{I \subseteq A}\{|I| \mid I \in \mathcal{I}\}.$$

Then define $\mathcal{I}'$ to be those subsets of $E$ satisfying $s(I) = |I|$. Show that $\mathcal{I} = \mathcal{I}'$. Conclude that this composition of cryptomorphisms always returns the original family, regardless of whether that family satisfies the independent set axioms.

(30) Let $E = \{a, b, c\}$ and define a function on subsets of $E$ as follows; $r(\emptyset) = 0, r(abc) = 2$ and $r(A) = 1$ for the remaining six subsets of $E$.

    (a) Show $r$ is not the rank function of a matroid. (Hint: check axiom (r3).)

    (b) In the composition part of the proof of the cryptomorphism given in Theorem 2.15, we used the rank function $r$ to create a family $\mathcal{I}$, then we used $\mathcal{I}$ to create a (potentially) new rank function $s$. Find the family $\mathcal{I}$ created from $r$ and the new function $s$, then show $s(abc) < r(abc)$. Conclude that the composition may not return the original function when axiom (r3) is violated.

(31) Prove Theorem 2.50, the cryptomorphism between the closure operator and the rank function.

(32) Prove Theorem 2.58, the cryptomorphism between hyperplanes and flats.

(33) Let $E$ be a finite set and let $\mathcal{D}$ be a family of subsets of $E$ called dependent sets satisfying:

    (D1) $\emptyset \notin \mathcal{D}$.

    (D2) If $D_1 \in \mathcal{D}$, and $D_1 \subseteq D_2$ then $D_2 \in \mathcal{D}$.

    (D3) If $D_1, D_2 \in \mathcal{D}$ with $D_1 \neq D_2$, then either $D_1 \cap D_2 \in \mathcal{D}$ or $D_1 \cup D_2 - x \in \mathcal{D}$ for any $x \in E$.

Show that a matroid can be defined in terms of dependent sets by finding a crytpomorphism between dependent sets and circuits.

(34) Let $E$ be a finite set and let $\mathcal{S}$ be a family of subsets of $E$ called spanning sets satisfying:

    (S1) $E \in \mathcal{S}$.

    (S2) If $S_1 \in \mathcal{S}$, and $S_1 \subseteq S_2$ then $S_2 \in \mathcal{S}$.

    (S3) If $S_1, S_2 \in \mathcal{S}$ and $|S_1| < |S_2|$, then there is an element $x \in S_2 - S_1$ so that $S_2 - x \in \mathcal{S}$.

Show that a matroid can be defined in terms of spanning sets by finding a crytpomorphism between spanning sets and bases.

### Posets and lattices

(35) Let $(P, \preceq)$ be a finite poset.

    (a) Show that $P$ must have at least one maximal element and one minimal element.

    (b) A poset $(P, \preceq_1)$ has an extension to a *total order* $(P, \preceq_2)$ if

        (i) every pair of elements of $P$ is related by $\preceq_2$: for all $x, y \in P$, either $x \preceq_2 y$ or $y \preceq_2 x$, and

        (ii) $\preceq_2$ *extends* $\preceq_1$: if $x \preceq_1 y$, then $x \preceq_2 y$.

*Dependent sets*

*Spanning sets*

Use mathematical induction and part (a) to prove that every finite poset has an extension to a total order.[43]

(36) For the poset $D_n$ of Example 2.30, show $a \vee b = lcm(a, b)$ and $a \wedge b = gcd(a, b)$.

(37) Baseball relations. Let $E$ be the set of all people who have ever played major league baseball, where $(a, b) \in R$ if players $a$ and $b$ were ever teammates at any point in either of their careers. Then (Lou Gehrig, Babe Ruth) $\in R$, but (Sandy Koufax, Joe DiMaggio) $\notin R$.

　(a) Show $R$ is reflexive and symmetric, but not antisymmetric and not transitive.

　(b) There is a game analogous to the "Kevin Bacon" movie game that uses this relation. In the baseball version of the game, Rogers Hornsby usually plays the role of Kevin Bacon.[44] To play this game, include managers. Find a path of length at most 3 connecting New York Yankee second baseman Joe Gordon and Cleveland Indian outfielder Pat McNulty.

　(c) A mathematical version of the baseball or movie relation goes like this: let $E$ be the set of all people who have authored or co-authored a paper in a mathematics journal. Define $(a, b) \in R$ if $a$ and $b$ are co-authors, and show $R$ is not an equivalence relation. The American Mathematics Society has a tool to allow one to calculate "collaboration distance," that is the length of a shortest path connecting two mathematicians. Find the collaboration distance between each of the authors and Hassler Whitney.

　(d) Your *Erdős number* is the collaboration distance between you and the famous mathematician Paul Erdős[45] using this relation. Interesting statistics about this relation appear on-line at the Erdős Number Project Site. Find the Erdős Number for Hassler Whitney.

(38) Show that the lattice $(D_n, |)$ is atomic if and only if $n$ is square-free, i.e., $n = p_1 p_2 \cdots p_k$ for distinct primes $p_1, p_2, \ldots, p_k$.

(39) (a) Show that each poset $L_1$, $L_2$ and $L_3$ in Figure 2.34 is a lattice.

　(b) Determine which properties each lattice in Figure 2.34 has: atomic, graded (if so, find the rank of each element), semimodular, modular and geometric.

---

[43] This is also true for infinite posets, but the proof requires the Axiom of Choice. This is often referred to as the *Szpilrajn* Extension Theorem, or the *Szpilrajn–Marczewski* Theorem. Edward Szpilrajn (1907–1976) was a Polish topologist who was called Edward Marczewski after 1940. After World War II, he founded the journal *Colloquium Mathematicum* in 1946 in Wroclaw, Poland, and also played a prominent role in the new university in that city.

[44] We would love to see Kevin Bacon play Rogers Hornsby in a biopic.

[45] Paul Erdős (1913–1996) was a Hungarian born mathematician who was a master of problem-solving. One of the greatest and most prolific mathematicians of all time, Erdős published over 1200 papers. For more on this legendary mathematician see [2].

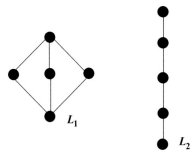 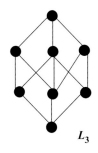

Figure 2.34. Three of our favorite lattices – Exercise 39.

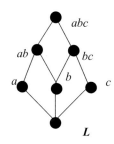

Figure 2.35. The lattice from Example 2.44.

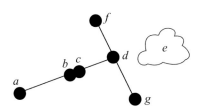

Figure 2.36. The matroid for Exercises 42(b), 43(b) and 48.

(c) Find the positive integers $n$ so that the Hasse diagram for $D_n$ is the same as that of $L_2$. Repeat for $L_1$ and $L_3$ if possible.

(40) Prove Proposition 2.37: A finite non-empty lattice $L$ has a least element $\hat{0}$ and a greatest element $\hat{1}$.

(41) (*If you know some group theory.*) Let $G$ be a finite group and $P$ be the set of normal subgroups of $G$. Show that $(P, \subseteq)$ is a lattice. Determine what properties the lattice has for your choice of $G$.

(42) (a) Consider the lattice $L$ in Figure 2.35. This lattice was studied in Example 2.44.

This is not the lattice of flats of a matroid, but the hyperplanes are those of a matroid. Show that if $\mathcal{H} = \{ab, bc\}$ is a set of hyperplanes on $E = \{a, b, c\}$, then the hyperplane axioms (H1), (H2) and (H3) are satisfied. Find the (unique) matroid on $E = \{a, b, c\}$ with hyperplanes $ab$ and $bc$.

(b) Find the lattice of flats of the matroid given in Figure 2.36. Verify that the lattice is geometric.

(43) Let $M$ be a matroid and let $F$ be a flat. Then we can understand the flat covering axiom (F3) via an *equivalence relation*. Define a

relation $R$ on $E - F$ as follows: for all $x$, $y \in E - F$, define $x R y$ if $x \in \overline{F \cup y}$.

(a) Show that $R$ is an equivalence relation.

(b) Consider the flat $F = \{b, c, e\}$ of the matroid in Figure 2.36. Find the equivalences classes for $R$.

(44) Let $(L, \preceq)$ be a lattice.

(a) Prove  $x \vee (y \wedge z) \preceq (x \vee y) \wedge (x \vee z)$  for all elements $x, y, z \in L$.

(b) Show that if $L$ is a semimodular lattice, then for all $x$, $y \in L$, if $x$ and $y$ cover $x \wedge y$, then $x \vee y$ must cover $x$ and $y$.

(c) For a challenge try the reverse implication of part (b).

Interval

(d) The interval $[x, y]$ is defined as $[x, y] = \{z : x \preceq z \preceq y\}$. Show that every interval of a geometric lattice is geometric.

(45) (a) Suppose $P$ is a finite poset so that $x \wedge y$ is defined for all $x, y \in P$ (this means $P$ is a meet-semilattice) and $P$ has a greatest element $\hat{1}$. Show that $P$ is a lattice.

(b) Use the axioms for flats $\mathcal{F}$ along with part (a) to show $(\mathcal{F}, \subseteq)$ forms a lattice. (You don't need (F3) to do this!)

(46) Let $F$ be a flat in a matroid $M$. Show there is a flat $G$ satisfying all of the following requirements:

- $r(F) + r(G) = r(F \cup G) + r(F \cap G)$;
- $r(F \cup G) = r(M)$; and
- $r(F \cap G) = 0$.

$G$ is called a *relative complement* of $F$. (Hint: if $x_1, x_2, \ldots, x_n$ are the atoms of the lattice of flats, show that $F = \bigvee_{x_i \in F} x_i$. Then form $G$ as the supremum of the remaining atoms.)

(47) A matroid is *modular* if $r(F_1 \cup F_2) + r(F_1 \cap F_2) = r(F_1) + r(F_2)$ for all flats $F_1$, $F_2$.

(a) Give an example of a matroid and two flats $F_1$ and $F_2$ where

$$r(F_1 \cup F_2) + r(F_1 \cap F_2) < r(F_1) + r(F_2),$$

i.e., give an example of a matroid that is not modular.

(b) Show that the matroid $M$ on the six-element set $E = \{a, b, c, d, e, f\}$ with bases $\mathcal{B} = \{ace, ade, bce, bde, cde\}$ (the matroid from Example 2.7) is modular.

(c) Show that any rank 2 matroid is modular.

(d) Show that inverting the lattice of flats of a matroid $M$ gives a geometric lattice if and only if $M$ is modular.

## Greedy algorithm

(48) Solve the maximum-weight basis matroid optimization problem for the matroids below.

(a) $M$ is the matroid in Figure 2.36, with weights: $w(a) = w(b) = 1$, $w(c) = w(d) = 2$, $w(e) = w(f) = 3$ and $w(g) = 8$.

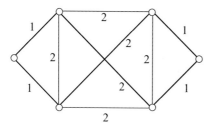

Figure 2.37. Graph for
Exercise 48.

(b) *M* is the graphic matroid $M = M(G)$ for the graph in Figure 2.37.

(c) *M* is the graphic matroid for the graph in Figure 2.30 with the new weight function $w(a) = w(e) = 3$, $w(b) = 5$ and $w(c) = w(d) = 4$.

(49) Let $M = (E, \mathcal{I})$ be a matroid with weight function $w : E \rightarrow \mathbb{R}$.

(a) Modify the greedy algorithm to find a basis $B \in \mathcal{B}$ of minimum weight. Prove that your modification produces a minimum-weight basis.

(b) Show that if the weights are distinct ($w(e) \neq w(f)$ for two distinct elements $e, f \in E$), then there is a unique solution to the matroid optimization problem.

# 3

# New matroids from old

Given a matroid, how can you use that matroid to create a new matroid? When does a given matroid (or a vector space, or a group, or a ring, ...) decompose as a union or a sum of two (or more) smaller matroids? These are important questions, and this chapter attempts to answer them. We begin with two very important operations on a matroid that are motivated by graph theory.

## 3.1 Matroid deletion and contraction

This section is devoted to two very important operations we can perform on a matroid: *deletion* and *contraction*. Both operations reduce the size of the matroid by removing an element from $M$; this is frequently useful for proofs involving mathematical induction. Recall that an element $e$ of a matroid is an *isthmus* if it's in every basis, and $e$ is a loop if it's in no basis. (For a quick review of the different ways to describe an isthmus and a loop, see Exercises 6 and 7 from Chapter 2.)

**Definition 3.1.** Let $M$ be a matroid on the ground set $E$ with independent sets $\mathcal{I}$.

Deletion (1) **Deletion** For $e \in E$ ($e$ not an isthmus), the matroid $M - e$ has ground set $E - \{e\}$ and independent sets that are those members of $\mathcal{I}$ that do not contain $e$:

> $I$ is independent in $M - e$ if and only if $e \notin I$ and $I$ is independent in $M$.

Contraction (2) **Contraction** For $e \in E$ ($e$ not a loop), the matroid $M/e$ has ground set $E - \{e\}$ and independent sets that are formed by choosing all those members of $\mathcal{I}$ that contain $e$, and then removing $e$ from each such set:

> $I - \{e\}$ is independent in $M/e$ if and only if $e \in I$ and $I$ is independent in $M$.

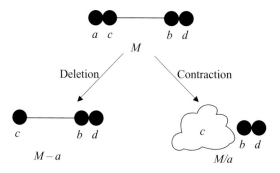

Figure 3.1. Three matroids: $M$, $M - a$ and $M/a$.

Equivalently, a subset $I$ is independent in $M/e$ if $e \notin I$, but $I \cup \{e\}$ is independent in $M$ (see Exercise 3). It's useful to think about all of the independent sets of the matroid $M$ being partitioned into two families: those independent sets that don't contain the element $e$ and those that do. The former collection becomes the independent sets of $M - e$, while the latter becomes the independent sets in $M/e$ (where $e$ is removed from each of these sets). In particular, if $i(M)$ denotes the number of independent sets of a matroid $M$, then $i(M) = i(M/e) + i(M - e)$.[1]

Proving the deletion $M - e$ and contraction $M/e$ are actually matroids on the ground set $E - e$ is a completely routine exercise,[2] but it's worth stating this as a proposition:

**Proposition 3.2.** *If $e$ is neither an isthmus nor a loop, then $M - e$ and $M/e$ are both matroids.*

Proving this proposition involves showing that the three independent set axioms (I1), (I2) and (I3) are satisfied by the families of sets we claimed were the independent sets of $M - e$ and $M/e$. We leave the proof to you in Exercise 8.

This seems like a good time for an example, don't you think?

**Example 3.3.** Let $M$ be the matroid on $E = \{a, b, c, d\}$ with independent sets $\{\emptyset, a, b, c, d, ab, ad, bc, cd\}$. Pictures of $M$, $M - a$ and $M/a$ are given in Figure 3.1. Note that the loop $c$ in $M/a$ is depicted in our customary, cloud-like illustration. Let's figure out the independent sets of $M - a$ and $M/a$:

- For $M - a$, we just find the independent sets of $M$ that do not contain the element $a$. Here they are: $\{\emptyset, b, c, d, bc, cd\}$. See Table 3.1. (Note that $c$ is an isthmus in $M - a$.)

---

[1] There are lots of important things we can count that obey a similar recursive deletion–contraction formula. See Chapter 9 for much more information about such *matroid invariants*.

[2] Routine proofs are inherently unsatisfying: when completed, they provide very little sense of accomplishment, and if you get stuck, you feel rather foolish.

Table 3.1. *Independent sets and bases in M − a and M/a – see Example 3.3.*

|  | $M$ | $M - a$ | $M/a$ |
|---|---|---|---|
| Independent sets | $\emptyset, a, b, c, d, ab, ad, bc, cd$ | $\emptyset, b, c, d, bc, cd$ | $\emptyset, b, d$ |
| Bases | $ab, ad, bc, cd$ | $bc, cd$ | $b, d$ |

- It's no harder to find the independent sets of $M/a$: the independent sets of $M$ that *do* contain $a$ are the ones we didn't write down before: $\{a, ab, ad\}$. Then removing $a$ from each of these sets gives the independent sets of $M/a$: $\{\emptyset, b, d\}$. (Note that $c$ is a loop in $M/a$.)

Note that deletion and contraction split the bases $\mathcal{B}$ of $M$ into two families precisely the same way we split up the independent sets:

**Proposition 3.4.**  *Let M be a matroid on the ground set E, with $e \in E$ neither an isthmus nor a loop.*

(1) **Deletion** *The bases of $M - e$ are those bases of $M$ that do not contain $e$.*

(2) **Contraction** *The bases of $M/e$ are those bases of $M$ that do contain $e$, with $e$ then removed from each such basis.*

The proof is left for you – see Exercise 4.

### 3.1.1  Drawing $M - e$ and $M/e$

Since $M - e$ and $M/e$ are matroids on the ground set $E - e$, it makes sense to ask how we can modify a picture of the matroid $M$ to produce a picture for $M - e$ or $M/e$. In each case, we will need to eliminate the point $e$ from the picture for $M$; deletion will be very easy, while contraction will involve the geometric operation of *projection*.

**Example 3.5.** Let $M = F_7$ be the Fano plane (Figure 3.2), and let's

*Drawing the deletion*   delete the point $g$. Then a picture for $F_7 - g$ is obtained from the picture for $F_7$ by simply erasing $g$ (and all three-point lines $g$ was on). See Figure 3.3.

We don't want to make too much of a fuss about why this works – it's so easy that an extensive discussion just clouds the issue. It should be clear that a set will be independent in our picture of $F_7 - g$ precisely when the same set is independent in $F_7$, and that's all we need.

The matroid $F_7$ possesses an unusual amount of symmetry[3] – every element "looks the same." (A more precise statement is: "The

---

[3] What is the "usual" amount of symmetry?

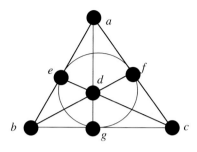

Figure 3.2. The Fano plane $F_7$.

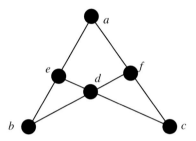

Figure 3.3. $F_7 - g$. Deletion of $g$ from the Fano plane $F_7$.

automorphism group of $F_7$ is transitive.") Thus, it should be no surprise that $F_7 - g$ is the same as (more precisely, *isomorphic to*) $F_7 - b$, for example.

The next example shows you how to draw $M/e$.

**Example 3.6.** Let $M = F_7$ be the Fano matroid again, and choose the element $a$ from $M$. Then recall the independent sets of the contraction $M/a$ are simply those independent sets that *do* contain $a$, with $a$ then removed from each such set. How can we get a picture for $F_7/a$ from a picture of $F_7$? This is harder than it was for deletion, and it involves a thought experiment.

Drawing the contraction

- Place your eye at the point to be contracted.
- Construct a "screen" of dimension one less than the original matroid picture, and position your screen so that each line in the *pencil of lines*[4] through $a$ meets the screen.
- Now use the pencil of lines through $a$ to project the rest of the points of the matroid onto your screen.

See Figure 3.4 for a depiction of this procedure, and see Figure 3.5 for a drawing of $F_7/a$. In this case, our screen is a matroid of rank 2 – a line.

In this example, it's easy to line things up so that projecting from the point $a$ onto a line is a snap. What if we want to contract a different

---

[4] *Pencil of lines* is the correct term for a group of lines, as is a *pride of lions*, a *murder of crows* or an *exaltation of larks*. James Lipton's book *An Exaltation of Larks* is an excellent resource for these terms [22].

Figure 3.4. Contract the element *a* by projecting onto a line.

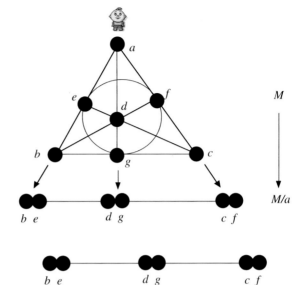

Figure 3.5. $F_7/a$. Contraction of *a* from the Fano plane $F_7$.

Figure 3.6. $F_7/e$. Contraction of *e* from the Fano plane $F_7$.

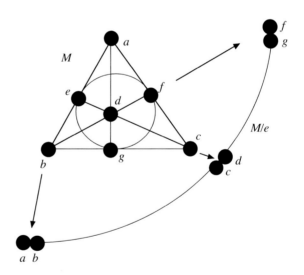

point, say *e*? We can still use the projection idea, but we may wish to bend the flat we project onto.[5] In Figure 3.6, we show how to contract *e*. Our pencil of lines through *e* still meets our projecting line, but we've curved the line to make it easier to capture the points of $F_7/e$ and we've made sure the line *efg* projects to the double point *fg* in $M/e$.

As with deletion, the symmetry of the Fano plane tells us that it doesn't matter which point we contract, e.g., $F_7/a$ is isomorphic to $F_7/e$.

---

[5] "Bent flat" is an oxymoron.

Definition 3.1 does not allow for the deletion of an isthmus or the contraction of a loop. Why do we insist on this prohibition? Well, let's talk about the contraction of loops first. If $e$ were in no independent set (that's what it means for $e$ to be a loop), then we would have no independent sets to remove $e$ from when creating $M/e$. So $\emptyset$ would not be independent in the contraction (and, in fact, no set would be independent in the contraction), which violates axiom (I1) from our definition of independent sets in a matroid. For that reason, we won't contract loops.

*Don't contract loops!*

Deletion of isthmuses is not as serious a problem when considering independent sets, but there is a problem when we consider the bases. If we were allowed to delete isthmuses, what would happen?[6] First, if $e$ is an isthmus of $M$, then $e$ is in every basis of $M$. Then we would find that $M - e$ had no bases (see Proposition 3.4), which is not allowed (since every matroid has at least one basis).

*Don't delete isthmuses!*

We should mention that some authors work around this prohibition by *defining* $M - e$ for an isthmus to be the contraction $M/e$. A similar definition allows you to "contract" loops, but we will never delete isthmuses or contract loops in this text.

Note also that *contraction reduces rank*:

$$r(M/e) = r(M) - 1.$$

*$r(M/e) = r(M) - 1$*

It's easy to prove this: the rank of the matroid is the number of elements in a basis. But, from Proposition 3.4(2), a basis for $M/e$ is formed by taking a basis for $M$ that contains $e$, then removing $e$ from that basis. Thus, the number of elements in a basis has dropped by 1 in moving from $M$ to $M/e$, so $r(M/e) = r(M) - 1$.

Can deletion reduce rank? No: if $r(M - e) < r(M)$, then it must be that $e$ was in every maximum-size independent set, i.e., $e$ is in every basis. But this means $e$ is an isthmus. This is one more reason to avoid deleting isthmuses.

*$r(M - e) = r(M)$*

We just proved the next proposition:

**Proposition 3.7.** *Let M be a matroid.*

(1) *If $e$ is not an isthmus, then $r(M - e) = r(M)$.*
(2) *If $e$ is not a loop, then $r(M/e) = r(M) - 1$.*

### 3.1.2 Commutativity of deletion and contraction

Does the order of these operations matter? Suppose Katie is given a matroid $M$ and is told to first contract the element $a$, then delete the

---

[6] Honestly? Not much.

element $b$, while her friend Jorge is told to do these operations in the reverse order: he first deletes $b$ then contracts $a$. Who wins?

Well, no one "wins." Let's try it for the matroid from Example 3.3. The bases of $M$ are $\{ab, ad, bc, cd\}$.

- First contract $a$, then delete $b$: the independent sets of $M/a$ are $\{\emptyset, b, d\}$. Deleting $b$ gives a new matroid $(M/a) - b$ on the ground set $\{c, d\}$ with independent sets $\{\emptyset, d\}$.
- This time, first delete $b$, then contract $a$: the independent sets of $M - b$ are $\{\emptyset, a, c, d, ad, cd\}$. Contracting $a$ now gives a new matroid $(M - b)/a$ on the ground set $\{c, d\}$ with independent sets $\{\emptyset, d\}$.

Katie and Jorge ended up with the same matroid! More succinctly, we can write: $(M/a) - b = (M - b)/a$. In fact, more is true – repeated deletion commutes with itself and with contraction, and repeated contraction. This is made precise in the next proposition.

**Proposition 3.8.**  *Let $a, b \in E$, the ground set of the matroid $M$. Assuming everything is well-defined, we have:*

(1) $(M - a) - b = (M - b) - a$;
(2) $(M/a)/b = (M/b)/a$;
(3) $(M/a) - b = (M - b)/a$.

Before the proof, we offer a brief comment on what we mean by "well-defined." For part (1), we need $(M - a) - b$ and $(M - b) - a$ to make sense. Since we are prohibiting the deletion of isthmuses, that means $b$ cannot be an isthmus in $M - a$ and $a$ cannot be an isthmus in $M - b$. For part (2), we require the set $\{a, b\}$ to be independent in $M$, and, for part (3), we need to ensure $a$ is not a loop and $b$ is not an isthmus.

*Proof Proposition 3.8.* We'll now prove one of these and leave the rest for the exercises (Exercise 12). Let's do (2): We need to show that the two matroids $(M/a)/b$ and $(M/b)/a$ are identical. We'll do this by showing that they have the same independent sets.

What are the independent sets of $(M/a)/b$? First, find all the independent sets of $M$ that contain the element $a$, then remove $a$ from each such set. This gives the independent sets in $M/a$. Now, among *these* sets, find all that contain $b$, then remove $b$ from each to get the family of subsets that are independent in $(M/a)/b$.

The net effect of this two-step process is that you found all of the independent sets in $M$ that contained both $a$ and $b$, then removed both $a$ and $b$. (Since $\{a, b\}$ is an independent set in $M$, we know neither $a$ nor $b$ is a loop in $M$, and we also know $a$ is not a loop in $M/b$ and $b$ is not a loop in $M/a$.) But this statement does not depend on the order you found the points: you could have first found the independent sets of $M$ containing $b$, then, among those, found the ones contain-ing $a$. These are the independent sets of $(M/b)/a$, which is what we wanted. □

Note that in our proof of (2), we could have just found all independent sets in $M$ containing the subset $\{a, b\}$ all at once, then remove both $a$ and $b$ from each of them simultaneously. This justifies writing $M/\{a, b\}$ for repeated contractions, or, more compactly, $M/ab$. Similarly, we'll write $M/a - b$ and $M - ab$ without worrying about parentheses or the order of operations, since the order doesn't matter.

We now know what the independent sets and bases for $M - e$ and $M/e$ are. What about circuits? What about the rank function, hyper-planes, the closure operator, and so on? For circuits, for example, we would like short, pithy[7] descriptions of the circuits of $M - e$ and $M/e$ in terms of the circuits of $M$.

**Proposition 3.9.** *Let $M$ be a matroid and $e$ an element that is neither an isthmus nor a loop. Then*

(1) **Circuits**
    (a) **Deletion** *$C$ is a circuit of $M - e$ if and only if $e \notin C$ and $C$ is a circuit of $M$.*
    (b) **Contraction** *$C$ is a circuit of $M/e$ if and only if*
        (i) *$C \cup \{e\}$ is a circuit of $M$, or*
        (ii) *$C$ is a circuit of $M$ and $C \cup \{e\}$ contains no circuits except $C$.*
(2) **Rank function** *Let $A \subseteq E$ with $e \notin A$. Then*
    (a) **Deletion** *$r_{M-e}(A) = r_M(A)$.*
    (b) **Contraction** *$r_{M/e}(A) = r_M(A \cup e) - 1$.*

We leave the proofs of these to you (Exercise 13), along with descrip-tions of the flats, hyperplanes and geometric lattice of $M - e$ and $M/e$. We point out that part (2) of Proposition 3.9 generalizes Proposition 3.7.

As a final example, we give a final example.

**Example 3.10.** Let $M$ be the rank 4 matroid at the top of Figure 3.7. Let's draw $M/x$ for various points $x$. Note that, up to symmetry, there are only two different candidates for the point $x$ to consider:

(1) $x = a$. Then the three-point line $abc$ becomes a double point $bc$ in the contraction $M/a$. Further, the plane determined by $bcef$ projects to a line, and so does the other non-trivial plane containing $a$, i.e., $bcdg$.
(2) $x = f$. Since $f$ is on no three-point lines, the contraction $M/f$ will have no double points. The non-trivial planes containing $f$

---

[7] The more pith, the better.

Figure 3.7. Contraction in a
rank 4 matroid.

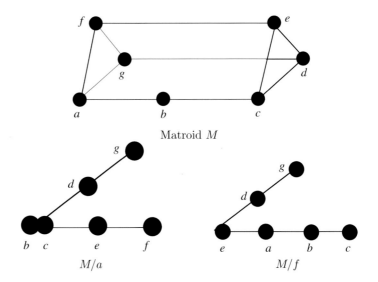

Figure 3.7. Contraction in a rank 4 matroid.

Matroid $M$

$M/a$          $M/f$

are *abcef* and *defg*. These project to a pair of intersecting lines;
the four-point line *abce* intersecting the three-point line *deg* at the
common point *e*.

$M/a$ and $M/f$ are all shown in Figure 3.7.

It's worth pointing out that *a* and *b* are interchangeable in the matroid:
swapping *a* and *b* does not change the collection of bases of $M$ (or
independent sets, or circuits, etc.). See Exercise 24.

## 3.2 Deletion and contraction in graphs and representable matroids

Much of matroid theory is motivated by graph theory. Recall that if $G$
is a graph, the *cycle matroid* $M(G)$ is defined on the edge-set $E$, and a
subset of edges is independent precisely when it contains no cycles in
the graph $G$. Now suppose you are given a graph $G$ and its cycle matroid
$M(G)$. If $e$ is an edge of the graph (and so a point in the matroid $M(G)$),
we can form the new matroids $M(G) - e$ and $M(G)/e$. This set-up begs
the following two questions:

- Are the deletion $M(G) - e$ and contraction $M(G)/e$ graphic matroids,
  i.e., are there graphs $G_d$ and $G_c$ with $M(G_d) = M(G) - e$ and
  $M(G_c) = M(G)/e$?
- If so, then how can we create graphs $G_d$ and $G_c$ whose cycle matroids
  are $M(G) - e$ and $M(G)/e$, respectively?

This is really easy, and it's really important. That's a good thing. The
answer to the first question is yes (we wouldn't have asked the second
question if it weren't). For the second question, here is the procedure for
deleting and contracting edges in a graph.

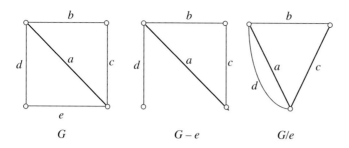

Figure 3.8. Deletion and contraction in a graph.

---

### Deletion and contraction in graphs

(1) **Deletion** Let $G - e$ be the graph obtained from $G$ by erasing the edge $e$. Then the cycle matroid of $G - e$ is the same as the matroid $M(G) - e$.

(2) **Contraction** Let $G/e$ be the graph obtained from $G$ by erasing the edge $e$ and then identifying its endpoints. Then the cycle matroid of $G/e$ is the same as the matroid $M(G)/e$.

---

You can think of the contraction $G/e$ as *shrinking* (or, more reaffirming with respect to the terminology, *contracting*) the edge $e$ to a point.

In Figure 3.8, we draw the graphs $G, G - e$ and $G/e$ for our old friend, the matroid $M$ on $E = \{a, b, c, d, e\}$ with circuits $abc$, $ade$ and $bcde$.

What's missing from our discussion? A proof that the cycle matroids of the two graphs $G - e$ and $G/e$ are the same matroids as $M(G) - e$ and $M(G)/e$, respectively. This appears as Proposition 4.4 in Chapter 4, along with the proof that $M(G)$ is a matroid, i.e., the acyclic subsets of edges satisfy the independent set axioms (Theorem 4.1).[8]

Both deletion and contraction reduce the number of edges of the graph, but contraction also reduces the number of vertices (by 1). If $G$ is a connected graph, then $G/e$ will still be connected. Now the rank $r(M(G))$ of the cycle matroid of a connected graph is $v - 1$ (where $v$ is the number of vertices) since a spanning tree has $v - 1$ edges – this is Theorem 1.17 from Chapter 1. Then we get $r(M(G/e)) = v - 2 = r(M(G)) - 1$, which agrees with Proposition 3.7(2).

### 3.2.1 Representable matroids

Recall that the columns of a matrix define a matroid: given a matrix $A$, let $E$ be the set of columns of $A$. Then the matroid $M[A]$ has

---

[8] One reason for proving fundamental connections between graphs and matroids in Chapter 4 (instead of earlier, when we first encounter the results) is that this allows Chapter 4 to be more self-contained. We hope you can wait.

ground set $E$, and a subset of $E$ is independent precisely when those columns are linearly independent. Matroids that arise in this way are called *representable* matroids, and we study these matroids in more detail in Chapter 6.[9]

We can interpret the matroid operations of deletion and contraction in terms of matrix operations, just as we did for graphs. We'll ask the same two questions we asked before, but you should guess the answer to the first immediately.

- If $M$ is a representable matroid, then are the deletion $M - e$ and contraction $M/e$ also always representable matroids?
- If so, then how can we create matrices $A_d$ and $A_c$ whose column vector matroids are $M - e$ and $M/e$?

As in the graphic case, we wouldn't have asked the first question if it weren't true. As before, deletion is very easy, but this time, contraction is a little bit more involved.

---

### Deletion and contraction in representable matroids

(1) **Deletion**  Suppose $e$ is not an isthmus of the matroid $M[A]$. Let $A - e$ be the matrix obtained from $A$ by erasing the column vector corresponding to $e$ in the matrix $A$. Then $A - e$ represents the matroid $M[A] - e$.

(2) **Contraction**  Suppose $e$ is not a loop of the matroid $M[A]$. Let $B$ be the matrix obtained from $A$ by performing elementary row operations on $A$ so that there is exactly one non-zero entry in the column corresponding to $e$, and row $r$ is the row of $B$ containing the one non-zero entry of that column. Then remove *both* column $c$ and row $r$ from the matrix $B$. This new matrix, which we'll (suggestively) call $A/e$, is the one we want: $A/e$ represents the matroid $M[A]/e$.

---

Example time!

**Example 3.11.**  Here's a lovely matrix:

$$A = \begin{array}{c} \begin{array}{ccccc} a & b & c & d & e \end{array} \\ \left[ \begin{array}{ccccc} 1 & 0 & 0 & 1 & 1 \\ 0 & 1 & 0 & 2 & -1 \\ 0 & 0 & 1 & 0 & 1 \end{array} \right]. \end{array}$$

[9]  As was the case with graphs, we have not yet *proven* this fact: we have not yet shown that subsets of linearly independent vectors satisfy (I1), (I2) and (I3). The proof is given in Theorem 6.1 of Chapter 6.

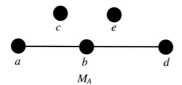

Figure 3.9. Deletion and contraction in a representable matroid.

$M_A$

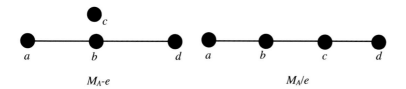

$M_A$-$e$                    $M_A/e$

We'll delete and contract $e$. For deletion, just kill the offending column:

$$A - e = \begin{array}{cccc} a & b & c & d \\ \left[\begin{array}{cccc} 1 & 0 & 0 & 1 \\ 0 & 1 & 0 & 2 \\ 0 & 0 & 1 & 0 \end{array}\right]. \end{array}$$

That was pleasant.

For contraction, we need to row reduce $A$ to produce a new matrix $B$ with exactly one non-zero entry in the last column. This is pretty easy:

$$A = \begin{array}{ccccc} a & b & c & d & e \\ \left[\begin{array}{ccccc} 1 & 0 & 0 & 1 & 1 \\ 0 & 1 & 0 & 2 & -1 \\ 0 & 0 & 1 & 0 & 1 \end{array}\right] \end{array} \rightarrow \begin{array}{ccccc} a & b & c & d & e \\ \left[\begin{array}{ccccc} 1 & 0 & 0 & 1 & 1 \\ 1 & 1 & 0 & 3 & 0 \\ 0 & 0 & 1 & 0 & 1 \end{array}\right] \end{array}$$

$$\rightarrow \begin{array}{ccccc} a & b & c & d & e \\ \left[\begin{array}{ccccc} 1 & 0 & 0 & 1 & \boxed{1} \\ 1 & 1 & 0 & 3 & 0 \\ -1 & 0 & 1 & -1 & 0 \end{array}\right]. \end{array}$$

Then $B$ is the last matrix in this sequence, and the single non-zero entry in column $e$ is boxed. Now cross out the first row of $B$ and column $e$. This gives

$$A/e = \begin{array}{cccc} a & b & c & d \\ \left[\begin{array}{cccc} 1 & 1 & 0 & 3 \\ -1 & 0 & 1 & -1 \end{array}\right]. \end{array}$$

The matroids $M[A]$, $M[A - e]$ and $M[A/e]$ are pictured in Figure 3.9.

The proof that this procedure always works is given in Proposition 6.5 in Chapter 6. It depends on the fact from linear algebra that the column dependences of a matrix are not changed by elementary row operations; see Proposition 6.2.

By the way, how could you tell if a column vector $e$ corresponds to an isthmus or a loop in our matroid? Well, loops are easy: $e$ is a loop in the matroid $M[A]$ if and only if $e$ corresponds to the zero vector.

Loops and isthmuses in representable matroids

For isthmuses in representable matroids, we need to look at our construction of the matrix $B$ in the procedure described above. Here's a how-to guide for recognizing an isthmus from a given matrix:

Isthmus recognition procedure

- Let $B$ be the matrix obtained from $A$ by performing elementary row operations on $A$ so that there is exactly one non-zero entry in column $e$.
- Let row $r$ of $B$ be the row containing the one non-zero entry of $e$.
- Then $e$ is an isthmus of $M[A]$ if and only if the single non-zero entry of $e$ is also the only non-zero entry in row $r$.

Proofs are deferred to Exercise 9 of Chapter 6. Finally, note that this procedure fails when the column corresponding to $e$ is the zero vector. This isn't a problem for us, since such an $e$ must be a loop, but it does give us another reason to avoid contracting loops.

## 3.3  Duality in matroids

One of the most important properties matroids enjoy[10] is a well-developed theory of duality. Duality is an important and ubiquitous concept throughout mathematics; for instance, vector spaces have duals, and so do *planar* graphs. The dual of a matroid $M$, written $M^*$, is defined on the same ground set as $M$, and the bases of $M^*$ are just the complements of the bases of $M$.

### 3.3.1  Motivation and examples

Planar graph

The motivation for duality comes from graph theory. First, a graph is *planar* if it can be drawn in the plane so that no edges cross each other. For example, the complete graph $K_4$ in Figure 3.10 is a planar graph. We've drawn $K_4$ twice, once "badly" (i.e., with an edge crossing), and once without any edge crossings. We emphasize that the property of a graph being planar is a property of the *graph*, not a property of the particular drawing of that graph (what a topologist or graph theorist would call an *embedding*).

Now suppose $G$ is a planar graph that is drawn in the plane with no edge crossings. Then the *dual* graph $G^*$ is formed by placing one vertex somewhere in the interior of each *region* of $G$ and joining two vertices with an edge $e^*$ precisely when the corresponding regions in $G$ share

---

[10] It's unclear how a matroid can "enjoy" anything, but we'll leave the philosophical implications of this to the reader.

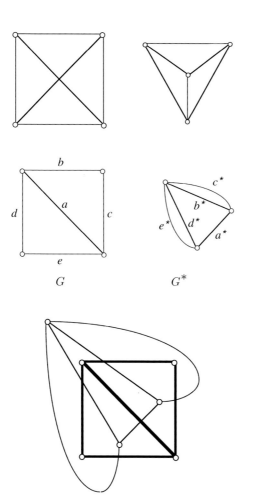

Figure 3.10. The complete graph $K_4$ is planar, regardless of how it is drawn: Left: $K_4$ drawn with an edge crossing. Right: $K_4$ without any edge crossings.

Figure 3.11. A graph $G$ and its dual $G^*$.

Figure 3.12. The connection between a planar graph and its dual.

the edge $e$. See Figure 3.11 for a graph and its dual, and see Figure 3.12 for a hint about how these two graphs are related. It is always possible to draw $G$ and $G^*$ simultaneously so that each edge $e$ in $G$ crosses exactly one edge of $G^*$, namely $e^*$.

For the pair of dual graphs in Figure 3.11, you can see that the edges $\{a, c, d\}$ form a spanning tree in the graph $G$, while the complementary edges $\{b^*, e^*\}$ form a spanning tree in the dual graph $G^*$. This fact about the complements[11] of spanning trees for planar graphs motivates our definition of matroid duality.

**Definition 3.12.** Let $M$ be a matroid on the ground set $E$. Then the dual matroid $M^*$ is a matroid on the same ground set $E$ so that

Dual matroid

$$\mathcal{B}(M^*) = \{E - B : B \in \mathcal{B}(M)\}.$$

---

[11] This is completely different from the operation of *graph complementation*, where the complement $G^C$ of a graph $G$ is formed by joining two vertices in $G^C$ if and only if those vertices are not joined in $G$.

Figure 3.13. Our favorite
matroid $M$ and its dual $M^*$.

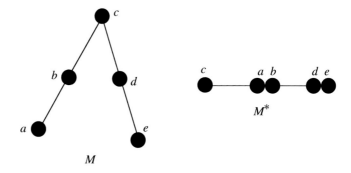

**Example 3.13.** Let $M$ be the matroid on the ground set $E = \{a, b, c, d, e\}$ whose circuits are $\{a, b, c\}, \{c, d, e\}$, and $\{a, b, d, e\}$. (This is the graphic matroid $M(G)$ for the graph $G$ in Figure 3.11.) We list the bases of $M$ alongside the bases for $M^*$:

| Bases of $M$ | | Bases of $M^*$ |
|---|---|---|
| $\{a, b, d\}$ | $\leftrightarrow$ | $\{c, e\}$ |
| $\{a, b, e\}$ | $\leftrightarrow$ | $\{c, d\}$ |
| $\{a, c, d\}$ | $\leftrightarrow$ | $\{b, e\}$ |
| $\{a, c, e\}$ | $\leftrightarrow$ | $\{b, d\}$ |
| $\{a, d, e\}$ | $\leftrightarrow$ | $\{b, c\}$ |
| $\{b, c, d\}$ | $\leftrightarrow$ | $\{a, e\}$ |
| $\{b, c, e\}$ | $\leftrightarrow$ | $\{a, d\}$ |
| $\{b, d, e\}$ | $\leftrightarrow$ | $\{a, c\}$ |

More concisely, since every subset of size 3 *except* $\{a, b, c\}$ and $\{c, d, e\}$ is a basis of $M$, every subset of size 2 *except* $\{d, e\}$ and $\{a, b\}$ will be a basis for $M^*$. Thus, $M^*$ is a rank two matroid on the same ground set $E = \{a, b, c, d, e\}$ with two multiple points $\{a, b\}$ and $\{d, e\}$. We draw the dual matroid $M^*$ in Figure 3.13.

### 3.3.2 Fundamental results on dual matroids

Why does our definition of duality work? More precisely, why do the complements of the bases of a matroid also satisfy the basis axioms from Chapter 2? Proving this always works is a fundamental result in matroid theory (Theorem 3.15).

To construct a proof, we have some freedom in choosing the axiom system for bases. Exercise 14 in Chapter 2 introduced *strong basis exchange* (B3') for the bases of a matroid. Then we will need the following cryptomorphic characterization of the bases of a matroid, proven in that exercise.[12]

**Lemma 3.14.** *Suppose a family $\mathcal{B}$ satisfies:*

(B1) $\mathcal{B} \neq \emptyset$.
(B2′) *If $B_1, B_2 \in \mathcal{B}$, then $|B_1| = |B_2|$.*
(B3′) *If $B_1, B_2 \in \mathcal{B}$ and $x \in B_1 - B_2$, then there is an element $y \in B_2 - B_1$ so that both $B_1 - x \cup \{y\} \in \mathcal{B}$ and $B_2 - y \cup \{x\} \in \mathcal{B}$.*   Strong basis exchange

*Then $\mathcal{B}$ is the family of bases of a matroid.*

We now prove the dual matroid $M^*$ is really a matroid.

**Theorem 3.15.** *Let $M$ be a matroid with bases $\mathcal{B}$. Then $\{E - B : B \in \mathcal{B}\}$ are the basis of a matroid.*

*Proof Theorem 3.15.* Let $\mathcal{B}^*$ for the set of all basis complements, i.e., $\mathcal{B}^* = \{E - B : B \in \mathcal{B}\}$. We will show $\mathcal{B}^*$ satisfies (B1), (B2′) and (B3′), which will complete the proof that $M^*$ is a matroid, by Lemma 3.14.

Now (B1) is easy to check: since $M$ is a matroid, there is a basis $B \in \mathcal{B}$, so there is a basis complement $E - B \in \mathcal{B}^*$. Axiom (B2′) is just as easy: every basis has $r(M)$ elements, so every basis complement has $|E| - r(M)$ elements, i.e., $\mathcal{B}^*$ satisfies (B2′) because $\mathcal{B}$ does.

As usual, the only non-trivial thing to check is the last axiom. We know $\mathcal{B}$ satisfies strong exchange (B3′); we must show $\mathcal{B}^*$ also satisfies strong exchange. So let $B_1^c = E - B_1$ and $B_2^c = E - B_2$ be two basis complements, where $B_1, B_2 \in \mathcal{B}$. Let $x \in B_1^c$; we must find $y \in B_2^c$ so that both $B_1^c - x \cup y$ and $B_2^c - y \cup x$ are in $\mathcal{B}^*$.

But this is pretty easy, too. The idea is to simply take complements, apply strong basis exchange in $M$, then take complements again. Now $x \in B_1^c$; if $x \in B_2^c$, too, then we can set $y = x$ and be done. So assume $x \notin B_2^c$. Translating this back to bases of $M$, we have $x \notin B_1$ and $x \in B_2$.

Now applying strong basis exchange (B3′) to the matroid $M$, there is an element $y \in B_1$ so that both $B_1 \cup x - y$ and $B_2 \cup y - x$ are bases of $M$. Taking complements again, we have $(B_1 \cup x - y)^c$ and $(B_2 \cup y - x)^c$ are both in $\mathcal{B}^*$. But $(B_1 \cup x - y)^c = B_1^c \cup y - x$ and $(B_2 \cup y - x)^c = B_2^* \cup x - y$, which is what we wanted.   □

Why did we need strong exchange (B3′) in the proof of Theorem 3.15? The problem is that if $x \in B_1^c - B_2^c$, after taking complements we get $x \in B_2 - B_1$. Using (weak) basis exchange (B3) in $M$ then produces $y \in B_1 - B_2$, as in the above proof, but we are only guaranteed that $B_2 \cup y - x$ is a basis of $M$. Taking complements then gives us $B_2^c \cup x - y \in \mathcal{B}^*$, but we need $B_1^c \cup y - x$ to be in $\mathcal{B}^*$ to prove property (B3) for $M^*$.

[12] This is an excellent time to do Exercise 14 from Chapter 2.

Moral: it's easier to prove a stronger property (B3′) for the family of basis complements $\mathcal{B}^*$ than it is for the weaker (B3) (under the correspondingly stronger assumption that (B3′) holds for the family $\mathcal{B}$).

Set complementation is an especially simple operation. It should be no surprise that this gives rise to some simple proofs. Since $r(M)$ is the size of a basis of $M$, the next result is obvious.

**Proposition 3.16.**  $r(M) + r(M^*) = |E|$.

Since the complement of the complement of a set is the original set, the next result is just as obvious.[13]

**Proposition 3.17.**  $(M^*)^* = M$.

We can also get descriptions of independent sets, circuits, spanning sets and hyperplanes of $M^*$ in terms of subsets of $M$.

**Proposition 3.18.**  *Let $M$ be a matroid on the ground set $E$. Then the bases, independent sets, spanning sets, circuits and hyperplanes of the dual matroid $M^*$ are determined as follows:*

| $M^*$ | $\Leftrightarrow$ | $M$ |
|---|---|---|
| *B is a basis* | $\Leftrightarrow$ | *E − B is a basis* |
| *I is independent* | $\Leftrightarrow$ | *E − I is spanning* |
| *S is spanning* | $\Leftrightarrow$ | *E − S is independent* |
| *C is a circuit* | $\Leftrightarrow$ | *E − C is a hyperplane* |
| *H is a hyperplane* | $\Leftrightarrow$ | *E − H is a circuit* |

Most of these relations are easy to check, but it's entirely possible you're saying "What's the deal with circuits and hyperplanes?"[14] We prove the circuit–hyperplane correspondence now.

*Proof Proposition 3.18.* We omit the proofs connecting independent sets in $M$ and spanning sets in $M^*$ (these follow immediately from the definition of matroid duality). The circuit–hyperplane story is a nice exercise in complementation. Recall the complement $E − H$ of the hyperplane $H$ is called a *cocircuit* in the matroid. Then:

---

[13]  It might be interesting to generate a debate about which of these two results, Propositions 3.16 or 3.17, is more obvious. The second of these results will be more useful, though.

[14]  We wish we had several million dollars for every time we've heard that question. We would then be several million dollars richer.

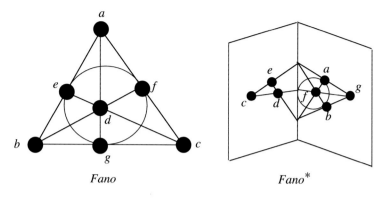

Figure 3.14. The Fano plane $F_7$ and its dual $F_7^*$.

*Fano*                    *Fano*\*

| | | |
|---|---|---|
| $C$ is a circuit in $M$ | $\Leftrightarrow$ | $C$ is a minimal dependent set in $M$ |
| | $\Leftrightarrow$ | $C$ is minimal with respect to not being *contained in* a basis $B$ in $M$ |
| | $\Leftrightarrow$ | $E - C$ is maximal with respect to not *containing* a basis complement |
| | $\Leftrightarrow$ | $E - C$ is maximal with respect to not *containing* a basis $B^*$ of $M^*$ |
| | $\Leftrightarrow$ | $E - C$ is a hyperplane of $M^*$ |
| | $\Leftrightarrow$ | $C$ is a cocircuit of $M^*$. |

☐

Our next result will relate the rank function $r^*$ in the dual matroid $M^*$ to the rank function $r$ of $M$. In order to develop a formula for $r^*(A)$ of a subset $A \subseteq E$, we first look at an example.

**Example 3.19.** Let $M$ be the Fano matroid $F_7$ on $E = \{a, b, c, d, e, f, g\}$ (see Figure 3.14). Since $r(M) = 3$, we know $r(M^*) = |E| - r(M) = 7 - 3 = 4$.

Let's compute the rank of $A = \{a, b, c, d\}$ in the dual matroid $M^*$. To do this, evidently we need to find a basis of $M^*$ whose intersection with $A$ is as large as possible. If $B'$ is this lucky[15] basis of $M^*$, then $B = E - B'$ will be a basis of $M$ whose intersection with $E - A$ is as large as possible, too.

Now $E - A = \{e, f, g\}$, so one candidate for $B$ is $\{a, e, f\}$ (there are many others that would also work). This gives $B' = \{b, c, d, g\}$. Then $A \cap B' = \{b, c, d\}$, and this is as large as possible, so $r^*(A) = 3$. This is depicted in the right-hand drawing in Figure 3.14 in the following way. Although the points $a, b, c, d$ appear to span three-dimensional space

---

[15] This seems to be a fortuitous opportunity for a joke. We leave this joke to the reader, however.

Figure 3.15. Computing dual rank: $r^*(A) = r(E - A) + |A| - r(M)$.

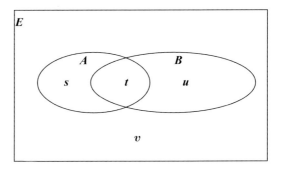

in the dual matroid, the "circular" line containing $a$ and $b$ intersects the line through $c$ and $d$, so we really do have $r^*(A) = 3$.

Sometimes math books like to mix the order of exposition up a bit. Instead of slavishly adhering to a Definition, Theorem, Proof, Example format, they might try to sneak in a proof before a theorem. Pay attention!

Our immediate goal is to get a general formula for $r^*(A)$ in terms of the rank function $r$ of the original matroid. Let $\mathcal{B}$ and $\mathcal{B}^*$ be the collections of bases for $M$ and $M^*$, respectively. Now $r^*(A)$ is the size of the largest independent set of $M^*$ residing inside of $A$. More consisely, we have

$$r^*(A) = \max\{|A \cap B| \mid B \in \mathcal{B}^*\}.$$

So choose $B \in \mathcal{B}^*$ with $|A \cap B|$ maximal. Complementing, that means that $E - A$ also meets the complementary basis $B' = E - B$ of $M$ in a maximum-size subset: $|(E - A) \cap B'|$ is also maximal among bases of $M$ that meet $E - A$. Thus $|(E - A) \cap B'| = r(E - A)$.

In Figure 3.15, we partition the set $E$ into four pieces: $A - B$, $A \cap B$, $B - A$, and $E - (A \cup B) = (E - A) \cap B'$. We label the sizes of these four sets $s$, $t$, $u$ and $v$, respectively. By our choice of $B$, we know $t = r^*(A)$ and $v = r(E - A)$. But we also have $u + v = |E| - |A|$ and $t + u = |B| = |E| - r(M)$. Eliminating $u$ from these last two equations gives $t = v - r(M) + |A|$. Replacing $t = r^*(A)$ and $v = r(E - A)$ finally gives the following theorem.

Dual rank $r^*(A)$

**Theorem 3.20.** *Let $M$ be a matroid on the ground set $E$ with rank function $r$, and let $r^*(A)$ be the rank of the set $A$ in the dual matroid $M^*$. Then*

$$r^*(A) = r(E - A) + |A| - r(M).$$

Let's use the formula to check our computation of $r^*(A)$ for $A = \{a, b, c, d\}$ in Example 3.19. Since $E - A = \{e, f, g\}$, we have $r(E - A) = 2$ in $M = F_7$. Further, $|A| = 4$ and $r(M) = 3$. Then the formula gives $r^*(A) = 2 + 4 - 3 = 3$, which agrees[16] with our previous computation.

Combining duality with the operations of deletion and contraction gives two very attractive formulas:

**Proposition 3.21.** *Let M be a matroid on the ground set E and suppose $e \in E$.*

(1) *If e is not an isthmus, then $(M - e)^* = M^*/e$.*
(2) *If e is not a loop, then $(M/e)^* = M^* - e$.*

*Proof Proposition 3.21.* (1) Let's keep track of the bases of $(M - e)^*$ and $M^*/e$. Assume $B$ is a basis for $(M - e)^*$. Then $B \subseteq (E - e)$ and $(E - e) - B$ is a basis for $M - e$. This means $(E - e) - B$ is also a basis for $M$, so its complement (in $E$) is a basis for $M^*$: $E - [(E - e) - B]$ is a basis for $M^*$. But $E - [(E - e) - B] = B \cup e$. Thus, $B \cup e$ is a basis for $M^*$, so $B$ is a basis for $M^*/e$.

We just showed that if $B$ is a basis for $(M - e)^*$, then $B$ is also a basis for $M^*/e$. To complete the proof, we need to show the reverse: if $B$ is a basis for $M^*/e$, then $B$ is also a basis for $(M - e)^*$. But it's easy to see all of the above implications can be reversed, so $(M - e)^* = M^*/e$.

(2) The proof of (2) comes along (more or less) for free from (1): to show $M^* - e = (M/e)^*$, let $N = M^*$, use the fact that $(N - e)^* = N^*/e$ (which we just proved) and simply take the duals:

$$((N - e)^*)^* = (N^*/e)^*.$$

Now replace $N$ by $M^*$ and use Proposition 3.17 twice:

$$M^* - e = (M/e)^*. \qquad \square$$

Some authors[17] sometimes write $r(E)$ for $r(M)$ in these formulas; they are (obviously) identical. Then we could have expressed the rank of the dual matroid $M^*$ as $r(M^*) = r(E - A) + |A| - r(E)$.

We conclude this discussion of duality with an easy result connecting isthmuses, loops and duality.

**Proposition 3.22.** *e is a loop of M if and only if e is an isthmus of $M^*$.*

*Proof Proposition 3.22.* $e$ is a loop of $M$ if and only if $e$ is in no basis of $M$. This is true if and only if $e$ is in every basis complement, i.e., $e$ is an isthmus of $M^*$. $\qquad \square$

Of course, we could also say $e$ is an isthmus of $M$ if and only if $e$ is a loop of $M^*$. For this reason, isthmuses are also called *coloops*.[18]

Coloop = isthmus

---

[16] Well, it better.
[17] Including the authors of this text.
[18] Are loops also called *co-isthmuses*? Don't be silly.

Figure 3.16. The icosahedron and the dodecahedron; dual Platonic solids that are dual planar graphs.

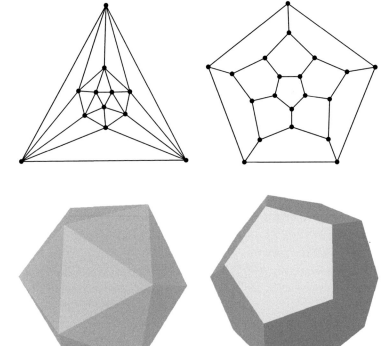

Figure 3.17. Two dual Platonic solids: the icosahedron and the dodecahedron.

## 3.4 Duality in graphic and representable matroids

We specialize the general discussion on matroid duality to concentrate on our two main matroid classes, graphic and representable matroids. First, we return to graphs, which motivated much of our general treatment.

### 3.4.1 Duals of graphic matroids

When is the dual of a graphic matroid graphic? Rephrasing this, if $G$ is a graph, when can we find a graph $G^*$ so that $M(G^*) = M(G)^*$? The surprising[19] answer depends on whether or not the graph is *planar*.

We begin with two examples.

**Example 3.23.** Consider the planar graphs $I$ and $D$ in Figure 3.16. $I$ is a projection of the icosahedron into the plane, and $D$ is a projection of its dual Platonic solid, the dodecahedron. More familiar projections of these solids are shown in Figure 3.17. Graph duality tells us we can create an icosahedron by placing a point at the center of each of the 12 faces[20] of a dodecahedron, then joining two points with an edge if the corresponding faces of the dodecahedron share an edge.

---

[19] It's not that surprising, but this is a useful plot device we employ to build suspense.

[20] When drawn in the plane, the *faces* of the solid are *regions* in the plane.

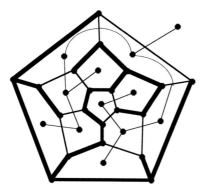

Figure 3.18. A spanning tree for the doecahedron and its complement, a spanning tree for the icosahedron.

From the matroid point of view, since $I$ and $D$ each have 30 edges, $M(I)$ and $M(D)$ each have 30 points. Furthermore, since the icosahedron has 12 vertices and the dodecahedron has 20 vertices, we get $r(M(I)) = 11$ and $r(M(D)) = 19$. These are dual matroids: $M(I)^* = M(D)$. Thus,

> The complement of any spanning tree of the dodecahedron is a spanning tree for the icosahedron.[21]

See Figure 3.18 for a spanning tree and its complement. To see why this is true, there are just two things to check:

- The complement of a spanning tree $T$ is connected in the dual.
- The complement of a spanning tree $T$ is acyclic in the dual.

The first fact follows because $T$ is acyclic, and the second follows from the fact that $T$ is spanning.[22]

One consequence of these facts is Euler's famous formula involving the number of vertices $v$, edges $e$ and faces (or planar regions) $f$ of a connected, planar graph.

**Theorem 3.24.** *[Euler's polyhedral formula] Let G be a connected planar graph with v vertices, e edges and f faces, including the unbounded face. Then*

$$v - e + f = 2.$$

*Proof Theorem 3.24.* We give a very short, matroid-inspired proof. A spanning tree for $G$ has $v - 1$ edges (see Theorem 1.15 in Chapter 1). The $e - v + 1$ complementary edges form a spanning tree for the dual graph, so $r(M(G)^*) = e - v + 1$. But there are $f$ vertices in the dual, so $r(M(G)^*) = f - 1$. The formula follows immediately. ☐

Euler discovered this formula around 1750, or about 180 years before matroids were invented.

---

[21] As we have previously commented, this is *set theoretic complementation*, not the *graph complement*.
[22] There are some technicalities to check, and a careful proof makes use of the *Jordan Curve Theorem*.

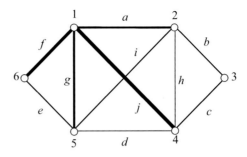

Figure 3.19. A vertex cut-set separating vertex 1 is the cocircuit $\{a, f, g, j\}$ in the cycle matroid $M(G)$.

It's easy to identify the circuits of the cycle matroid $M(G)$ directly from the graphs: $C$ is a circuit of $M(G)$ if and only if $C$ is a cycle in the graph $G$.[23]

How can we identify the *cocircuits* of the cycle matroid $M(G)$ from the graph $G$? Well, cocircuits are hyperplane complements. For graphs, this means cocircuits are cut-sets, i.e., minimal subsets of edges in a connected graph whose removal disconnects the graph. If we take all the edges in a connected graph $G$ that are incident to a given vertex (and that vertex is not a *cut-vertex*), then we get a cocircuit in $M(G)$ – the complement of a maximal flat in $M(G)$ (since adding another edge gives a spanning set).[24] See Figure 3.19.

*Cocircuits are cut-sets*

*Cut-vertex*

When $G$ is not planar, the dual matroid $M(G)^*$ still exists, but $M(G)^*$ won't be graphic. That's the point of the next example.

**Example 3.25.** Let $K_5$ be the complete graph on five vertices. This graph is not planar – there is no way to redraw it so that we eliminate all edge crossings. (A short proof of this based on Theorem 3.24 is given in Proposition 3.26 below.) Since $K_5$ has five vertices, a spanning tree for $K_5$ includes four edges. Thus, $M(K_5)$ is a rank 4 matroid on 10 points. The graph is pictured in Figure 3.20, and the matroid $M(K_5)$ is pictured in Figure 3.21. The geometric picture is a three-dimensional Desargues configuration, which we'll see again in Chapter 6.

Now the dual matroid $M(K_5)^*$ also has 10 points, but has rank $10 - 4 = 6$. Drawing a picture of a rank 6 matroid would involve projection from five-dimensional space. In theory, one could place a basis of size 6 at the vertices of a simplex in $\mathbb{R}^5$, then figure out where to place the remaining four points on the flats determined by those six points. We won't attempt this here, though.

But, if $M(K_5)^*$ were graphic, we would be able to represent this matroid graphically. So, our question is simply:

Is $M(K_5)^*$ graphic?

---

[23] This is why $M(G)$ is called the *cycle matroid* on $G$.

[24] A more detailed treatment of flats, hyperplanes and cocircuits in graphs appears in Example 4.2 in Chapter 4. The vertex cocircuits generate all cocircuits – see Theorem 4.13 in Chapter 4.

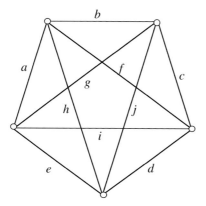

Figure 3.20. The complete graph $K_5$.

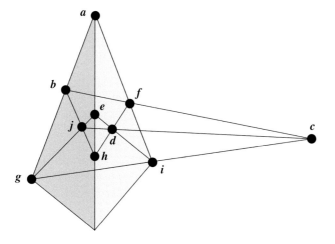

Figure 3.21. The rank 4 matroid $M(K_5)$ is the three-dimensional version of Desargues configuration.

The answer is no. This is a classic proof by contradiction – we suppose $M(K_5)^*$ is graphic, and hope something goes wrong. Now if $M(K_5)^*$ were graphic, the corresponding graph would have 10 edges, and, since a spanning tree would have six edges, it would have seven vertices.[25]

It's probably easiest to use the circuit–cocircuit correspondence to finish this argument.[26] We identify the cocircuits of $M(K_5)^*$, and then argue the corresponding circuits of $M(K_5)$ cannot exist. By the argument given above, we know the set of edges incident to a vertex is a cocircuit.[27]

Now our supposed graph representing $M(K_5)^*$ has seven vertices and 10 edges, so the average degree of a vertex is 20/7, which is less than 3. That means at least one vertex has degree at most 2. It's clear any vertex of degree 1 in our graph would give us an isthmus in $M(G)^*$,

---

[25] This assumes that if $M$ is a graphic matroid, there is a connected graph $G$ that can be found with $M = M(G)$. See Proposition 4.3 in Chapter 4.

[26] Even if it isn't, it's good practice.

[27] This is true provided that our vertex $v$ is not a cut-vertex. In that case, the edges incident to $v$ can be partitioned into more than one cocircuit.

and so would correspond to a loop in $M(G)$, which is impossible. But if a vertex has degree 2, then the two edges incident to that vertex form a two-element cocircuit in $M(K_5)^*$, so they must form a two-element circuit in $M(K_5)$. But the smallest circuit in $M(K_5)$ has size 3, so we get our contradiction.

As we hinted at prior to the example, the problem with $M(K_5)^*$ failing to be graphic is essentially the same problem that keeps $K_5$ from being planar. If you haven't studied graph theory, here is a quick argument why $K_5$ is not a planar graph.[28]

$K_5$ is not planar

**Proposition 3.26.** $K_5$ *is not a planar graph.*

*Proof Proposition 3.26.* From Euler's formula (Theorem 3.24), we know that a connected planar graph $G$ with $v$ vertices, $e$ edges and $f$ faces satisfies $v - e + f = 2$. If the graph has no multiple edges, then every cycle in the graph has at least three edges and every edge bounds exactly two regions. Then $3f \leq 2e$, and we get $e \leq 3v - 6$. But $e = 10$ and $v = 5$, so $K_5$ is not planar.    □

The moral of these two examples is that planar graphs are precisely the graphs whose dual matroids are graphic. When $G$ is a planar graph, then the dual graph gives the dual matroid: $M(G^*) = M(G)^*$. This is the point of the next theorem.

**Theorem 3.27.** *Let $G$ be a connected graph. Then the dual matroid $M(G)^*$ is graphic if and only if $G$ is planar. Further, if $G$ is planar, then $M(G^*) = M(G)^*$.*

We omit the proof of Theorem 3.27, but give an outline of the main ideas involved. You are asked to fill in these details in Exercise 29. First, suppose $G$ is planar. Then the dual graph $G^*$ exists.

- Observe that the complements of spanning trees in $G$ are spanning trees of $G^*$, as in Figures 3.22 and 3.23. (We assumed this in our proof of Euler's polyhedral formula – Theorem 3.24.)
- Then $M(G^*) = M(G)^*$, since the bases of $M^*$ are the complements of the bases of $M$.

For the converse, we suppose $G$ is a graph and $M(G)^*$ is graphic.

- Since $M(G)^*$ is graphic, we know $G$ cannot contain a copy of $K_5$ or $K_{3,3}$ as a *topological minor*.[29] This takes a bit of work, but $G \neq K_5$ by Example 3.25 and $G \neq K_{3,3}$ by Exercise 28.
- Now use Kuratowski's Theorem: $G$ is planar if and only if it contains neither $K_5$ nor $K_{3,3}$ as a topological minor.

---

[28]  The existence of this argument does not depend on your previous studies, however.

[29]  A subgraph $H$ is a topological minor of a graph $G$ if you obtain $H$ by removing edges of $G$ and also "removing" vertices of degree 2 – this latter operation is equivalent to contracting edges having a vertex of degree 2.

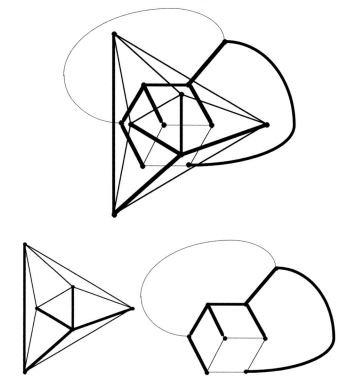

Figure 3.22. The complement of a spanning tree for a cube is a spanning tree for the dual octahedron.

Figure 3.23. The complementary spanning trees of a cube and an octahedron.

The duals of graphic matroids are called *cographic* matroids. If $G$ denotes the class of all graphic matroids, then we write $G^*$ for the class of cographic matroids:

Cographic matroids

**Definition 3.28.** The class of cographic matroids is defined by

$$G^* = \{M \mid M^* \in G\}.$$

Thus, a matroid $M$ is cographic if and only if $M^*$ is graphic. So, for instance, $M(K_5)^*$ is cographic since its dual, $(M(K_5)^*)^* = M(K_5)$ is graphic. Then we can restate Theorem 3.27 in terms of the classes $G$ and $G^*$.

**Corollary 3.29.** *Let $G$ be the class of all graphic matroids, $G^*$ the class of cographic matroids, and $P$ the class of all cycle matroids of planar graphs. Then*

$$G \cap G^* = P.$$

We conclude this brief treatment of duality in graphs with a provocative statement:

Matroids were invented to define duals of non-planar graphs.

Were you provoked?

Figure 3.24. The matroid for
Example 3.30 and its dual.

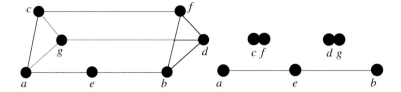

### 3.4.2  Duality in representable matroids

The dual of a graphic matroid is graphic precisely when the given graph
is planar. We now ask a similar question for representable matroids:
given a matrix $A$ and the associated matroid $M[A]$ we get by looking
at column dependences, when can we find a matrix $A^*$ so that the dual
matroid $M[A]^*$ is the same as the column dependences of $A^*$?

**Example 3.30.**  Let $M$ be the rank 4 matroid in Figure 3.24 (this is the
same matroid we saw in Figure 3.7, but we've relabeled the points).
You can check that $M$ is represented by the following matrix $A$:

$$A = \begin{array}{c c c c c c c c} & a & b & c & d & e & f & g \\ & \begin{bmatrix} 1 & 0 & 0 & 0 & 1 & 1 & 1 \\ 0 & 1 & 0 & 0 & 1 & -1 & -1 \\ 0 & 0 & 1 & 0 & 0 & 1 & 0 \\ 0 & 0 & 0 & 1 & 0 & 0 & 1 \end{bmatrix} \end{array}.$$

Our immediate goal is to get a matrix (that we'll call $A^*$) representing
$M^*$. To do this, first note that the set $\{a, b, c, d\}$ is a basis for $M$, so the
complement $\{e, f, g\}$ must be a basis for $M^*$. We'll use a $3 \times 3$ identity
matrix[30] to represent the columns $e$, $f$ and $g$ in $A^*$:

$$A^* = \begin{array}{c c c c c c c} & a & b & c & d & e & f & g \\ & \begin{bmatrix} & & & & 1 & 0 & 0 \\ & & & & 0 & 1 & 0 \\ & & & & 0 & 0 & 1 \end{bmatrix} \end{array}.$$

To find coordinates for columns $a, b, c$ and $d$, we'll choose entries
that make *every row of A orthogonal to every row of $A^*$*. (Recall that
two vectors are *orthogonal* if their dot product is zero.) There's a nice
explanation for why this always works, but don't worry about that yet.[31]
To do this, do the following two steps:

---

[30]  In fact, if $B$ is a basis of a representable matroid, we can always assume the columns
corresponding to $B$ form an identity matrix. This follows from Proposition 6.9 of
Chapter 6.

[31]  We'll worry about this in Chapter 6, when we prove Theorem 6.6.

- Let $D$ be the submatrix formed by the last three columns of $A$:

$$D = \begin{array}{c} \begin{array}{ccc} e & f & g \end{array} \\ \begin{bmatrix} 1 & 1 & 1 \\ 1 & -1 & -1 \\ 0 & 1 & 0 \\ 0 & 0 & 1 \end{bmatrix} \end{array}.$$

- Form $D^T$, the transpose of $D$, and label the columns with the initial basis $\{a, b, c, d\}$ of $M$:

$$D^T = \begin{array}{c} \begin{array}{cccc} a & b & c & d \end{array} \\ \begin{bmatrix} 1 & 1 & 0 & 0 \\ 1 & -1 & 1 & 0 \\ 1 & -1 & 0 & 1 \end{bmatrix} \end{array}.$$

- Now negate each entry in $D^T$ and append these four columns to $e$, $f$ and $g$ to form $A^*$:

$$A^* = \begin{array}{c} \begin{array}{ccccccc} a & b & c & d & e & f & g \end{array} \\ \begin{bmatrix} -1 & -1 & 0 & 0 & 1 & 0 & 0 \\ -1 & 1 & -1 & 0 & 0 & 1 & 0 \\ -1 & 1 & 0 & -1 & 0 & 0 & 1 \end{bmatrix} \end{array}.$$

You can check the rows of $A$ and $A^*$ are orthogonal. For example, row 2 of $A$ is $(0, 1, 0, 0, 1, -1, -1)$, and row 3 of $A^*$ is $(-1, 1, 0, -1, 0, 0, 1)$. Then the dot product is 0:

$$(0, 1, 0, 0, 1, -1, -1) \cdot (-1, 1, 0, -1, 0, 0, 1) = 0.$$

You can check the other 11 combinations[32] if you like. In each case, you should get a dot product of 0. In fact, this is easy to see by the way the matrix $A^*$ was constructed: each dot product $v \cdot w$ will have either zero or two non-zero terms, and these two terms will always cancel each other out.

Finally, multiplying the columns of $A^*$ by non-zero constants doesn't change the column dependences. This means we can represent $M^*$ by a "cleaner" matrix – one with fewer negative entries, if we like (which we do):

$$M^* \text{ is also represented by } \begin{array}{c} \begin{array}{ccccccc} a & b & c & d & e & f & g \end{array} \\ \begin{bmatrix} 1 & -1 & 0 & 0 & 1 & 0 & 0 \\ 1 & 1 & 1 & 0 & 0 & 1 & 0 \\ 1 & 1 & 0 & 1 & 0 & 0 & 1 \end{bmatrix} \end{array}.$$

Note that the points $c$ and $f$ are represented by the same column vector in this matrix. This means these two points form a double point in $M^*$; this makes sense since $E - \{c, f\}$ is a hyperplane in $M$, so $\{c, f\}$

---

[32] Although we wouldn't recommend it.

is a cocircuit of $M$, and therefore a circuit of $M^*$. The same thing is true for the pair of points $d$ and $g$.

We can summarize this procedure in a theorem. The proof is given in Chapter 6.

Duals of representable matroids are representable

**Theorem 6.6.** *The matrix $A = [I_{r \times r} \mid D]$ represents the matroid $M$ precisely when the matrix $A^* = [-D^T \mid I_{(n-r) \times (n-r)}]$ represents the dual matroid $M^*$.*

The moral of the story is simple:

The dual of a graphic matroid is graphic precisely when the graph is planar. The dual of a representable matroid is always representable.

For representable matroids, we have more: Theorem 6.6 gives us a recipe for creating the matrix $A^*$ that represents $M^*$. We'll return to this story in Chapter 6 when we prove Theorem 6.6.

## 3.5  Direct sums and connectivity

Direct sums appear throughout mathematics; for instance, you may have seen the direct sum of two groups, or vector spaces, or modules. One nice feature of direct sums is that they are easy, important and ubiquitous.[33]

### 3.5.1  Direct sums

Given two matroids $M_1$ and $M_2$ on *disjoint* ground sets $E_1$ and $E_2$, it's easy to put them together to get a new matroid, the *direct sum $M_1 \oplus M_2$* on the ground set $E_1 \cup E_2$.

Direct sum

**Definition 3.31.** Let $M_1$ and $M_2$ be matroids on disjoint ground sets $E_1$ and $E_2$, respectively. Define the *direct sum $M_1 \oplus M_2$* to be the matroid on the ground set $E = E_1 \cup E_2$ with independent sets $I_1 \cup I_2$, where $I_1 \subseteq E_1$ is independent in $M_1$ and $I_2 \subseteq E_2$ is independent in $M_2$.

As usual, we[34] need to show $M_1 \oplus M_2$ is a matroid, i.e., $\{I_1 \cup I_2 \mid I_1 \in \mathcal{I}_1 \text{ and } I_2 \in \mathcal{I}_2\}$ satisfies the independent set axioms. This is left for the exercises, along with descriptions of the bases, circuits, flats, etc. – see Proposition 3.33 and Exercise 35.

---

[33] That's really three features, isn't it?
[34] You.

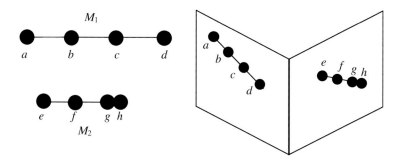

**Example 3.32.** Let $M_1 = U_{2,4}$, the four-point line on the ground set $\{a, b, c, d\}$, and let $M_2$ be the matroid on the ground set $\{e, f, g, h\}$ with circuits $\{efg, efh, gh\}$. Then an independent set of $M_1 \oplus M_2$ is formed by taking the union of an independent set in $M_1$ with one from $M_2$. For example, the set $\{a, c, e\}$ is independent in $M_1 \oplus M_2$.

Drawing a picture that represents the direct sum can be challenging, but the entire challenge has to do with rank, because $r(M_1 \oplus M_2) = r(M_1) + r(M_2)$ (since a basis for $M_1 \oplus M_2$ is simply the union of a basis of $M_1$ with a basis of $M_2$ – see Proposition 3.33(2)). We indicate how to put pictures representing $M_1$ and $M_2$ together to get one for $M_1 \oplus M_2$ in Figure 3.25.

**Proposition 3.33.** *Let $M_1$, $M_2$ be matroids defined on disjoint ground sets $E_1$, $E_2$ with independent sets $\mathcal{I}(M_1), \mathcal{I}(M_2)$, respectively. Then,*

(1) *Independent sets:* $\mathcal{I}(M_1 \oplus M_2) = \{I_1 \cup I_2 \mid I_i \in \mathcal{I}(M_i)\}$.
(2) *Bases:* $\mathcal{B}(M_1 \oplus M_2) = \{B_1 \cup B_2 \mid B_i \in \mathcal{B}(M_i)\}$.
(3) *Rank function:* $r_{M_1 \oplus M_2}(X) = r_{M_1}(X \cap E_1) + r_{M_2}(X \cap E_2)$.
(4) *Flats:* $\mathcal{F}(M_1 \oplus M_2) = \{F_1 \cup F_2 \mid F_i \in \mathcal{F}(M_i)\}$.
(5) *Hyperplanes:* $\mathcal{H}(M_1 \oplus M_2) = \{H_1 \cup E_2 \mid H_1 \in \mathcal{H}(M_1)\} \cup \{E_1 \cup H_2 \mid H_2 \in \mathcal{H}(M_2)\}$.
(6) *Spanning sets:* $\mathcal{S}(M_1 \oplus M_2) = \{S_1 \cup S_2 \mid S_i \in \mathcal{S}(M_i)\}$.
(7) *Circuits:* $\mathcal{C}(M_1 \oplus M_2) = \mathcal{C}(M_1) \cup \mathcal{C}(M_2)$.
(8) *Cocircuits:* $\mathcal{C}^*(M_1 \oplus M_2) = \mathcal{C}^*(M_1) \cup \mathcal{C}^*(M_2)$.
(9) *Duals:* $(M_1 \oplus M_2)^* = M_1^* \oplus M_2^*$.

Virtually all of the proofs here are completely straightforward. We comment briefly on hyperplanes now, and not-so-briefly on circuits shortly. If $H$ is to be a hyperplane in $M_1 \oplus M_2$, $H$ must be a maximal flat that isn't the entire matroid $M_1 \oplus M_2$. Such a flat is formed by taking a hyperplane $H_1$ of $M_1$, for instance, together with the *entire* matroid $M_2$. Thus $H_1 \cup E_2$ is a hyperplane of $M_1 \oplus M_2$. Similarly, a hyperplane $H_2$ of $M_2$ gives rise to the unique hyperplane $E_1 \cup H_2$ in $M_1 \oplus M_2$. This explains why the hyperplanes of $M_1 \oplus M_2$ are *not* the unions $H_1 \cup H_2$, with $H_i$ a hyperplane in $M_i$. In general, $r(H_1 \cup H_2) = r(M_1 \oplus M_2) - 2$.

Figure 3.26. The matroid $M$ is the direct sum of $U_{1,3}$ and $M_2$ from Figure 3.25.

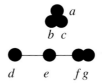

By the way, we can (and will) extend direct sums to include more than two matroids. It should be clear how to define $M_1 \oplus M_2 \oplus M_3$, for example. You can also verify $(M_1 \oplus M_2) \oplus M_3 = M_1 \oplus (M_2 \oplus M_3)$ and $M_1 \oplus M_2 = M_2 \oplus M_1$.

**Definition 3.34.** Let $M_1, M_2, \ldots, M_k$ be matroids on disjoint ground sets $E_1, E_2, \ldots, E_k$, respectively. Define the *direct sum* $M_1 \oplus M_2 \oplus \cdots \oplus M_k$ to be the matroid on the ground set $E = E_1 \cup E_2 \cup \cdots \cup E_k$ with independent sets $I_1 \cup I_2 \cup \cdots \cup I_k$, where $I_j \subseteq E_j$ is independent in $M_j$ for $1 \le j \le k$.

At this point in your life, you may have a favorite matroid $M$. Maybe it's the Fano plane, or a uniform matroid, or the matroid associated to one of the graphs or matrices you've met. Is your matroid $M$ the direct sum of two smaller matroids $M_1$ and $M_2$? How can you tell?

The key to answering this question is the behavior of the circuits of $M$.

**Example 3.35.** Let $M$ be a matroid on $E = \{a, b, c, d, e, f, g\}$ with the following circuits: $ab, ac, bc, def, deg,$ and $fg$. These circuits suggest a *partition* of the ground set $E$ into two pieces: $E_1 = \{a, b, c\}$ and $E_2 = \{d, e, f, g\}$. Note that no circuit of $M$ uses points from both $E_1$ and $E_2$.

What's up with the bases of $M$? Every basis has a Chinese menu flavor:[35] choose $B_1$ from $E_1$ and $B_2$ from $E_2$ and form $B = B_1 \cup B_2$. For instance, we could take $B_1 = \{b\}$ and $B_2 = \{e, g\}$, giving the basis $B = \{b, e, g\}$. You can (easily) check that every basis of $M$ is formed this way.

Then $M = M_1 \oplus M_2$, where $M_1$ is the triple point $U_{1,3}$ and $M_2$ is the matroid pictured on the lower left of Figure 3.25. See Figure 3.26 for a picture of $M$.

Our immediate goal is to figure out when a matroid breaks up as a direct sum, as in Example 3.35. The following relation on the ground set of a matroid plays a key role. Let $M$ be a matroid on $E$ with circuits $\mathcal{C}$. Define a relation $R$ on $E$ as follows:

> Relation $R$: $x$ is related to $y$ if $x = y$ or if there is a circuit $C$ containing both $x$ and $y$.

---

[35] The menu has very little flavor.

Recall that a relation is an *equivalence relation* if it is reflexive, symmetric and transitive. Of special significance, any equivalence relation on a set $E$ induces a *partition* of the set $E$ into disjoint subsets: $E = E_1 \cup E_2 \cup \cdots \cup E_k$, where each $E_i$ is an equivalence class. See Definition 2.28 for a quick review.

**Theorem 3.36.** *Let $M$ be a matroid on $E$ with circuits $C$ and relation $R$ given above. Then $R$ is an equivalence relation on $E$.*

The difficult part of the proof is transitivity. We state this as a separate lemma.

**Lemma 3.37.** *Let $x, y, z$ be distinct elements of $E$ and $C_1, C_2$ be circuits of $M$ with $x, y \in C_1$ and $y, z \in C_2$. Then there is a circuit $C_3$ containing both $x$ and $z$.*   Circuit transitivity

To prove Lemma 3.37, we need to assume *strong circuit elimination* (see Exercise 46):

(C3′) If $C_1$ and $C_2$ are circuits with $x \in C_1 - C_2$ and $y \in C_1 \cap C_2$, then there is a circuit $C_3 \subseteq C_1 \cup C_2$ such that $x \in C_3$ and $y \notin C_3$.   Strong circuit elimination

*Proof Lemma 3.37.* Assume we have circuits $C_1$ and $C_2$ with $x \in C_1 - C_2$ and $z \in C_2 - C_1$, with $y \in C_1 \cap C_2 \neq \emptyset$. Our goal is to produce a circuit $C'$ that contains both $x$ and $z$.

We do this by induction, and the key idea involved is figuring out what parameter in the problem we should use for the induction. One parameter that works is $|C_2 - C_1|$.

If $|C_2 - C_1| = 1$, then, since $z \in C_2 - C_1$, we must have $C_2 - C_1 = \{z\}$. Now use strong circuit exchange to get a circuit $C' \subseteq C_1 \cup C_2$ with $x \in C'$ and $y \notin C'$. Then $z \in C'$ because, otherwise, $C' \subsetneq C_1$, which can't happen with two circuits.

Now assume $|C_2 - C_1| = n > 1$ and the result is true whenever $|C_2 - C_1| < n$. We'll use strong circuit elimination twice to produce a new circuit $C_4 \subseteq C_1 \cup C_2$ satisfying $x \in C_1 - C_4$ and $z \in C_4 - C_1$ with $|C_4 - C_1| < |C_2 - C_1| = n$.

To that end, we first use (C3′) to produce a circuit $C_3 \subseteq C_1 \cup C_2$ with $x \in C_3$ and $y \notin C_3$. Now if $z \in C_3$, we're done – we have our circuit containing both $x$ and $z$. So assume $z \notin C_3$. We also know there is some $z^* \in C_2 - C_1$ with $z^* \in C_3$ (or else $C_3 \subsetneq C_1$, a contradiction).

It will be helpful to use a Venn diagram to keep track of the players in this game – see Figure 3.27. Now $z \in C_2 - C_3$ and $z^* \in C_2 \cap C_3$, so we can apply (C3′) to $C_2$ and $C_3$. This produces a circuit $C_4$ such that $C_4 \subseteq C_2 \cup C_3$ where $z \in C_4$ and $z^* \notin C_4$. Further, if $x \in C_4$, we would be done since $C_4$ would then contain both $x$ and $z$. So assume $x \notin C_4$.

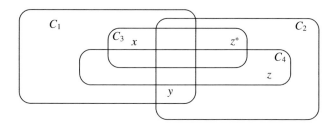

Now let's look at what we know about the circuits $C_1$ and $C_4$. First, note $C_1 \cap C_4 \neq \emptyset$ (or else $C_4 \subsetneq C_2$, a contradiction). We also have $x \in C_1 - C_4$, $z \in C_4 - C_1$ and $z^* \notin C_1 \cup C_4$ (see Figure 3.27). By the way we've constructed these circuits, we also know $C_4 - C_1 \subseteq C_2 - C_1$. But this containment is strict because $z^* \in C_2 - C_1$ but $z^* \notin C_4 - C_1$. Thus $|C_4 - C_1| < n$, so, by our induction hypothesis,[36] we know there is a circuit $C^* \subseteq C_1 \cup C_4$ containing both $x$ and $z$. □

*Proof Theorem 3.36.* It follows immediately from the definition that $R$ is reflexive and symmetric. Transitivity is proven in Lemma 3.37, so we're done.

Cocircuits and circuits satisfy many of the same theorems. In particular, cocircuits also generate an equivalence relation, partitioning the ground set. You are asked to prove *cocircuit transitivity* in Exercise 45.

You might wonder if circuit transitivity can be combined with strong exchange (C3′) in the following way: given circuits $C_1$ and $C_2$ with $x \in C_1 - C_2$, $y \in C_1 \cap C_2$ and $z \in C_2 - C_1$, is there a circuit $C_3$ containing both $x$ and $z$ but *also avoiding* $y$? The next example shows that this is not true in general.

**Example 3.38.** Consider the matroid from Figure 3.28. Let $C_1 = bxy$ and $C_2 = ayz$. Then $x \in C_1 - C_2$, $z \in C_2 - C_1$ and $y \in C_1 \cap C_2$. But the circuit $C_3 = xyz$ is the *only* circuit containing both $x$ and $z$, so there is no circuit containing $x$ and $z$, but avoiding $y$.

Theorem 3.36 gives us a recipe for how to take a matroid and write it as a direct sum, if possible. First, we need the following: if $M$ is a matroid on $E$ and $S \subseteq E$, then the *restriction* $M|_S$ is the matroid on $S$ obtained by simply restricting the ground set to $S$.

---

[36] At last!

**Definition 3.39.** Let $M$ be a matroid on the set $E$ and let $S \subseteq E$. Then    Matroid restriction
the restriction $M|_S$ has ground set $S$, and $I \subseteq S$ is independent in $M|_S$
if $I$ is independent in $M$.

When $r(M|_S) = r(M)$, then restriction is just deletion of the comple-
ment: $M|_S = M - (E - S)$. See Proposition 8.3 of Chapter 8.[37] Then
Theorem 3.40 uses restriction to break up the ground set $E$ of a matroid
into circuit equivalence classes in order to write $M$ as a direct sum. The
proof of Theorem 3.40 is straightforward, and we leave it for Exercise 36.

**Theorem 3.40.** *Let $M$ be a matroid on $E$ and suppose $E = E_1 \cup E_2 \cup$*
*$\cdots \cup E_k$ is a partition of $E$, where each $E_i$ is a circuit equivalence class*
*under relation $R$. Then*

$$M = M|_{E_1} \oplus M|_{E_2} \oplus \cdots \oplus M|_{E_k},$$

*where $M|_{E_i}$ is the restriction of $M$ to the elements of $E_i$.*

In general, a loop or an isthmus is only related to itself by our circuit
equivalence relation $R$. This means isthmuses and loops are especially
well-behaved under direct sums. The next result is a special case of
Theorem 3.40.

**Proposition 3.41.** *Let $M$ be a matroid on $E$ with $|E| \geq 2$. If $I$ is an*
*isthmus of $M$, then $M = (M/I) \oplus I$. If $L$ is a loop of $M$, then $M =$*
*$(M - L) \oplus L$.*

Usually in mathematics, it's easier to put things together than to
pull them apart.[38] For an apt analogy, think about the difference
between multiplying two large numbers, say $p = 179{,}424{,}673$ and
$q = 179{,}424{,}691$ (an easy task) compared with the task of factoring
a single number, say $s = 32{,}193{,}216{,}510{,}801{,}043$ (much harder). In
fact, $p$ and $q$ are the 10,000,000th and 10,000,001st primes, and $s$ is
their product. The security of credit card numbers (and other personal
information) entered in commercial Web sites on the Internet rests in
large part on the difficulty in factoring the product of two very large
primes.

## 3.5.2 Connected matroids

Connectivity is another important idea that appears throughout mathematics. It is (almost always) defined in a negative way – an object is
*connected* if it doesn't somehow break apart. For matroids, we use our
notion of direct sum to define connectivity.

---

[37] Chapter 8 contains much more information on matroid *minors* (matroids obtained
from a given matroid by repeated deletion and contraction).
[38] Unlike everything else in life.

Connected matroid    **Definition 3.42.**  A matroid $M$ is *connected* if $M$ cannot be written as a direct sum of smaller matroids, i.e., if, for all $x, y \in E$, there is a circuit $C$ containing both $x$ and $y$. If $M = M_1 \oplus N$, where $M_1$ is connected, then $M_1$ is a *connected component* of $M$.

Connectivity is important because it is frequently possible to prove a theorem for connected matroids, then extend it (more or less trivially) to all matroids. For instance, a matroid is graphic if and only if each of its components is graphic, and similar theorems hold when "graphic" is replaced by "transversal," "representable over the field $F$," and so on. Thus, the class of connected matroids plays a central role in many proofs.

**Example 3.43.**  Let's look at the connectivity of some small matroids. To simplify notation, we write $I$ or $L$ for the one-element matroids $U_{1,1}$ and $U_{0,1}$ formed by an isthmus or a loop, respectively.

(1)  If $|E| = 1$, then $M$ is either an isthmus or a loop. In either case, $M$ is connected.
(2)  There are four matroids on two points:
 - $M_1$ consists of two isthmuses.
 - $M_2$ is a double point.
 - $M_3$ consists of an isthmus and a loop.
 - $M_4$ is just two loops.

   Then $M_1 = U_{2,2} = I \oplus I$ is disconnected, $M_2 = U_{1,2}$ is connected, $M_3 = I \oplus L$ is disconnected and $M_4 = L \oplus L$ is also disconnected. Note that $M_1^* = M_4$, while $M_2$ and $M_3$ are self-dual, i.e., $M_2 \cong M_2^*$ and $M_3 \cong M_3^*$.
(3)  There are eight matroids on three points. Of these, only two are connected: the three-point line $U_{2,3}$ and the triple point $U_{1,3}$. The remaining six matroids all have an isthmus or a loop (or both).
(4)  There are 17 matroids on four points. Listing them all is a bit tedious,[39] but you can check that only four of these matroids are connected: see Figure 3.29.

The data for the number of connected matroids on nine and fewer points is given in Table 3.2.[40] Note that the percentage of connected matroids rises as the number of points increases, and they seem to predominate (99.1% of all matroids on nine points are connected). One might conjecture that the percentage of connected matroids approaches 100% i.e., $c_n/m_n \to 1$ as $n \to \infty$, where $c_n$ is the number of connected matroids on $n$ points and $m_n$ is the total number of matroids on $n$ points. This conjecture is true, but the proof (by Oxley, Semple, Warshauer and Welsh) is in press, as of this writing. Problems about the asymptotic behavior of matroids seem rather difficult.

---

[39]  "A bit tedious" is a synonym for "a good exercise."
[40]  We thank Gordon Royle for providing these data.

Table 3.2. *The number of connected matroids on k ≤ 9 points.*

| $n$ | 1 | 2 | 3 | 4 | 5 | 6 | 7 | 8 | 9 |
|---|---|---|---|---|---|---|---|---|---|
| Connected | 2 | 1 | 2 | 4 | 10 | 31 | 128 | 1,141 | 379,776 |
| Total | 2 | 4 | 8 | 17 | 38 | 98 | 306 | 1,724 | 383,172 |
| Percent conn. | 100% | 25% | 25% | 24% | 26% | 32% | 42% | 66% | 99% |

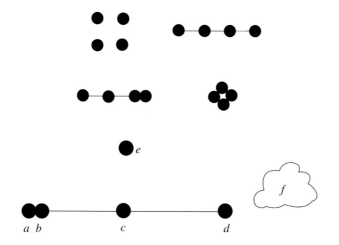

Figure 3.29. The connected matroids on four points.

Figure 3.30. A matroid with three components.

**Example 3.44.** Let's find the connected components of the matroid $M$ depicted in Figure 3.30.

We'll start by listing the circuits of $M$. Recall the cloud represents a loop, so $\{f\}$ is the only one-element circuit. The multiple point $\{a, b\}$ is a two-element circuit. Each set of three distinct points on a line also gives a circuit; so $C = \{f, ab, acd, bcd\}$. Then $M = M' \oplus L \oplus I$, where $M'$ is the matroid $M$ restricted to $\{a, b, c, d\}$, $L$ is a loop and $I$ is an isthmus. ($M'$ appears as one of the connected matroids on four points in Figure 3.29.)

As we noted at the beginning of this chapter, deletion and contraction play a central role in matroid theory because they facilitate inductive arguments. Here's a very important result that relates matroid connectivity to these operations. The following theorem is due to Tutte.

**Theorem 3.45.** *Let $M$ be a connected matroid. For all $e \in E$, either $M - e$ or $M/e$ (or both) is connected.*

*Proof Theorem 3.45.* Suppose $M - e$ is not connected (and note this implies $M$ has at least three points). We need to prove $M/e$ must be connected. So let $x, y \in E - e$. We need to find a circuit in $M/e$ containing both $x$ and $y$.

Now since $M$ is connected, we know there is a circuit $C$ in $M$ containing both $x$ and $y$. Since we are assuming $M - e$ is not connected,

Figure 3.31. Our favorite
matroid is connected.

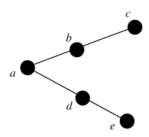

we can partition $E - e$ into connected components using our relation
$R$: this is the direct sum decomposition of $M - e$. Now if $x$ and $y$ are in
different components of $M - e$, then no circuit of $M - e$ contains both
$x$ and $y$. But, by Proposition 3.9, the circuits of $M - e$ are simply the
circuits of $M$ that avoid $e$. Thus $e \in C$. By the same proposition, $C - e$
is then a circuit of $M/e$, which is what we were after.

What if $x$ and $y$ are in the same component of $M - e$? Then all we
need to do is find some $z$ not in that component and use the fact that the
relation $R$ is transitive on the matroid $M/e$. Then, by the above argument,
there are circuits $C_x$ and $C_y$ in $M/e$ with $x, z \in C_x$ and $y, z \in C_y$. Circuit
transitivity (Lemma 3.37) in $M/e$ then gives us the circuit we need.  □

Time for another example.

**Example 3.46.**   Let $M$ be the matroid on $E = \{a, b, c, d, e\}$ in Figure 3.31. Then $M$ has precisely three circuits: $\{a, b, c\}$, $\{a, d, e\}$ and
$\{b, c, d, e\}$. This means every pair of elements of $E$ is in some circuit,
so $M$ is connected.

Let's see what happens to connectivity when we delete and contract
various points. Starting with the point $a$, note that $M/a$ has two circuits,
$\{b, c\}$ and $\{d, e\}$, so $M/a$ is disconnected. Then, by Theorem 3.45, we
are guaranteed that $M - a$ is connected. We can check this easily: the
only circuit of $M - a$ is $\{b, c, d, e\}$.

If we delete $b$ from $M$, then the point $c$ becomes an isthmus. Thus
$M - b$ is disconnected as a matroid, so, by Theorem 3.45, $M/b$ must be
connected. You can check that the circuits of $M/b$ are $\{a, c\}$, $\{a, d, e\}$
and $\{c, d, e\}$, so $M/b$ is connected, as promised. (By symmetry, $c$, $d$ and
$e$ behave the same way as $b$ in the matroid.)

Summing up:

$$M - a \cong U_{3,4}, \qquad M/a \cong U_{1,2} \oplus U_{1,2},$$
$$M - b \cong I \oplus U_{2,3}, \qquad M/b \cong M',$$

where $M'$ is the rank 2 matroid on four points consisting of a double
point and two other points pictured in the lower left of Figure 3.29.

Since $(M_1 \oplus M_2)^* = M_1^* \oplus M_2^*$, we immediately get the following
corollary.

**Corollary 3.47.** *A matroid M is connected if and only if its dual matroid $M^*$ is connected.*

The next result, which follows from Corollary 3.47 and has a one-word-proof,[41] relates connectivity to cocircuits.

**Theorem 3.48.** *M is a connected matroid if and only if for every pair of elements x and y, there is a cocircuit $C^*$ containing both x and y, i.e., there is a hyperplane H avoiding both x and y.*

*Proof Theorem 3.48.* Dualize! ☐

So, a matroid $M$ is connected if and only if for all $x, y \in E$, there is a circuit *and* a cocircuit of $M$ containing both. We can reverse the logical dependence[42] between Theorem 3.48 and Corollary 3.47. It is not hard to prove Theorem 3.48 directly; see Exercise 50. Then Corollary 3.47 follows as an immediate consequence.

### 3.5.3 Connectivity and rank

If $M = M_1 \oplus M_2$ with $M_1$ defined on the ground set $E_1$ and $M_2$ defined on the disjoint set $E_2$, then $r(E_1) + r(E_2) = r(M)$, where the rank is computed in $M$. This follows from part (2) or part (3) of Proposition 3.33. The nice thing about this result is that the converse is true: If $M$ is a matroid and $A \subseteq E$ is a proper subset of the ground set $E$, then if $r(A) + r(E - A) = r(M)$, we must have $M = M_1 \oplus M_2$. We state this result in terms of connectivity:

**Proposition 3.49.** *Let M be a matroid on E and suppose $A \subseteq E$ is a proper subset satisfying $r(A) + r(E - A) = r(M)$. Then M is disconnected, i.e., $M = M|_A \oplus M|_{E-A}$.*

*Proof Proposition 3.49.* Suppose $C$ is a circuit of $M$. We need to show $C \subseteq A$ or $C \subseteq E - A$; the result will then follow from our definition of connectivity via circuits (Definition 3.42). So suppose $C_1 := C \cap A$ and $C_2 := C \cap (E - A)$ are both non-empty. We plan to arrive at a contradiction.

Now $C_1$ is an independent set (since it's a proper subset of the circuit $C$), so we can extend $C_1$ to a basis $B_1$ for $M|_A$, i.e., $C_1 \subseteq B_1 \subseteq A$ and $r(A) = |B_1|$. We can do the same thing for $C_2$ in $E - A$, creating a basis $B_2$ for $M|_{E-A}$ with $C_2 \subseteq B_2 \subseteq (E - A)$ and $r(E - A) = |B_2|$. Then the hypothesis $r(A) + r(E - A) = r(M)$ implies $|B_1| + |B_2| = r(M)$.

Now we know $C = C_1 \cup C_2 \subseteq B_1 \cup B_2$. If we could show $B_1 \cup B_2$ is an independent set, we would have our contradiction because the

---

[41] This claim is debatable.
[42] Whatever that means.

independent set $B_1 \cup B_2$ would then contain the circuit $C$, which is impossible.

Now since $B_1$ and $B_2$ are bases for $M|_A$ and $M|_{E-A}$, respectively, we also know $\overline{B_1} \supseteq A$ and $\overline{B_2} \supseteq E - A$. But $E = \overline{B_1} \cup \overline{B_2} \subseteq \overline{B_1 \cup B_2}$, so $B_1 \cup B_2$ spans $M$. Since $|B_1| + |B_2| = r(M)$, we must have $B_1 \cup B_2$ independent. This contradiction finishes the argument.    □

Note $r(A) + r(E - A) \geq r(M)$ is always true – this follows immediately from the semimodularity of rank. We can "use"[43] Proposition 3.49 to determine if a matroid is connected. For example, the matroid in Figure 3.30 from Example 3.44 has $r(\{a, b, c, d\}) = 2$, $r(\{e, f\}) = 1$ and $r(M) = 3$, so $A = \{a, b, c, d\}$ will give $r(A) + r(E - A) = r(M)$. Since $M$ has three components, this choice of $A$ is not unique: for instance, if $A = \{a, b, c, d, e\}$, then $r(A) = 3$ and $r(E - A) = 0$, so $r(A) + r(E - A) = r(M)$ is still satisfied.

Using the rank function to discuss connectivity is important – it allows us to extend the definition to *higher* connectivity. Let $A$, $B$ be a non-trivial partition of the ground set $E$, i.e., $A \cup B = E$, $A \cap B = \emptyset$, and $|A|, |B| \geq 1$. Then we summarize our discussion in an attractive box:

> $M$ is *not* connected if and only if $r(A) + r(B) - r(M) = 0$ for some non-trivial partition $A$, $B$.

### 3.5.4 Connectivity motivation from graphs

Connectivity is a natural idea in graph theory – a graph is connected if any pair of vertices are joined by some edge-path. To interpret this from a matroid viewpoint, consider the graphs in Figure 3.32. Among the four graphs $G_1, G_2, G_3$ and $G_4$, only $G_1$ is disconnected. From the matroid viewpoint, it's easy to see $M(G_1) \cong M(G_2)$: the circuits of $G_1$ and $G_2$ are the same.

When does a given graph $G$ give rise to a connected matroid $M(G)$? From our definition of connectivity (Definition 3.42), we need every pair of edges of $G$ to be contained in some circuit. For the four graphs in Figure 3.32, you can check that $M(G_1)$ and $M(G_2)$ are disconnected matroids: there is no circuit containing the pair of edges $ab$ and $cd$.

*Connected graph $\neq$ connected matroid*

On the other hand, it's easy to see $M(G_3)$ and $M(G_4)$ are connected matroids.

In Figure 3.32, note that removing vertex $v$ (together with the four edges incident to $v$) from the connected graph $G_2$ disconnects the graph. We call the vertex $v$ a *cut-vertex*. The presence of cut-vertices is the key

---

[43] We use "use" loosely, here, since we would need to check all $2^{|E|}$ subsets $A$ for the rank condition.

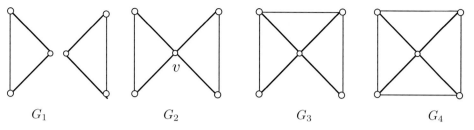

$G_1$          $G_2$          $G_3$          $G_4$

Figure 3.32. $M(G_1)$ and $M(G_2)$ are isomorphic, disconnected matroids, while $M(G_3)$ and $M(G_4)$ are connected matroids.

for figuring out if a connected graph gives a connected matroid. A graph with no cut-vertices is called a *block*.

**Theorem 3.50.** *[Whitney] Let G be a connected graph. Then the following are equivalent.*

(1) $M(G)$ *is a connected matroid.*
(2) *G has no cut vertices, i.e., G is a block.*
(3) *Every pair of vertices in G are joined by at least two vertex disjoint edge-paths.*

The central idea for the proof is as follows. If $G$ is a block with edges $e_1$ and $e_2$, then we can find two vertex-disjoint paths between the endpoints of $e_1$ and $e_2$ that can be glued together to form the cycle we need. On the other hand, if $M(G)$ is a connected matroid, then the existence of cycles containing any pair of edges prevents $G$ from having any cut-vertices. See Exercise 22 of Chapter 4 for an outline of the proof of Theorem 3.50.

The equivalence between (2) and (3) in Theorem 3.50 is due to Whitney (1932). A deeper result of Whitney that describes precisely when two distinct graphs can have the same cycle matroids is given in Theorem 4.21 in Chapter 4.

More generally, a connected graph is said to be *k-connected* if it remains connected whenever any $k - 1$ vertices are removed from the graph. So a block is 2-connected, and we can rephrase Theorem 3.50 in this language:

$M(G)$ is a connected matroid if and only if $G$ is a 2-connected graph.

The generalization of the equivalence between (2) and (3) to $k$-connectivity in graphs is known as Menger's Theorem.

## Exercises

### Section 3.1 – Deletion and contraction
(1) Let $M_1$, $M_2$ and $M_3$ be the three matroids pictured in Figure 3.33.
  (a) Draw a geometric picture for each of the following: $M_1/a$, $M_1/d$, $M_1 - d$, $M_2/a$ and $M_3/b$.

Figure 3.33. Three matroids for
Exercises 1, 17, 21 and 22.

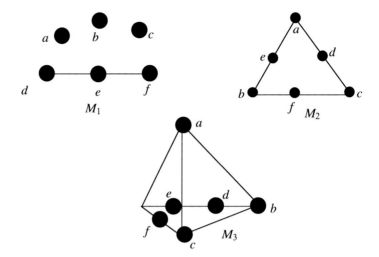

(b) For each of these three matroids, find the smallest number of
points you need to delete and/or contract to produce a uniform
matroid. (For instance, for $M_1$, the deletion $M_1 - d$ is the
uniform matroid $U_{3,5}$, so we need only delete one element.)

(2) Suppose $a, b \in E$, the ground set of a matroid $M$. True or false:
   (a) If $a$ is a loop of $M/b$, then $b$ is a loop of $M/a$.
   (b) If $a$ is a loop of $M - b$, then $a$ is a loop of $M$.
   (c) If $a$ is an isthmus of $M/b$, then $b$ is an isthmus of $M/a$.
   (d) If $a$ is an isthmus of $M - b$, then $a$ is an isthmus of $M$.

(3) Prove the following:

   $I$ is independent in $M/e$ if $e \notin I$, but $I \cup \{e\}$ is independent
   in $M$.

(4) Let $M$ be a matroid and suppose $e$ is neither an isthmus nor a loop.
   Prove Proposition 3.4 characterizing the bases of $M/e$ and $M - e$.

(5) Let $M = U_{r,n}$, the uniform matroid of rank $r$ on $n$ points.
   (a) Let $x$ be any point of $M$. Show that $M/x$ is a uniform matroid
      $U_{a,b}$ and figure out the values of $a$ and $b$ (assuming $r > 0$).
   (b) Repeat part (a) for $M - x$ (assuming $r < n$).
   (c) Suppose $M$ has no loops and $M/x$ is a uniform matroid for all
      $x$. Must $M$ be a uniform matroid? If true, find a proof; if false,
      find a specific counterexample; if neither true nor false, find
      another subject to study.

(6) Let's do some counting. Let $i(M)$ be the number of independent
   sets in $M$ and $b(M)$ the number of bases.
   (a) Suppose $x$ is neither a loop nor an isthmus. Show $i(M) =$
      $i(M - x) + i(M/x)$ and $b(M) = b(M - x) + b(M/x)$.
   (b) For $M = U_{r,n}$, compute $b(M), b(M - x)$ and $b(M/x)$. What
      does the equation $b(M) = b(M - x) + b(M/x)$ tell you?
   (c) Repeat part (b) for $i(M), i(M - x)$ and $i(M/x)$.
   (d) Let $s(M)$ be the number of spanning sets of $M$. Repeat
      parts (a) and (b) for $s(M)$.

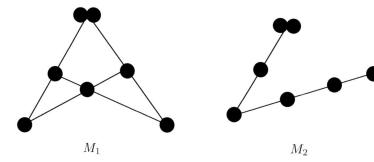

Figure 3.34. The two matroids for Exercise 7.

$M_1$                                    $M_2$

The deletion-contraction formulas obeyed by $i(M), b(M)$ and $s(M)$ are special cases of a more general result on the *Tutte polynomial*. See Chapter 9 for much more on the Tutte polynomial.

(7) It's possible for two different matroids to have isomorphic deletions and contractions. Let $M_1$ and $M_2$ be the matroids in Figure 3.34. Find points $x$ and $y$ with $M_1 - x \cong M_2 - y$ and $M_1/x \cong M_2/y$.

(8) This exercise asks you to prove Proposition 3.2: $M - e$ and $M/e$ are both matroids, provided $e$ is not a loop or an isthmus. Suppose $M$ is a matroid on the ground set $E$ and $e \in E$. Let $\mathcal{I}_1$ be the family of independent sets that don't contain $e$ and let $\mathcal{I}_2$ be the independent sets that do contain $e$, with $e$ removed from each set.

   (a) If $e$ is not an isthmus, show that the family $\mathcal{I}_1$ satisfies the independent set axioms (I1), (I2) and (I3). This shows $M - e$ is a matroid.

   (b) If $e$ is not a loop, show that $\mathcal{I}_2$ also satisfies the independent set axioms (I1), (I2) and (I3), so $M/e$ is also a matroid.

(9) Let $e$ be a non-isthmus. Then the bases of $M - e$ are defined as $\mathcal{B}(M - e) = \max\{B \subseteq E - e \mid B \in \mathcal{B}\}$. Define $\mathcal{B}_i$ for $1 \leq i \leq 4$ as follows:

   • $\mathcal{B}_1 = \mathcal{B}(M - e)$.
   • $\mathcal{B}_2 = \max\{B - e \mid B \in \mathcal{B}\}$.
   • $\mathcal{B}_3 = \max\{I \subseteq E - e \mid I \in \mathcal{I}\}$.
   • $\mathcal{B}_4 = \max\{I \subseteq E - e \mid I \in \mathcal{I}(M - e)\}$.

   Show these sets all describe the bases of the $M - e$, i.e., show $\mathcal{B}_1 = \mathcal{B}_2 = \mathcal{B}_3 = \mathcal{B}_4$.

(10) For this problem, assume $M$ is a matroid on the ground set $E$ with $n$ elements and no isthmuses or loops.

   (a) Suppose $r(M - e) = n - 1$ for every point $e \in E$. Show that $M$ is the uniform matroid $U_{n-1,n}$, i.e. the elements of $M$ form a single circuit.

   (b) Suppose $M/e$ is a circuit for every point $e \in E$. Show that the same conclusion holds as in part (a), i.e., the elements of $M$ form a single circuit.

(11) Give an example of a rank 3 matroid having *no* symmetry in the following sense. For any two distinct points $x$ and $y$, the two

Figure 3.35. $K_5$ and its
realization as a geometry – see
Exercise 14.

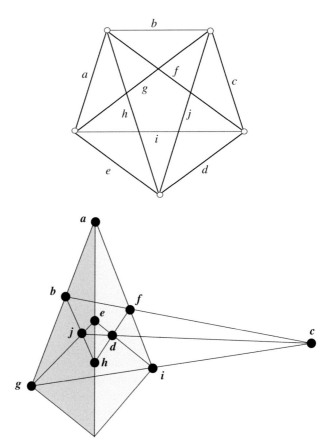

matroids $M - x$ and $M - y$ are not isomorphic and $M/x$ and $M/y$
are also not isomorphic.

(12) Prove the rest of Proposition 3.8.

(13) Prove Proposition 3.9.

## Section 3.2 – Deletion and contraction in graphs and matrices

(14) Verify the claim in the text that $M(K_5)$ is represented geometrically by the three-dimensional Desargues configuration – see
Figure 3.35. For instance, the triangle $abg$ in $K_5$ corresponds to a
three-point line in the geometry.

(15) Let

$$
A = \begin{array}{c} \begin{array}{ccccccc} a & b & c & d & e & f & g \end{array} \\ \left[ \begin{array}{ccccccc} 1 & 0 & 0 & 0 & 1 & 1 & 1 \\ 0 & 1 & 0 & 0 & 1 & -1 & -1 \\ 0 & 0 & 1 & 0 & 0 & 1 & 0 \\ 0 & 0 & 0 & 1 & 0 & 0 & 1 \end{array} \right]. \end{array}
$$

(a) Verify that $M[A]$ is the matroid $M$ pictured on the left in
Figure 3.36.

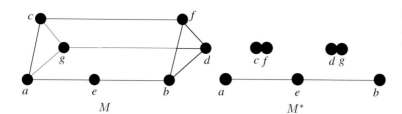

Figure 3.36. A matroid and its dual: see Exercises 15 and 27.

(b) Find a matrix $A_f$ that represents the contraction $M/f$. (Hint: apply the contraction procedure outlined in Section 3.2.1.)

(c) From symmetry considerations, you can see $M/c \cong M/f$. Find a matrix $A_c$ that represents $M/c$ and show $M[A_c]$ and $M[A_f]$ (from part (b)) are isomorphic matroids. (This is asking you to show the two matrices have the same column dependences, where column $c$ in the matrix $A_f$ corresponds to column $f$ in $M_c$ under a matroid isomorphism.)

(16) $U_{n-1,n}$ is the uniform matroid of rank $n - 1$ on $n$ points, i.e., a circuit of size $n$. Assume the ground set is $E$ and $n \geq 2$.

(a) Find a matrix $A$ that represents $U_{n-1,n}$. (Hint: assume your matrix has $n - 1$ rows and $n$ columns and the first $n - 1$ columns form an identity matrix.)

(b) Show that removing any column of your matrix $A$ gives a new matrix $A'$ with $M[A'] \cong U_{n-1,n-1}$, the Boolean algebra $B_{n-1}$ on $n - 1$ points.

(c) Let $e \in E$. Apply the contraction procedure outlined in Section 3.2.1 to the matrix $A$ to produce a matrix $A''$ that represents $M[A]/e$, and show $M[A''] \cong U_{n-2,n-1}$.

## Section 3.3 – Matroid duality

(17) For the matroid $M_3$ in Figure 3.33, show that $(M_3 - b)^* \cong M_3^*/b$ and $(M_3/b)^* \cong M_3^* - b$.

(18) Show that $U_{r,n}^* = U_{n-r,n}$. Conclude that $U_{r,n}$ is self-dual if and only if $r = n/2$. (A matroid is *self-dual* if it's isomorphic to its dual.)

(19) $F_7$ is the Fano matroid shown on the left in Figure 3.14. Show that the dual $F_7^*$ is represented by the matroid pictured on the right in Figure 3.14, verifying Example 3.19. (It's probably easiest to check the seven circuits of $F_7$ correspond to seven cocircuits in $F_7^*$, but a listing of all the bases of the two matroids pictured is not too onerous.)

(20) Let $F_7$ be the Fano matroid (Figure 3.14 shows $F_7$ and $F_7^*$). Draw the matroid $(F_7 - x)^*$ and show you get the same matroid as $F_7^*/x$.

(21) (a) For each of the matroids in Figure 3.33, draw the dual. Which of these matroids are self-dual?

(b) Draw the dual of the matroid $M_1 - a$. Then use your drawing and the drawing of $M_1^*$ from part (a) to show directly that $(M_1 - a)^* = M_1^*/a$. Repeat this procedure for the point $d$.

(c) Now draw the dual of the matroid $M_2/a$ and use that drawing and your picture of $M_2^*$ from part (a) to show directly $(M_2/a)^* = M_2^* - a$.

(22) In Theorem 3.20, the rank of a subset in the dual $M^*$ of a matroid $M$ is given in terms of the rank function of $M$:

$$r^*(A) = |A| + r(E - A) - r(E).$$

(a) Use the formula to find the rank of the set $\{c, d, e, f\}$ in the matroid $M_3^*$ in Figure 3.33, checking that the picture of $M_3^*$ you drew in Exercise 21 matches your answer.

(b) Find $r^*(\emptyset)$ and $r^*(E)$ using the formula.

(c) For all $A \subseteq E$ show $r^{**}(A) = r(A)$. (In this problem, you should interpret $r^{**}(A)$ as the rank of $A$ in the matroid $(M^*)^*$. Your proof should *not* use the fact that $(M^*)^* = M$.)

(23) Let $A \subseteq E$, where $E$ is the ground set of the matroid $M$. Show that every point in $A$ is an isthmus of $M$ if and only if $|A| + r(E - A) = r(M)$. (Hint: this is easier if you switch to the dual and use Theorem 3.20.)

(24) Show that the number of rank $k$ matroids on $n$ points is the same as the number of rank $n - k$ matroids on $n$ points.

(25) Define a set $C \subseteq E$ to be a *cycle* if $C$ is the union of circuits. Show that $C$ is a cycle in $M$ if and only if $E - C$ is a flat in $M^*$.

(26) You can *define* contraction from deletion and duality.[44] Let $M$ be a matroid and assume $x$ is not a loop. Then define $M/x = (M^* - x)^*$. Show this definition agrees with our definition of contraction by showing that $B$ is a basis of $(M^* - x)^*$ precisely when $B \cup \{x\}$ is a basis of $M$.

(27) (*If you know some group theory.*)

(a) Show that the automorphism groups $Aut(M)$ and $Aut(M^*)$ are equal. (Note: this is stronger than saying the two groups are isomorphic.)

(b) Consider the dual pair of matroids from Figure 3.36. Show $Aut(M^*) = S_3 \times (\mathbb{Z}_2)^3$. Conclude $Aut(M) = S_3 \times (\mathbb{Z}_2)^3$, as in Exercise 24 of Chapter 1. (The point of this exercise is that even though $Aut(M) = Aut(M^*)$, it is sometimes easier to work with the dual.)

### Section 3.4 – Duality in graphic and representable matroids

(28) Prove directly that $M(K_{3,3})^*$ is not graphic by modifying the circuit–cocircuit argument given in Example 3.25. (An indirect

---

[44] It's another question as to whether you *should* do this.

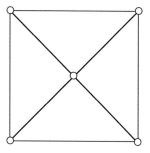

Figure 3.37. The wheel $W_4$.
See Exercise 30.

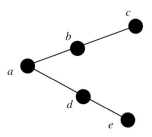

Figure 3.38. The matroid
for Exercise 31.

proof would use the two facts that $K_{3,3}$ is not planar and Theorem 3.27 about duals of non-planar graphs.)

(29) Fill in the details in the proof of Theorem 3.27: let $G$ be a connected graph. Then the dual matroid $M(G)^*$ is graphic if and only if $G$ is planar. Further, if $G$ is planar, then $M(G^*) = M(G)^*$.

(30) Let $n \geq 3$. The *wheel* $W_n$ is formed by adding one vertex to the $n$-cycle and then joining this new vertex to each vertex in the cycle (see Figure 3.37 for a picture of $W_4$). So $M(W_n)$ is a rank $n$ matroid on $2n$ points.

  (a) Show that $W_n$ is self-dual.

Wheels and whirls

  (b) Let $\mathcal{W}_n$ be the matroid derived from $W_n$ by changing the $n$-element circuit corresponding to the rim of the wheel from a dependent set to an independent set. This means the ground set of $\mathcal{W}_n$ is the set of edges of $W_n$, and the bases of $\mathcal{W}_n$ are the bases of $W_n$ with one additional basis; the new basis is the collection of edges corresponding to the rim. ($\mathcal{W}_n$ is called a *whirl*, and the process of changing a set that is both a circuit and a hyperplane to a basis always produces a new matroid. This is called *relaxing* the circuit-hyperplane – see Project P.3.)

    (i) Show $\mathcal{W}_n$ is always a matroid. (Hint: you can use any of the matroid characterizations you like, but showing the independent sets satisfy (I1), (I2) and (I3) is probably easiest.)

    (ii) Show $\mathcal{W}_n$ is not graphic for all $n \geq 3$.

    (iii) Is $\mathcal{W}_n$ self-dual?

(31) Let $M$ be the matroid of Figure 3.38.

(a) Show that $M$ is represented by the matrix

$$
A = \begin{array}{c} \\ \\ \end{array}
\begin{array}{ccccc} a & b & c & d & e \end{array}
\left[\begin{array}{ccccc}
1 & 0 & 1 & 0 & 1 \\
0 & 1 & 1 & 0 & 0 \\
0 & 0 & 0 & 1 & 1
\end{array}\right].
$$

(b) Use Theorem 6.6 to find a matrix that represents $M^*$. (See Figure 3.13 for a picture of $M^*$. You may wish to reorder the columns of $A$.)

(c) Check your answer in part (b) by showing that the matrix you found represents $M^*$.

(32) Suppose $M$ is a self-dual matroid on the ground set $E$. Call a bijection $\phi : E \to E$ a *dual isomorphism* if, for all $B \subseteq M$, $B$ is a basis of $M$ if and only if $\phi(B)$ is a basis of $M^*$.

(a) Let $M_1$ be the self-dual matroid in Figure 3.33, and suppose $\phi$ is a dual isomorphism. Show that $\{\phi(a), \phi(b), \phi(c)\} = \{d, e, f\}$ and $\{\phi(d), \phi(e), \phi(f)\} = \{a, b, c\}$. Conclude that there are 36 dual isomorphisms.

(b) (*If you know some group theory.*) Let $M$ be a matroid on an $n$-element set $E$.

(i) Show that the automorphism group $Aut(M)$ is a subgroup of $S_n$, the symmetric group on $n$ symbols. Use the matroid $M_1$ in Figure 3.33 to show this subgroup need not be normal in $S_n$.

(ii) Now suppose $M$ is self-dual. Show that the collection of all dual isomorphisms is a coset of $Aut(M)$ in $S_n$. Conclude that the number of dual isomorphisms equals the number of automorphisms of $M$.

(33) We can use Theorem 6.6 and graphs to create self-dual matroids. Here's the recipe:

- Let $G$ be a graph on $r$ vertices.
- Let $D$ be the vertex–vertex adjacency matrix for $G$ – i.e., $D$ is an $r \times r$ matrix in which the entry $d_{ij}$ is 1 if there is an edge joining vertices $i$ and $j$ and 0 otherwise.
- Form the new matrix $A = [I_{r \times r} \mid D]$.
- Then the matroid $M[A]$ is self-dual. How about that?

For example, let $G$ be the graph on the left in Figure 3.39. Then

$$
A = \begin{array}{c} \\ \\ \\ \end{array}
\begin{array}{cccccccc} a & b & c & d & e & f & g & h \end{array}
\left[\begin{array}{cccccccc}
1 & 0 & 0 & 0 & 0 & 1 & 0 & 0 \\
0 & 1 & 0 & 0 & 1 & 0 & 1 & 1 \\
0 & 0 & 1 & 0 & 0 & 1 & 0 & 1 \\
0 & 0 & 0 & 1 & 0 & 1 & 1 & 0
\end{array}\right].
$$

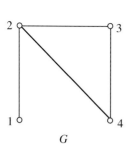

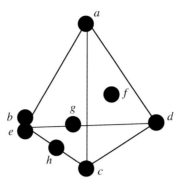

Figure 3.39. The graph $G$ is used to create a self-dual matroid – Exercise 33.

$G$

The matroid $M[A]$ is drawn on the right in Figure 3.39. Prove this always produces a self-dual matroid. What is the smallest self-dual matroid that does not arise in this way? (Hint: use Theorem 6.6 and the fact that $D$ is symmetric.)

## Section 3.5 – Direct sums

(34) Let $M$ be a matroid on $E = \{a, b, \ldots, h\}$ with circuits $C = \{ab, ac, bc, de, dfg, efg\}$. Show that $M$ is a matroid, and $M$ is a direct sum of three smaller matroids: $M = M_1 \oplus M_2 \oplus M_3$. Draw a picture of $M$ that shows these three components.

(35) Prove that the direct sum of two matroids is a matroid. Then prove as many of the characterizations of Proposition 3.33 as you can.

(36) Prove Theorem 3.40: let $M$ be a matroid on $E$ and suppose $E = E_1 \cup E_2 \cup \cdots \cup E_k$ is a partition of $E$, where each $E_i$ is a circuit equivalence class under relation $R$. Then $M = M|_{E_1} \oplus M|_{E_2} \oplus \cdots \oplus M|_{E_k}$, where $M|_{E_i}$ is the restriction of $M$ to the elements of $E_i$.

(37) Suppose $M_1$ and $M_2$ are matroids that are represented by matrices $A_1$ and $A_2$, respectively. Show that the direct sum $M_1 \oplus M_2$ is represented by the matrix $\left( \begin{array}{c|c} A_1 & 0 \\ \hline 0 & A_2 \end{array} \right)$.

(38) Suppose the rank $r$ matroid $M$ on $r + s$ points has exactly one basis. True/false: $M = U_{r,r} \oplus U_{0,s}$, i.e., $M$ must be a direct sum of $r$ isthmuses and $s$ loops. (See Exercise 8b of Chapter 2.)

### Two new constructions: Matroid union and intersection

Matroid Union

(39) [(39)] Let $M_1$ be a matroid on the ground set $E_1$ and $M_2$ a matroid on $E_2$. Then define the *matroid union* $M_1 \vee M_2$ as follows: $I$ is independent in $M_1 \vee M_2$ if $I = I_1 \cup I_2$, where $I_1$ is independent in $M_1$ and $I_2$ is independent in $M_2$. (Note that there is no assumption on the ground sets this time; they do not need to be disjoint, as they do for direct sums.)

Figure 3.40. Matroid union. The matroid $M_1 \vee M_2$ on the right is the union of $M_1$ and $M_2$.

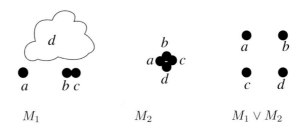

$M_1$ $\qquad\qquad$ $M_2$ $\qquad\qquad$ $M_1 \vee M_2$

(a) Let $M_1$ and $M_2$ be matroids with the same ground set $E = \{a, b, c, d\}$, as pictured in Figure 3.40. Then, for instance, the set $abd$ is independent in $M_1 \vee M_2$ since $ab$ is independent in $M_1$ and $d$ is independent in $M_2$. Show that every set of size three is independent in $M_1 \vee M_2$, so $M_1 \vee M_2 = U_{3,4}$, as in the figure.

(b) Prove $M_1 \vee M_2$ is always a matroid by showing $\mathcal{I}(M_1 \vee M_2)$ satisfies the independent set axioms (I1), (I2) and (I3). (There are no assumptions on the ground sets $E_1$ and $E_2$ for this part.)

(c) Show that $\mathcal{I}(M_1 \vee M_2) \supseteq \mathcal{I}_1 \cup \mathcal{I}_2$, and give an example to show this containment is strict, in general. Then show that $\mathcal{I}_1 \cup \mathcal{I}_2$ may not be the family of independent sets of a matroid.

(d) Give an example of two graphic matroids $M(G_1)$ and $M(G_2)$ in which the matroid union $M(G_1) \vee M(G_2)$ is not graphic. (Hint: Choose graphs with only two vertices, but having multiple edges and loops, so $r(M(G_1)) = r(M(G_2)) = 1$.)

(e) Show that if $E_1 \cap E_2 = \emptyset$, then $M_1 \vee M_2 = M_1 \oplus M_2$.

(f) Suppose $C$ is a circuit of $M$, with $|C| > 1$. Show that $C$ is independent in $M \vee M$. Conclude that if $M$ is a rank $r$ matroid on $n$ points with $M \vee M = M$, then $M = U_{r,r} \oplus U_{0,n-r}$.

Matroid Intersection    (40) Let $M_1$ and $M_2$ be matroids on the same ground set $E$. We attempt to define a new matroid $M_1 \wedge M_2$ as follows: $I \subseteq E$ is independent in $M_1 \wedge M_2$ if $I = I_1 \cap I_2$, where $I_1$ is independent in $M_1$ and $I_2$ is independent in $M_2$.

(a) Show that $\mathcal{I}_1 \cap \mathcal{I}_2 = \{I_1 \cap I_2 \mid I_1 \in \mathcal{I}_1 \text{ and } I_2 \in \mathcal{I}_2\}$, where $\mathcal{I}_1$ and $\mathcal{I}_2$ are the families of independent sets of $M_1$ and $M_2$, respectively.

(b) Suppose $E = \{a, b, c\}$, let $M_1$ be a matroid with bases $ab$ and $ac$, and let $M_2$ be a matroid with bases $ab$ and $bc$. Show that $M_1 \wedge M_2$ is not a matroid.

(c) Although $M_1 \wedge M_2$ is not a matroid, in general, it is an important problem to find the largest independent set in $\mathcal{I}_1 \cap \mathcal{I}_2$. Let $I \in \mathcal{I}_1 \cap \mathcal{I}_2$ and $A \subseteq E$. Show

$$|I| \leq r_1(A) + r_2(E - A),$$

where $r_1$ is the rank function for $M_1$ and $r_2$ is the rank function for $M_2$. Conclude that

$$\max_{I \in \mathcal{I}_1 \cap \mathcal{I}_2} |I| \le \min_{A \subseteq E} (r_1(A) + r_2(E - A)).$$

(In fact, this inequality is actually an equality: $\max_{I \in \mathcal{I}_1 \cap \mathcal{I}_2} |I| = \min_{A \subseteq E} (r_1(A) + r_2(E - A))$. This is Edmonds' *Matroid Intersection Theorem*.)

## Section 3.5 – Connectivity

(41) Show that the Fano matroid $F_7$ is connected (see Figure 3.2).

(42) For this problem, we define the following classes of matroids:
- $\mathcal{C}_1$ = all matroids on 10 elements.
- $\mathcal{C}_2$ = all matroids on 10 or fewer elements.
- $\mathcal{C}_3$ = all rank 4 matroids on 10 elements.
- $\mathcal{C}_4$ = all matroids of rank at most 5.
- $\mathcal{C}_5$ = all graphic matroids.
- $\mathcal{C}_6$ = all connected matroids.
- $\mathcal{C}_7$ = all matroids having no isthmuses.
- $\mathcal{C}_8$ = all simple matroids, i.e., matroids having no loops or multiple points.
  - (a) For each class, determine whether or not the class is closed under deletion, closed under contraction, and closed under duality (so there are 24 things to check). Note: although there are eight classes and eight potential yes/no answers to three questions, not all eight yes/no categories are represented.
  - (b) For each set of yes/no answers that did not appear in part (a), either construct a class with those properties or prove no class can be in that yes/no category.

(43) Let $r > 1$ be fixed. Show that there are an infinite number of simple, connected, rank $r$ matroids.

(44) Let $B$ be a basis of a matroid $M$ and suppose $x \in B$. This exercise is a dual version of Exercise 13 from Chapter 2. Prove the following directly, or use Exercise 13 from Chapter 2 and duality.
  - (a) Show that $B - x$ contains a *unique* cocircuit. This is called the *basic cocircuit* determined by $B$ and $x$.  Basic cocircuit
  - (b) Suppose $C^*$ is a cocircuit in $M$. Show that there is a basis $B$ and an element $x \in B$ so that $C^*$ is the basic cocircuit determined by $B$ and $x$. Thus, *every* cocircuit is a basic cocircuit.

(45) Prove transitivity for cocircuits: if $x, y \in C_1^*$ and $y, z \in C_2^*$ for cocircuits $C_1^*, C_2^*$, then there is a cocircuit $C_3^*$ containing both $x$ and $z$.

(46) The goal of this problem is to build a proof of strong circuit elimination (C3′): if $C_1$ and $C_2$ are circuits with $x \in C_1 - C_2$ and  Strong circuit elimination

$y \in C_1 \cap C_2$, then there is a circuit $C_3 \subseteq C_1 \cup C_2$ such that $x \in C_3$ and $y \notin C_3$. First, let $C = C_1 \cup C_2 - \{x, y\}$.

(a) Show $y \in \overline{C_2 - y}$ and $\overline{C_2 - y} \subseteq \overline{C}$, so $y \in \overline{C}$.

(b) Show $x \in \overline{C}$. Conclude (with part (a)) $\overline{C} = C_1 \cup C_2$.

(c) Now let $B \subseteq C$ be a basis for $C$ and show $B \cup \{x\}$ contains a circuit $C_3$ where $x \in C_3$ and $y \notin C_3$.

The interested reader is invited to construct a proof of (C3′) that only uses the circuit axioms (C1), (C2) and (C3). (One approach would be to modify the proof outlined here by using the crypto-morphism between closure and circuits.)

(47) Is the uniform matroid $U_{r,n}$ always connected? Find all values of $r$ and $n$ so that $U_{r,n}$ is connected.

(48) Show that if $M$ is connected and $M - e$ is not connected for some non-isthmus $e$, then $M^* - e$ must be connected.

(49) Suppose $M$ is a rank $r$ matroid with no loops, and suppose $C$ is a circuit with $|C| = r + 1$.

(a) Show $M$ has no isthmuses.

(b) Show $M$ must be connected.

(c) Show the converse to (b) is false: Find a connected rank $r$ matroid $M$ with no circuits of size $r + 1$. (Hint: find an example using graphic matroids.)

(d) Show that if $r(M) \leq 3$ and $M$ is connected, then $M$ has a circuit of size $r(M) + 1$. What is the smallest $r$ that produces an example for part (c)?

(50) (a) Prove Theorem 3.48 directly: $M$ is a connected matroid if and only if there is a cocircuit $C^*$ containing both $x$ and $y$.

(b) Use part (a) to prove Corollary 3.47: $M$ is connected if and only if $M^*$ is connected.

Extended hint:

$\Rightarrow$ First let $x, y \in E$ and let $C$ be a circuit containing $x$ and $y$. Next, find a basis $B \supseteq C - y$ (why is there such a basis?) Finally, let $H = \overline{B - x}$. To finish this direction, you need to show $H$ is a hyperplane, and that $x \notin H$ and $y \notin H$.

$\Leftarrow$ Let $x, y \in E$ and suppose $x \in C^*$ and $y \in C^*$ for some cocircuit $C^*$. Then $H = E - C^*$ is a hyperplane. Let $I$ be a basis for $H$. First show $B_1 = I \cup x$ and $B_2 = I \cup y$ are bases of $M$. Then there is a unique circuit $C_1 \subseteq B_1 \cup y$ (explain!). Do the same thing for $B_2 \cup x$. Then show these two circuits $C_1$ and $C_2$ coincide and contain $x$ and $y$.

(51) Suppose $G$ is a 2-connected graph (a *block*) with at least two edges. Show that, for any edge $e$ of $G$, either $G - e$ or $G/e$ is a block.

(52) There's an interesting connection between matroid connectivity and the connectivity of a certain bipartite graph associated with the matroid. Let $M$ be a matroid and let $B$ be a basis for $M$. Define a bipartite graph $G_B(M)$ in which the elements of the basis $B$ form

one side of the bipartite graph and the rest of the matroid $E - B$ forms the other side of the bipartite graph. Draw an edge between $x \in B$ and $y \in E - B$ if the unique basic circuit $C_y$ contained in $B \cup y$ also contains $x$.

(a) Draw $G_B(M)$ for the matroid $M_2$ of Figure 3.33 for two different bases $B$: *abc* and *def*.

(b) Show that $x$ and $y$ are joined by an edge if and only if $B \cup y - x$ is a basis.

(c) Show that if $G_B(M)$ is a connected graph, then $M$ is a connected matroid.

(d) Prove the converse of (c) is also true: if $M$ is a connected matroid, then $G_B(M)$ is connected as a graph.

# 4

# Graphic matroids

Matroids are important for lots of reasons, but the fact that they arise from graphs, vectors, matchings in bipartite graphs and arrangements of hyperplanes is a powerful argument for their study. We will focus our attention on these four specific classes:

- graphic matroids (this chapter).
- representable matroids (Chapter 6).
- transversal matroids (Sections 7.1 and 7.2 of Chapter 7).
- hyperplane arrangements (Sections 7.3 and 7.4 of Chapter 7).

Studying these classes in more depth from the matroid viewpoint is rewarding. The real payoff is the following:

---

**Why matroids matter**

If a certain class of objects is a matroid, then all the theorems that are true for matroids are true for this class.

---

Said in a different way, every theorem we prove for matroids gives us, for free, a theorem about graphs, vectors in a vector space, transversals in bipartite graphs, arrangements of hyperplanes, and so on.[1] Matroids are the "right" level of generalization for us because

- there are so many equivalent ways to axiomatize matroids (see Chapter 2), and
- so many important classes in combinatorics and finite geometry satisfy the matroid axioms.[2]

For a specific example, suppose we've proven that the spanning trees of a connected graph are the bases of a matroid. Then we immediately

---

[1] This is a broad argument for generalization. It's also a smart way to shop: you can get several theorems for the price of one.

[2] This is really an advertisement for matroids. If you're reading this text – which presumably you are – it probably isn't necessary. But now you can tell your friends why matroids matter.

know that the cycles in the graph are circuits of a matroid from the cryptomorphism between circuits and bases from Chapter 2. This allows us to interpret the circuit axioms for graphs, so, for instance, circuit elimination (C3) can be stated as a *theorem* about the cycles of a graph. We will highlight the theorems that arise in this way throughout this chapter. We'll also give examples to show this relationship is usually a one way street, that is, theorems that hold for the edges of a graph need not be true for (non-graphic) matroids.

## 4.1  Graphs are matroids

Throughout the first three chapters of this text, we have been using the fact that the edges of a graph form a matroid (technically, the family of acyclic subsets of edges satisfy the independent set axioms). But we haven't proven this yet! We fill in this gap now by proving that the *cycle matroid* $M(G) = (E, \mathcal{I})$ is in fact a matroid; we show the collection $\mathcal{I}$ of acyclic subsets of edges satisfy the independent set axioms of a matroid. Several alternative proofs are outlined in the exercises – see Exercise 4 for three different approaches.

**Theorem 4.1.** *Let $G$ be a graph with edge set $E$, and let $\mathcal{I}$ be the collection of all subsets of $E$ that do not contain a cycle. Then $\mathcal{I}$ forms the independent sets of a matroid on the ground set $E$, called the cycle matroid $M(G)$.*

Before giving the proof, we will need to recall the following fact from Chapter 1. Recall a tree $T$ is defined as an acyclic, connected graph.

**Theorem 1.15.** *If $T$ is a tree with n vertices, then $T$ has $n - 1$ edges.*

*Proof Theorem 1.15.* This theorem appears as Exercise 17 from Chapter 1, but we include the proof here because this result is of central importance. First, $T$ must contain a *leaf*, an edge incident to a vertex of degree[3] 1. If not, then every edge in $T$ would have degree at least 2. Then we could choose a vertex $v_1$ and we could create a list of adjacent vertices $v_1 \rightarrow v_2 \rightarrow v_3 \rightarrow \cdots$, at each stage choosing a new vertex adjacent to the previous vertex. Since each vertex has degree at least 2, there must be some vertex $v_i$ that appears twice in our list of vertices visited (since there are a finite number of vertices in $T$). But this gives a cycle, contradicting the fact that $T$ is a tree.

We now finish the proof by induction on the number of vertices. If $n = 1$, then $T$ has no edges, so the result is true.

Now suppose $T$ is a tree with $n$ vertices and let $e$ be an edge incident to a vertex $v$ of degree 1, i.e., $v$ is a leaf vertex. Then remove the vertex

---

[3] The *degree* of a vertex is the number of edges it is adjacent to.

$v$ and the edge $e$ from $T$. The result is a tree $T'$ (why?) with $n-1$ vertices. Therefore, by induction, $T'$ has $n-2$ edges. But $T$ has exactly one more edge than $T'$, namely the edge $e$. Thus, the number of edges in $T$ is $(n-2)+1 = n-1$, and we are done.    □

We also recall (see Section 1.4) that an acyclic subset of edges in a graph is simply a disjoint union of trees (called a *forest*). One consequence of Theorem 1.15 is that if $A$ is a forest composed of $v$ vertices, $e$ edges and $c$ disjoint trees, then $e = v - c$. The number of disjoint trees $c$ is called the number of components of $A$. We'll use this count in our proof.

*Proof Theorem 4.1.* We show the three conditions (I1), (I2) and (I3) are all satisfied by the acyclic subsets of edges of a graph. (I1) is trivial: $\emptyset$ is certainly acyclic. (I2) is also trivial: if $B$ is a subset of edges that do not contain a cycle and $A \subseteq B$, then $A$ is also acyclic.

Proving (I3)      As usual, the hardest part of the proof is showing that (I3) is always satisfied. So let $A$ and $B$ be acyclic subsets with $|A| < |B|$. We need to find an edge from $B$ that is not in $A$ which can be added to $A$ without creating a cycle.

First assume that $A$ is a tree (this is the easier case). Then the edges of $A$ meet only $|A|+1$ vertices of the graph $G$. We need to find an edge $b \in B$ to add to $A$ without creating a cycle. Since $B$ is a forest, it meets $|B|+k$ vertices, where $k \geq 1$ is the number of components of $B$. Since $|B| > |A|$, we have $|B|+k > |A|+1$ and so the edges of $B$ meet some vertex of $G$ that is not met by any edge of $A$. Call this vertex $v$ and let $b \in B$ be any edge that meets $v$ (there may be several such edges). Then $A \cup \{b\}$ must be acyclic and (I3) holds.

What if $A$ is not a tree? Then we can't guarantee that $B$ will hit a new vertex, but we can still make the proof work by looking at each tree within $A$. Suppose $A$ is composed of $c$ disjoint trees, which we'll call $T_1, T_2, \ldots, T_c$. If we're lucky and the edges of $B$ do meet a new vertex of $G$, then we're done; so we'll suppose otherwise that $B$ doesn't meet any new vertices of $G$. Let $e_i$ denote the number of edges in the tree $T_i$, so that $e_1 + e_2 + \cdots e_c = |A|$. In this case, we make the following claim.

*Claim.* There is an edge of $B$ that joins up two of the trees $T_i$ and $T_j$ from $A$.

Why is this true? Well, if not, then how many edges could $B$ have? Since $B$ is acyclic, it would have at most $e_1$ edges of $T_1$ (otherwise we'd get a cycle among the vertices of $T_1$). Similarly, $B$ could have at most $e_2$ edges of $T_2$, and so on. Then $|B| \leq e_1 + e_2 + \cdots e_c = |A|$, which contradicts $|A| < |B|$. Thus, $B$ *must* include an edge that joins the vertices of some $T_i$ to some other $T_j$. This edge can now be safely added to $A$ without creating any cycles, so (I3) is satisfied and we're done. See Figure 4.1.    □

Table 4.1. *Matroid terms interpreted for graphs.*

| Matroid term | Symbol | Graph interpretation |
|---|---|---|
| Independent sets | $\mathcal{I}$ | Edges of forests. |
| Bases | $\mathcal{B}$ | Edges of spanning forests. (Spanning trees if $G$ is connected.) |
| Circuits | $\mathcal{C}$ | Edges of cycles. |
| Rank | $r$ | $r(A)$ is the number of edges of a spanning forest in $A$. |
| Flats | $\mathcal{F}$ | Edges $F$ for which there is a partition $\Pi$ of the vertices so that $e \in F$ whenever $e$ joins two vertices of the same block of $\Pi$. |
| Cocircuits | $\mathcal{C}^*$ | Minimal edge cut-sets. |
| Hyperplanes | $\mathcal{H}$ | Cocircuit complements or maximal flats. |
| Closure | $^-$ | $\bar{A}$ contains all edges whose endpoints are connected by a path in $A$. |

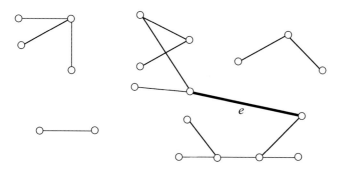

Figure 4.1. When $A$ and $B$ are forests with $|A| < |B|$, either $B$ hits a new vertex or it contains an edge $e$ joining two trees in the $A$ forest. Here we show the forest $A$ and the single edge $e \in B$.

The proof of this theorem doesn't really need the two separate cases, but we think it's sometimes a good idea to do an easy case just to get started on a proof.[4]

Now that we have shown $M(G) = (E, \mathcal{I})$ is a matroid, we can use our cryptomorphisms to describe bases, circuits, cocircuits, the rank function, flats, hyperplanes and closure of $M(G)$ in purely graph theory terms. We summarize this information in Table 4.1, and the next example illustrates the connections between the matroid concepts and graphs.

**Example 4.2.** Let $G$ be the graph drawn in Figure 4.2. We give examples of all the interpretations in Table 4.1.

- Independent sets ↔ forests. For example, $I = \{a, b, c, d, e, j, o, l\}$ is    Independent sets
  a forest composed of three trees: $T_1 = \{a, b, c, d, e\}$, $T_2 = \{j\}$, $T_3 = \{l, o\}$. These three trees induce a partition of the vertices of $G$:

  $$\{1, 2, 3, 4, 5, 6\}, \{7, 8\}, \{9, 10, 11\}.$$

  Vertex partitions play an important role in describing flats – see below.
- Bases ↔ spanning trees. Since $G$ is connected, maximal acyclic sub-    Bases
  sets of edges are trees that meet every vertex, i.e., spanning trees. For

---

[4] "If you can't solve a problem, then there is an easier problem you can solve: find it." G. Polya.

Figure 4.2. The graph for
Example 4.2.

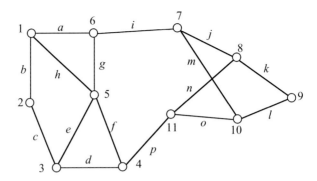

instance, $B = \{b, c, e, f, g, k, l, m, o, p\}$ is a spanning tree. Note that
every spanning tree has 10 edges.

Circuits   • Circuits ↔ cycles. For instance, the cycle $\{a, b, c, e, g\}$ is a circuit in
the matroid.

Rank   • Rank ↔ $r(A)$ is the size of the largest spanning forest contained in $A$.
For example, for $A = \{a, c, d, e, f, g, h, j, k, l, m\}$, we get a forest
composed of the two trees $\{c, d, f, g, h\}$ and $\{j, k, l\}$. This forest
includes eight edges, so $r(A) = 8$.

Flats   • Flats ↔ edge-sets corresponding to vertex partitions.[5] While we usually don't pay much attention to the vertices of a graph when thinking
about the matroid structure,[6] the vertices are of interest to us here.
Here's how this works. First, take a vertex partition.

$$\Pi = \{1, 3, 4, 5\}, \{2\}, \{6\}, \{7, 8, 10, 11\}, \{9\}.$$

Now find all edges that join vertices within each of the blocks of $\Pi$. For
the block $\{1, 3, 4, 5\}$, we have edges $d, e, f$ and $h$. For the singleton
blocks $\{2\}, \{6\}$ and $\{9\}$, we pick up no additional edges. Finally, for
the block $\{7, 8, 10, 11\}$, we have edges $j, m, n$ and $o$. So the flat $F_\Pi$
associated to the vertex partition $\Pi$ is $\{d, e, f, h, j, m, n, o\}$.

Why is this a flat? Adding any edge not in $F_\Pi$ to this set increases
the rank by either hitting a new vertex (for instance, edge $c$ hits vertex
2), or joining two components of $F_\Pi$ (for instance, edge $p$ joins the
two components $\{d, e, f, h\}$ and $\{j, m, n, o\}$).

Note we can get the same flat from different vertex partitions. The
partition $\Pi' = \{1, 3, 4, 5\}, \{2, 6, 9\}, \{7, 8, 10, 11\}$ induces the same
flat as before: $F_\Pi = F_{\Pi'}$.

Cocircuits   • Cocircuits ↔ minimal edge cut-sets. This is important and it has some
Cut-set   surprising implications. Recall a *cut-set* is a collection of edges whose
removal from the graph breaks a component into two or more pieces.[7]

---

[5] These were introduced in Example 2.18 in Chapter 2.
[6] Think of this as looking at the graph through matroid glasses, glasses that show the
independent and dependent subsets of edges. These glasses (more or less) ignore the
vertices.
[7] These were introduced in Chapter 3 – see Figure 3.19.

For instance, the set $C^* = \{j, m, n, o\}$ is a cocircuit since $E - C^*$ has two components. So is $\{e, f, g, h\}$. This latter cocircuit separates vertex 5 from the rest of the graph.

By the way, matroid cocircuits are also called *bonds*. This terminology comes directly from graph theory.

- Hyperplanes $\leftrightarrow$ maximal flat. Hyperplanes are cocircuit complements, Hyperplanes so $H = E - \{j, m, n, o\}$ is a hyperplane, for example. Since $G$ is connected, the set of edges that form a hyperplane is maximal with respect to not containing a spanning tree. It is then straightforward to see this happens precisely when $E - H$ is a minimal cut-set, i.e., a cocircuit.
- Closure $\leftrightarrow \overline{A}$ contains all edges that don't increase the rank of Closure $A$. For instance, if $A = \{c, d, e, h, j, k, l\}$, then $\overline{A} = \{b, c, d, e, f, h, j, k, l, m\}$, i.e., $\overline{A} = A \cup \{b, f, m\}$. Note the flat $\overline{A}$ can be realized by the vertex partition $\Pi = \{1, 2, 3, 4, 5\}, \{6\}, \{7, 8, 9, 10\}, \{11\}$.

Connected graphs are nicer than disconnected graphs.[8] From the matroid viewpoint, we can always assume the cycle matroid $M(G)$ comes from a connected graph $G$. An outline for the proof of this appears back in Section 1.4 (see Figure 1.26). We'll have more to say about connectivity in graphs and matroids in Section 4.4 (also see Section 3.5.4 in Chapter 3).

**Proposition 4.3.** *Let $G$ be a graph with cycle matroid $M(G)$. Then there is a connected graph $G'$ with $M(G) = M(G')$.*

In Chapter 3, we defined deletion and contraction as matroid operations, but also gave interpretations for graphs and representable matroids. For graphs, here is what we claimed (see Figure 3.8 of Chapter 3):

(1) **Deletion** Let $G - e$ be the graph obtained from $G$ by erasing the edge $e$.
(2) **Contraction** Let $G/e$ be the graph obtained from $G$ by erasing the edge $e$ and then identifying its endpoints.

Then these two graphs allow us to interpret matroid deletion and contraction for graphs.

**Proposition 4.4.** *Let $G$ be a graph and let $e$ be an edge that is neither a loop nor an isthmus in the cycle matroid $M(G)$. Then*

$$M(G) - e = M(G - e) \quad and \quad M(G)/e = M(G/e).$$

*Proof Proposition 4.4.* For deletion, $I$ is an independent set in $M(G) - e$ if and only if $I$ is a subset of edges satisfying two conditions:

(1) $e \notin I$, and
(2) $I$ contains no cycles.

<hr>
[8] Whatever that means.

Then the edges of $I$ remain acyclic in the graph $G - e$, so $I$ is an independent set in $M(G - e)$, and every independent set in $M(G - e)$ arises in this way. Since the two matroids have the same independent sets, we conclude $M(G) - e = M(G - e)$.

For contraction, a subset of edges $I$ is independent in the matroid $M(G)/e$ if $I \cup e$ is independent in the matroid $M(G)$, i.e., $I \cup e$ is an acyclic set in the graph $G$. Such a set can contain no path (besides $e$ itself) in $G$ joining the endpoints of $e$. Thus, $I$ remains acyclic in the contracted graph $G/e$, so the independent sets of $M(G)/e$ are independent in the matroid $M(G/e)$. But if $I'$ is an acyclic subset of edges in $G/e$, then $I' \cup e$ remains acyclic in $G$. So again, the two matroids $M(G)/e$ and $M(G/e)$ have the same independent sets, and we conclude $M(G)/e = M(G/e)$. $\qquad\square$

We conclude this section with a brief comment about isthmuses (coloops). If $G$ is a connected graph, an isthmus in the graphic matroid $M(G)$ is a cut-set of size 1, i.e., an edge whose erasure disconnects the graph. Graph theorists also use the term *bridge* instead of isthmus, and this term can occasionally lead students astray. For instance, the graph $G$ from Figure 4.2 has no bridges, but it is tempting to view the two edges $i$ and $p$ as bridges joining two pieces of the graph. (You could make the same argument for the pair of edges $k$ and $l$, or any other cut-set of size 2.) **Don't make this mistake!**[9]

## 4.2  Graph versions of matroid theorems

In Chapter 2, we introduced many of the axiom systems you can use to define a matroid. Stronger versions of some of these axioms hold for graphs.

### 4.2.1  Circuit elimination for graphs

We have seen two versions of circuit elimination axioms for matroids so far. Weak circuit elimination says the following:

(C3)  Let $C_1$ and $C_2$ be circuits in a matroid $M$ with $e \in C_1 \cap C_2$. Then there is a circuit $C_3 \subseteq C_1 \cup C_2$ that avoids $e$.

*Strong* circuit elimination requires the circuit $C_3$ to do a little bit more:

(C3′)  If $C_1$ and $C_2$ are circuits with $x \in C_1 - C_2$ and $y \in C_1 \cap C_2$, then there is a circuit $C_3 \subseteq C_1 \cup C_2$ such that $x \in C_3$ and $y \notin C_3$.

---

[9]  As a general rule (for matroids, math or life), it would be nice to know in advance which mistakes to avoid.

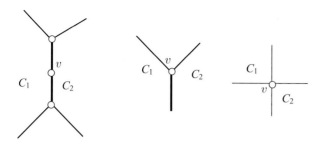

Figure 4.3. A vertex $v$ of $C_1 \cup C_2$ has three possibilities. The bold edges are in $C_1 \cap C_2$, and the lighter edges are in $C_1 \triangle C_2$.

Strong circuit elimination appears in the proof of circuit transitivity (Lemma 3.37), and you were asked to prove it in Exercise 46 in Chapter 3.

For graphs, these are both special cases of a more general result. We remind you that $(A - B) \cup (B - A)$ is called the *symmetric difference* of the sets $A$ and $B$, and is usually written $A \triangle B$.

Symmetric difference

**Theorem 4.5.** *Let $C_1$ and $C_2$ be circuits in the cycle matroid $M(G)$ of a graph $G$. Then $C_1 \triangle C_2$ is a disjoint union of circuits.*

Before proving the theorem, note that it immediately implies strong circuit elimination (C3′): every edge $e \in C_1 - C_2$ is in a circuit that completely avoids all of the edges of $C_1 \cap C_2$.

*Proof Theorem 4.5.* Every vertex in the circuit $C_1$ has degree 2, as does every vertex in $C_2$. What are the degrees of the vertices of $C_1 \triangle C_2$? If the vertex $v$ is not incident to any edge of $C_2$, then it clearly has degree 2 in $C_1 \triangle C_2$ (and, of course, the same argument is true if $v$ is only incident to edges of $C_2$).

What if a vertex $v$ is incident to edges in both $C_1$ and $C_2$? There are three possibilities: $v$ has degree 2, 3 or 4 in $C_1 \cup C_2$ – see Figure 4.3. If $v$ has degree 2, then both edges incident to $v$ are in $C_1 \cap C_2$, so $v$ will have degree 0 in $C_1 \triangle C_2$. If $v$ is incident to exactly one edge of $C_1 \cap C_2$, then $v$ will have degree 2 in $C_1 \triangle C_2$. Finally, if $v$ is incident to no edges of $C_1 \cap C_2$, then $v$ will have degree 4 in $C_1 \triangle C_2$.

Thus, every vertex of $C_1 \triangle C_2$ will have even degree. By a well-known theorem of Euler,[10] this means $C_1 \triangle C_2$ is a disjoint union of cycles. Thus, in the cycle matroid $M(G)$, we get $C_1 \triangle C_2$ is a disjoint union of circuits. □

For a quick example, consider the graph $G$ from Figure 4.2. For circuits $C_1 = \{a, b, c, d, p, o, l, k, j, i\}$ and $C_2 = \{f, g, i, j, n, p\}$, we get

---

[10] If every vertex in a graph has even degree, the entire graph can be broken into an edge-disjoint union of cycles. Euler proved this in 1735, solving the Königsberg bridge problem and more or less inventing graph theory. A graph is *Eulerian* if you can decompose all its edges into disjoint cycles. If you haven't seen this before, take time out to look it up!

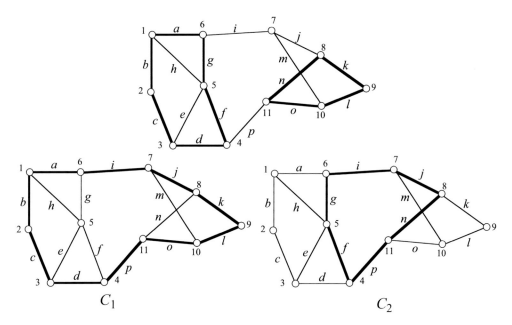

Figure 4.4. The symmetric difference of two circuits in a graph is a disjoint union of circuits: $C_1 \triangle C_2 = C_3 \cup C_4$. The top picture shows $C_1 \triangle C_2$.

$C_1 \triangle C_2 = C_3 \cup C_4$, where $C_3 = \{a, b, c, d, f, g\}$ and $C_4 = \{k, l, o, n\}$. See Figure 4.4.

Theorem 4.5 is not true for all matroids – see Exercise 5. But it remains true for *binary* matroids, i.e., matroids representable by matrices with entries from the field $\{0, 1\}$. (In fact, this theorem is one of many that characterizes binary matroids, and graphic matroids are binary. See Section 8.2.1 of Chapter 8 for more on this connection.)

### 4.2.2　Circuits, cocircuits and strong basis exchange

Theorem 4.5 tells us graphs are special within the class of matroids, at least as far as the circuits are concerned. What about bases? Let's look at an example illustrating the difference between weak and strong basis exchange in a graph. (Strong basis exchange was introduced in Exercise 14 in Chapter 2, and it was used in proving Theorem 3.15 in Chapter 3.)

Weak basis exchange:

(B3) If $B_1, B_2 \in \mathcal{B}$ and $x \in B_1 - B_2$, then there is an element $y \in B_2 - B_1$ so that $B_1 - x \cup \{y\} \in \mathcal{B}$.

Strong basis exchange:

(B3′) If $B_1, B_2 \in \mathcal{B}$ and $x \in B_1 - B_2$, then there is an element $y \in B_2 - B_1$ so that *both* $B_1 - x \cup \{y\} \in \mathcal{B}$ and $B_2 - y \cup \{x\} \in \mathcal{B}$.

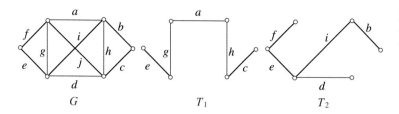

Figure 4.5. A graph $G$ (left) with two of its spanning trees $T_1$ and $T_2$.

From Exercise 14 in Chapter 2, we know the stronger property (B3′) holds for all matroids, but let's examine this more carefully in an example.

**Example 4.6.** Let $G$ be the graph on the left in Figure 4.5, and let $T_1$ and $T_2$ be the spanning trees of $G$ shown in the figure. Note $a \in T_1 - T_2$, and (B3) (weak exchange) ensures the existence of $y \in T_2 - T_1$ so that $T_1 - a \cup y$ is also a spanning tree.

Here's one way to find such an edge $y$ in our graph: we note that $T_1 - a$ has two components, so to complete $T_1 - a$ to a basis, we need to choose an edge from $T_2$ that connects these two pieces. The edges of $G$ that join these two pieces form the minimal cut-set (cocircuit) $C^* = \{a, d, i, j\}$. Then $T_1 - a \cup y$ will be a spanning tree precisely when $y \in C^* \cap T_2$. Now $C^* \cap T_2 = \{d, i\}$, so we can choose either $y = d$ or $y = i$. This verifies (B3), and we can go home happy now. (Note: $C^* = E - \overline{T_1 - a}$ is the complement of the hyperplane $\overline{T_1 - a}$.)

Verifying (B3′) will take a little more work. Note that although both $T_1 - a \cup i$ and $T_1 - a \cup d$ are spanning trees, only $T_2 - i \cup a$ is a spanning tree, i.e., $T_2 - d \cup a$ is not. How could we have figured this out?

Well, note that $T_2 \cup a$ contains a unique cycle $C = \{a, e, f, i\}$ – the *basic circuit* determined by $T_2$ and $a \notin T_2$. To ensure that $T_2 - y \cup a$ is also a spanning tree, we need to make sure $y \in C$ (otherwise $T_2 - y \cup a$ will still contain a circuit). So we finally conclude that $y \in C \cap C^*$. Now $C \cap C^* = ai$, so the edge $i$ will make both $T_1 - a \cup i$ and $T_2 - i \cup a$ spanning trees. This verifies (B3′) and we can go home absolutely ecstatic.[11]

Our proof that strong basis exchange holds for all matroids (Exercise 14 in Chapter 2) depended on the following facts. All of these are true *for all matroids*:

(1) (Exercise 13(a) in Chapter 2) For every basis $B$ and $x \notin B$, there is a unique circuit contained in $B \cup x$. This is the *basic* or *fundamental* circuit determined by $B$ and $x$.
(2) For every basis $B$ and $y \in B$, there is a unique cocircuit contained in $E - (B - y)$. This is the *basic cocircuit* determined by $B$ and $y$.
(3) (Exercise 9 in Chapter 2) If $C$ is a circuit and $C^*$ is a cocircuit, then $|C \cap C^*| \neq 1$.

---

[11] Our standards are pretty low.

It seems a bit unfair to expect you to have done all of these problems, so here are a few proofs.[12] The next collection of results, Propositions 4.7, 4.8, 4.9 and 4.10, are true *for all matroids*.

**Proposition 4.7.**  *Let B be a basis of a matroid M* $= (E, \mathcal{B})$.

(1)  *(Exercise 13(a) in Chapter 2) For all x* $\notin$ *B, there is a unique circuit* $C \subseteq B \cup x$ *which contains x.*

Basic circuit

Basic cocircuit

(2)  *For all y* $\in$ *B, there is a unique cocircuit* $C^* \subseteq E - (B - y)$.

*Proof Proposition 4.7.* (1) For a basis $B$ with $x \notin B$, $B \cup x$ is not an independent set, hence it must contain a circuit. Moreover, since $B$ itself contains no circuits, $x$ is contained in every such circuit. Suppose $C_1$ and $C_2$ are distinct circuits contained in $B \cup x$. Since $x \in C_1 \cap C_2$, we can use the circuit elimination axiom (C3) to get a circuit $C_3 \subseteq C_1 \cup C_2 - x$. But $C_3$ does not contain $x$, so $C_3 \subseteq B$, a contradiction.

(2) Dualize the above argument.    □

**Proposition 4.8.**  *[Exercise 9 in Chapter 2] For every circuit C and cocircuit C\* of the matroid M,* $|C \cap C^*| \neq 1$.

*Proof Proposition 4.8.* Suppose $C \cap C^* = \{x\}$. Let $H$ be the hyperplane $H = E - C^*$, and note $x \in C^*$ implies $x \notin H$. Now $C - x \subseteq H$ and $x \in \overline{C - x}$ gives $x \in \overline{C - x} \subseteq \overline{H}$. But $\overline{H} = H$, since $H$ is closed. So $x \notin C^*$, a contradiction.    □

In Example 4.6, we verified weak and strong exchange for the graph in Figure 4.5. The general situation for weak exchange is given in the next proposition. Given bases $B_1$, $B_2$ with $x \in B_1 - B_2$, what elements $y \in B_2 - B_1$ can be chosen so that $B_1 - x \cup \{y\}$ is a basis?

**Proposition 4.9.**  *Let* $B_1$, $B_2$ *be bases of the matroid M* $= (E, \mathcal{B})$ *with* $x \in B_1 - B_2$ *and* $y \in B_2 - B_1$. *The following are equivalent:*

(1)  $B_1 - x \cup \{y\}$ *is a basis.*
(2)  *x is in the basic circuit C of y with respect to* $B_1$.
(3)  *y is contained in the cocircuit* $C^* = E - \overline{B_1 - x}$.

*Proof Proposition 4.9.* (1) $\Rightarrow$ (2) If $B_1 - x \cup \{y\}$ is a basis and $x \notin C$ then $C \subseteq B_1 \cup \{y\} - x$; a contradiction.

(2) $\Rightarrow$ (3) We'll prove the contrapositive. If $y \notin C^*$, then $y \in \overline{B_1 - x}$, and since $B_1 - x$ is a basis of $\overline{B_1 - x}$, $B_1 - x \cup y$ is dependent. Let $C$ be a circuit contained in $B_1 - x \cup \{y\}$. Then this circuit must contain $y$ and hence must be the *unique* basic circuit $C_1$ of $y$ with respect to $B_1$ (Proposition 4.7(1)). But by construction, $x \notin C$.

---

[12] Did you know that the solutions to some of the homework in Chapter 2 would be presented in Chapter 4? Would it have mattered?

(3) $\Rightarrow$ (1) Again, we'll use the contrapositive. If $B_1 - x \cup \{y\}$ is a not a basis, then $r(B_1 - x \cup \{y\}) = r(B_1 - x) = r(B_1) - 1$. Thus, $\overline{B_1 - x \cup \{y\}}$ and $\overline{B_1 - x}$ are both hyperplanes, so they are equal. Hence $y \in \overline{B_1 - x}$, so $y \notin C^*$.   □

We now give a quick proof of strong basis exchange.   Strong basis exchange

**Proposition 4.10.**   *[Exercise 14 in Chapter 2] Let $M = (E, \mathcal{B})$ be a matroid. Then*

(B3$'$) *If $B_1, B_2 \in \mathcal{B}$ with $x \in B_1 - B_2$, then there is a $y \in B_2 - B_1$, so that $B_1 - x \cup y$ and $B_2 - y \cup x$ are bases.*

*Proof Proposition 4.10.* Let $B_1$ and $B_2$ be bases of $M$ with $x \in B_1 - B_2$. Let $C$ denote the basic circuit of $x$ with respect to $B_2$ and let $C^*$ denote the basic cocircuit $E - \overline{B_1 - x}$. Now $x \in C \cap C^*$, so, by Proposition 4.8, there is some $y \in C \cap C^*$ with $x \neq y$. We claim that $y \in B_2 - B_1$, and $B_1 - x \cup y$ and $B_2 - y \cup x$ are both bases.

First, $y \in B_2$, since $y \in C \subseteq B_2 \cup x$ and $y \neq x$. If $y \in B_1$, then since $y \neq x$, $y \in B_1 - x \subseteq \overline{B_1 - x}$ and $y \notin C^*$, a contradiction. Thus, $y \in B_2 - B_1$.

But now both results follow from Proposition 4.9: $B_1 - x \cup y$ is a basis since $y \in C^*$ (by Proposition 4.9(3)) and $B_2 - y \cup x$ is a basis since $y \in C$ (by Proposition 4.9(2)).   □

As we mentioned above, Propositions 4.7, 4.8, 4.9 and 4.10 are true for all matroids. In the spirit of this section, we now give a stronger version of Proposition 4.8 that is true for graphic matroids, but not general matroids (Exercise 12).

**Proposition 4.11.**   *Let $M(G)$ be a graphic matroid with circuit $C$ and cocircuit $C^*$. Then $|C \cap C^*|$ is even.*

*Proof Proposition 4.11.* We assume $G$ is connected – see Proposition 4.3. Now if $C^*$ is a cocircuit of $M(G)$, then $C^*$ gives a partition of the vertices of $G$ into two parts, say $S$ and $T$. Then a circuit $C$ in the matroid $M(G)$ corresponds to a cycle in the graph, and this cycle either uses only vertices in $S$, only vertices in $T$, or vertices in both $S$ and $T$. In either of the former cases, $C \cap C^* = \emptyset$.

If the circuit uses vertices in both $S$ and $T$, we *order* the vertices of $C$. Then, as we traverse our cycle $C$ in order, we keep track of which component of $E - C^*$ we are in. If we begin in $S$, then, for each edge of $C^*$ we take *from $S$ to $T$*, we must take a different edge back *from $T$ to $S$*. Since we end up back where we started,[13] $|C \cap C^*|$ is even. See Figure 4.6.   □

---

[13] Life is a journey, but it's always nice to come home.

Figure 4.6. Top: a graph $G$.
Bottom left: the circuit $abcj$.
Bottom right: the cocircuit
$achi$. The dashed line shows
the partition of the vertices:
$V = S \cup T$. Then
$C \cap C^* = \{a, c\}$ contains an
even number of edges. See the
proof of Proposition 4.11.

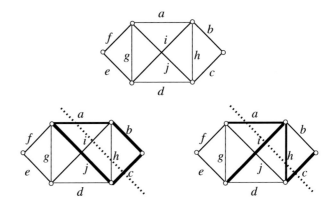

## 4.3 Duality and cocircuits in graphs

Matroid duality is motivated by planar graph duality – this was the
central message of Section 3.3.1 in Chapter 3. In Theorem 3.27, we saw
a close connection between planarity and duality. The proof is outlined
in Exercise 29 of Chapter 3.

**Theorem 3.27.** *Let $G$ be a connected graph. Then the dual matroid
$M(G)^*$ is graphic if and only if $G$ is planar. Further, if $G$ is planar, then
$M(G^*) = M(G)^*$.*

Although the dual $M(G)^*$ of the cycle matroid $M(G)$ is not graphic
when $G$ is not planar, we can still study the duals of cycle matroids
for non-planar graphs. One consequence of abstract matroid duality is
that every theorem involving circuits gives you a dual theorem about
cocircuits for free. For example, if $B$ is a basis and $x \in B$, then the fact
that there is a unique cocircuit contained in $E - (B - x)$ follows from
the corresponding fact about basic circuits – see Proposition 4.7.

We examine this connection for graphs in more detail in this section.
This allows us to study the duals of arbitrary graphs, including non-
planar graphs. These are the *cographic matroids* $\mathcal{G}^*$ from Definition 3.28
in Chapter 3. Let's start with an example.

**Example 4.12.** Let $G$ be the graph from Figure 4.2. We saw in Section 4.1 that cocircuits in the cycle matroid $M(G)$ are minimal edge
cut-sets. Let's investigate this in a bit more detail now. Consider the
four edges incident to the vertex 5: $efgh$. These edges form a cut-
set – we call such a cut-set a *vertex cut-set*.[14] Call this cocircuit
$C_5^* = \{e, f, g, h\}$. Similarly, $C_6^* = \{a, g, i\}$. Then the symmetric difference $C_5^* \triangle C_6^* = \{a, e, f, h, i\}$ is also a cocircuit – removing these
edges partitions the vertex set into two pieces. See Figure 4.7.

---

[14] We used vertex cut-sets in Example 3.25 in Chapter 3, where they were used in
proving the matroid $M(K_5)^*$ is not a graphic matroid.

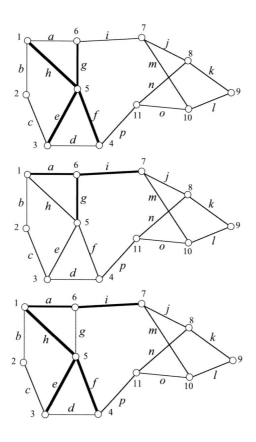

Figure 4.7. The symmetric difference $C_5^* \triangle C_6^*$ of two cocircuits can produce another cocircuit.

Let's turn this around. Given a cocircuit $C^*$, can we find vertices whose vertex cut-sets give $C^*$ as their symmetric difference? Try to create the cut-set $\{i, n, o\}$ like this. (Pause to draw.)

The cut-set $C^* = \{i, n, o\}$ separates vertices $\{7, 8, 9, 10\}$ from the rest of the vertices. To generate $C^*$, a good first try is to simply take the symmetric difference of these four vertices: $C_7^* \triangle C_8^* \triangle C_9^* \triangle C_{10}^*$. (Taking the symmetric difference of more than two sets is easy to do – just take all the elements that are in an odd number of the sets. See Exercise 14.) In this case, we get $C_7^* \triangle C_8^* \triangle C_9^* \triangle C_{10}^* = C^*$, so our first try was successful.

What about the symmetric difference of the vertex cut-sets of the remaining vertices? In our example, we would get $C_1^* \triangle C_2^* \triangle \cdots$, where we use the seven vertices $\{1, 2, 3, 4, 5, 6, 11\}$. This also works – the symmetric difference of the vertices in either one of the two components of the graph gives the same cut-set.

It's tempting to try to use the example to create a theorem, but a single example can sometimes lead you astray. Consider the graph $G$ from Figure 4.8. Then the vertex $v$ is a *cut-vertex*. This means that removing that vertex, along with all the edges incident to it, disconnects the graph. Then the edges incident to $v$ are not a *minimal* cut-set; in fact, these edges can be partitioned into two cocircuits (and, in general, the

Cut-vertex

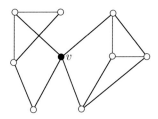

Figure 4.8. $v$ is a cut-vertex, and the edges incident to $v$ do not form a cocircuit in $M(G)$.

edges incident to a cut-vertex can always be partitioned into at least two cocircuits).

The next theorem puts all of this information in a nice, tidy package.

**Theorem 4.13.** *Let G be a graph with no cut-vertices.*

(1) *The symmetric difference of any two minimal cut-sets is a disjoint union of minimal cut-sets.*

(2) *Every minimal cut-set is a symmetric difference of vertex cut-sets, i.e., every cocircuit of the cycle matroid $M(G)$ is a symmetric difference of the vertex cocircuits.*

(3) *Let $S \subseteq V$ and let $C_S^*$ be the symmetric difference of the vertex cut-sets over the vertices of S. Then $C_S^* = C_{V-S}^*$.*

We leave the proof of Theorem 4.13 for you – see Exercise 15. Graphs with no cut-veritces are blocks. Blocks appeared in the discussion of Theorem 3.50 at the end of Chapter 3, and they will be important in Section 4.4.

A generalization of Theorem 4.13 to *binary matroids* (matroids representable by a matrix of 0's and 1's, using mod 2 arithmetic) appears as Corollary 8.20 in Chapter 8. In that context, Theorem 4.13 is a corollary of this more general result in Chapter 8.[15]

Part (1) is true for planar graphs by duality and Theorem 4.5.[16] When people[17] talk about the matroid associated with a graph, they are almost always referring to the cycle matroid $M(G)$. But that isn't the only matroid we can define on $G$. The circuits of the dual matroid $M(G)^*$ are the cocircuits of $M(G)$. Since the cocircuits obey the same axioms as the circuits, we could *define* a new matroid on a graph $G$ whose circuits are the cutsets of $G$.

**Definition 4.14.** Let $G$ be a graph. The *cocircuit* matroid $M^*(G)$ is the matroid whose circuits are the cocircuits of the cycle matroid $M(G)$.

Blocks

Cocircuit (bond) matroid

---

[15] But don't worry about that – be happy with what you have right now. This approach to life will serve you well.

[16] In fact, duality immediately shows this is true for all *cographic* matroids, where "minimal cut-set" should be replaced with "cocircuit" when the cographic matroid is not graphic.

[17] Almost all of these "people" are matroid theorists.

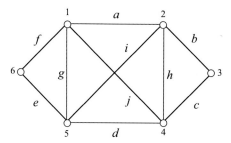

Figure 4.9. The graph for Example 4.17.

$M^*(G)$ is called the *corcicuit* or *bond* matroid. The next result follows immediately from the fact that $C$ is a circuit in $M$ if and only if $C$ is a cocircuit in $M^*$.

**Theorem 4.15.** *Let $G$ be a graph with cycle matroid $M(G)$. Then $M^*(G) = M(G)^*$, i.e., the cocircuit matroid associated to $G$ is the dual of the cycle matroid $M(G)$.*

These two matroids are both closely associated with the *vertex–edge incidence matrix*.

**Definition 4.16.** If $G$ has $m$ vertices and $n$ edges, the *vertex–edge incidence matrix* $A_G$ is an $m \times n$ matrix constructed as follows:

Incidence matrix

- Label the rows by the $m$ vertices and the columns by the $n$ edges.
- Set $a_{i,j} = 1$ if vertex $i$ is incident to edge $j$, and set $a_{ij} = 0$ otherwise.

Let's look at an example, shall we?

**Example 4.17.** Let $G$ be the graph of Figure 4.9. Here's the vertex–edge incidence matrix $A_G$:

$$A_G = \begin{array}{c} \\ 1 \\ 2 \\ 3 \\ 4 \\ 5 \\ 6 \end{array} \begin{array}{c} a\ \ b\ \ c\ \ d\ \ e\ \ f\ \ g\ \ h\ \ i\ \ j \\ \left[ \begin{array}{cccccccccc} 1 & 0 & 0 & 0 & 0 & 1 & 1 & 0 & 0 & 1 \\ 1 & 1 & 0 & 0 & 0 & 0 & 0 & 1 & 1 & 0 \\ 0 & 1 & 1 & 0 & 0 & 0 & 0 & 0 & 0 & 0 \\ 0 & 0 & 1 & 1 & 0 & 0 & 0 & 1 & 0 & 1 \\ 0 & 0 & 0 & 1 & 1 & 0 & 1 & 0 & 1 & 0 \\ 0 & 0 & 0 & 0 & 1 & 1 & 0 & 0 & 0 & 0 \end{array} \right] \end{array}.$$

The matrix has lots of nice properties: each column has exactly two non-zero entries, and the row sums are the degrees of the corresponding vertices. So, summing up all the entries in the matrix, row by row, gives the sum of all the degrees of all the vertices. Summing column by column gives $2n$, twice the number of edges. Thus, the incidence matrix gives a rapid proof of the familiar theorem that the sum of the degrees in a graph is twice the number of edges.[18]

---

[18] The fact that $\sum deg(v) = 2e$ is sometimes called the *first* theorem in graph theory.

We are interested in how the matrix $A_G$ relates to the two matroids associated to $G$, the cycle matroid $M(G)$ and the dual cocircuit matroid $M^*(G)$. First, note the circuits in the cycle matroid are easy to find in $A_G$: restricting ourselves to the columns corresponding to a circuit, each row will have exactly two 1's, and these 1's form a cycle among the rows. For instance, the circuit $\{a, b, c, j\}$ is represented by

$$
\begin{array}{c}
 \\ 1 \\ 2 \\ 3 \\ 4
\end{array}
\begin{array}{cccc}
a & b & c & j \\
\end{array}
\left[
\begin{array}{cccc}
1 & 0 & 0 & 1 \\
1 & 1 & 0 & 0 \\
0 & 1 & 1 & 0 \\
0 & 0 & 1 & 1
\end{array}
\right].
$$

How can we spot the cocircuits? From Theorem 4.13, we know all the cocircuits are generated from the vertex cocircuits (assuming $G$ has no cut-vertices). But it's also easy to identify these in $A_G$: look at a row of $A_G$ and select the columns corresponding to the 1's in that row. For example, row 4 of $A_G$ has 1's in columns $c, d, h$ and $j$, so we must have $\{c, d, h, j\}$ is a vertex cut-set of $G$, i.e., a cocircuit in $M(G)$. Then as in Theorem 4.13, we can get the rest of the cocircuits by selecting some rows of $A_G$ and taking the mod 2 sum of the selected rows.

The incidence matrix of a graph allows us to represent a graphic matroid by a matrix: graphic matroids are representable (see Exercise 4(c)).

Moreover, Example 4.17 shows us that both the cycle matroid and the cocircuit matroid of a graph $G$ are representable by matrices using only 0's and 1's, with arithmetic mod 2. (To get the representation for the cocircuit matroid, use Theorem 6.6.) Matroids representable by matrices in this way are called *binary* matroids. The fact that graphic matroids are binary appears as Theorem 8.13 in Chapter 8, where we discuss the class of binary matroids in more detail.

## 4.4 Connectivity and 2-isomorphism

When does the cycle matroid $M(G)$ uniquely determine a graph $G$? The answer: not always[19] – we have already seen several examples of non-isomorphic graphs with the same cycle matroid. The goal of this section is to determine how much freedom there is in constructing different graphs with the same matroid structure.

How bad can this get? At one extreme, note that every forest on $n$ edges gives the same matroid, namely the Boolean algebra $B_n$ in which

---

[19] This isn't much of an answer, but be patient.

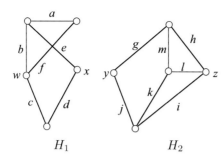

Figure 4.10. $H$ is the disconnected graph composed of the two components $H_1$ and $H_2$. $H'$ is formed by identifying $x$ with $y$, while $H''$ is formed by identifying $w$ with $z$. Then $M(H) = M(H') = M(H'')$.

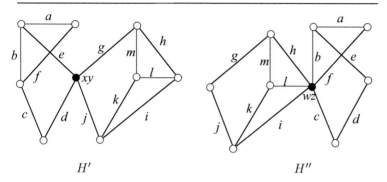

every subset is independent (and every point is an isthmus). So all the different $n$-edge forests give the same cycle matroid. The next example shows other ways non-isomorphic graphs can produce the same cycle matroid.

**Example 4.18.**

- **Vertex identification and splitting** First consider the disconnected graph $H$ composed of the two components $H_1$ and $H_2$ at the top of Figure 4.10. These two pieces can be joined to create a connected graph by identifying any vertex from $H_1$ with a vertex from $H_2$. We do this in two different ways in Figure 4.10; on the bottom left, the graph $H'$ is created by identifying the vertex $x$ of $H_1$ with $y$ from $H_2$. $H''$ is formed by identifying $w$ from $H_1$ with $z$ from $H_2$. Call this operation *vertex identification*.

  But all three graphs have precisely the same cycles! This is really important, and it's obvious. Since a matroid is completely determined by its circuits, which are the cycles of $G$, we immediately get $M(H) = M(H') = M(H'')$. Note that $H'$ and $H''$ each have a cut-vertex.

  We point out that vertex identification and the reverse procedure, *vertex splitting*, which we introduced in Example 1.19 in Chapter 1, are *graph theoretic operations*. Even though identification and splitting

Vertex identification

Vertex splitting

Figure 4.11. Two graphs that will be glued together in two different ways.

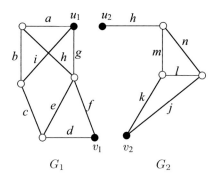

Figure 4.12. *G* is formed by identifying $u_1$ with $u_2$ and $v_1$ with $v_2$.

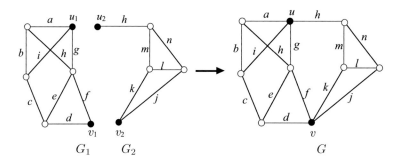

Figure 4.13. *G'* is formed by identifying $u_1$ with $v_2$ and $v_1$ with $u_2$.

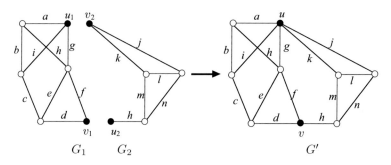

of vertices may look like they depend on the particular drawings of our graphs, they don't.

Twisting    • **Twisting** There is another way to get non-isomorphic graphs to have the same cycle matroid. Let $G_1$ and $G_2$ be the two graphs in Figure 4.11. Then we can glue $G_1$ and $G_2$ together to get a graph $G$ by identifying the vertex $u_1$ in $G_1$ with $u_2$ in $G_2$ and $v_1$ in $G_1$ with $v_2$ in $G_2$. See Figure 4.12.

But this is not the only way to put $G_1$ and $G_2$ together without changing our cycles. For instance, we could have identified $u_1$ with $v_2$ and $v_1$ with $u_2$, producing the graph $G'$ of Figure 4.13. As was the case above, this is a *graph theoretic operation* – it does *not* depend on the drawings!

Now the key observation is that $G$ and $G'$ have the same cycles, so $M(G)$ and $M(G')$ have identical circuits. Thus, as before, we have

$M(G) = M(G')$. For instance, the circuit $\{f, g, h, m, k\}$ in $M(G)$ is also a circuit in $M(G')$. Although the edges appear in a different order in the two graphs, the cycle matroid cannot detect this.

How do we know the graphs $G$ and $G'$ are not isomorphic, as graphs? A quick way to distinguish them is to note that $G'$ has a vertex of degree 5 (the vertex $u$), but $G$ does not. That does it.

Both of the operations in Example 4.18 can be reversed. We already saw that vertex identification and vertex splitting are inverses of each other. For instance, given the connected graph $H'$ on the bottom left in Figure 4.10, we note that vertex labeled $xy$ is a cut-vertex. Then we can split $H'$ into the two pieces $H_1$ and $H_2$, "cloning" the vertex $xy$ to get the two vertices $x$ in $H_1$ and $y$ in $H_2$.

We can also reverse the second operation from Example 4.18 when we find a pair of vertices $u$ and $v$ in a graph whose removal disconnects the graph. For instance, the graph $G$ on the right in Figure 4.12 has two such vertices. Then we do the following:

- Break $G$ into two pieces $G_1$ and $G_2$ of Figure 4.11, forming (temporary) "clones" $u_1$ and $u_2$ of the vertex $u$ of $G$ and $v_1$ and $v_2$ of the vertex $v$.
- Reattach $G_1$ and $G_2$ by identifying $u_1$ with $v_2$ and $v_1$ with $u_2$.

We call this operation *twisting* about the vertices $u$ and $v$. These three operations (vertex identification, vertex splitting and twisting) all preserve the cycle matroid. We give a definition to collect these operations.

*Twisting*

**Definition 4.19.** Two graphs $G$ and $G'$ are *2-isomorphic* if $G$ can be transformed to $G'$ by a sequence of the three operations of vertex identification, vertex splitting and twisting.

*2-isomorphic graphs*

In Example 4.18, we see $H$, $H'$ and $H''$ are 2-isomorphic; so are $G$ and $G'$. You can check that 2-isomorphism is an equivalence relation on the class of all graphs – see Exercise 20.

The next result follows immediately from the fact that 2-isomorphism preserves the circuits of the cycle matroid.

**Proposition 4.20.** *Suppose $G$ is 2-isomorphic to $G'$. Then $M(G) \cong M(G')$.*

The remarkable fact about Proposition 4.20 is that the converse is true. This is Whitney's 2-isomorphism theorem.

**Theorem 4.21.** *[Whitney 1933] Suppose $G$ and $G'$ are graphs with no isolated vertices, and $M(G) \cong M(G')$. Then $G$ and $G'$ are 2-isomorphic.*

We won't prove this theorem; we recommend Oxley's text [26] for a proof. The theorem tells us precisely how non-isomorphic graphs can give the same cycle matroid; the two graphs can be obtained from one another through the three operations of vertex splitting, vertex identification and twisting.

The operations of vertex splitting and twisting are only present when there are small subsets of vertices whose removal disconnects the graph. This motivates the next definition.

*k-connected graph* **Definition 4.22.** A graph $G$ is *k-connected* if $G$ remains connected after removing any $k$ vertices (along with all the incident edges). $G$ is *minimally k-connected* if it is $k$-connected, but not $(k + 1)$-connected.

So, if a connected graph $G$ has a cut-vertex, it is 1-connected, but not 2-connected. As we remarked earlier, graphs that are 2-connected are blocks. If $G$ has a pair of vertices where we can perform a twist, then $G$ is not 3-connected.

Graphs that are 3-connected play an important role in geometry. A classical theorem due to Steinitz[20] characterizes the graphs that can arise as the vertices and edges of a three-dimensional polytope: they are precisely the simple, planar, 3-connected graphs.

For our purposes, 3-connected graphs are especially nice: if $G$ is 3-connected (with no loops), then the cycle matroid $M(G)$ uniquely determines the graph $G$.

**Theorem 4.23.** *Suppose G is a 3-connected graph with no loops. Then G can be uniquely reconstructed from its cycle matroid $M(G)$.*

We can rephrase the theorem as follows. If $G$ is a loopless 3-connected graph and $M(G) = M(H)$ for some graph $H$ (where $H$ has no isolated vertices), then $G$ and $H$ are isomorphic as graphs. We point out that this follows immediately from Whitney's 2-isomorphism Theorem (Theorem 4.21), since we can't perform any of the 2-isomorphism operations on a loopless 3-connected graph. This answers the question we asked at the beginning of this section, telling us precisely when the cycle matroid $M(G)$ completely determines the graph $G$.

We give a proof of Theorem 4.23 that does not rely on Whitney's Theorem. The reason is two-fold. First, our proof uses the vertex-cocircuit ideas developed in Example 4.12 and Theorem 4.13. The other reason has to do with logic: Theorem 4.23 is often used as a lemma in the proof of Whitney's Theorem, so using Whitney's Theorem to prove Theorem 4.23 should set off alarm bells in the part of your brain that processes logic.

*Proof Theorem 4.23.* Suppose $G$ is loopless and 3-connected. We show how to reconstruct $G$ uniquely from the cocircuits of the cycle matroid $M(G)$. First, two observations:

---

[20] See [44] for a discussion of this theorem in geometry.

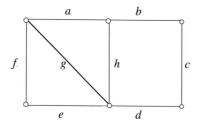

Figure 4.14. The graph for
Exercises 1 and 17.

- Since $G$ is 3-connected, $G - v$ is 2-connected for all vertices $v$, so the hyperplane consisting of the edges of $G - v$ is connected as a matroid (Exercise 22).
- If $H$ is a hyperplane in $M(G)$ that is *not* obtained by removing a vertex $v$, then $H$ is disconnected as a matroid. This is true because such a hyperplane must correspond to a non-trivial partition of the vertex set: $A \cup B$, where $|A| \geq 2$ and $|B| \geq 2$. Then $H$ is a direct sum of the submatroids on $A$ and $B$, and these are both non-empty.

Putting these two facts together tells us the connected hyperplanes of $M(G)$ correspond precisely to the complements of the vertex cut-sets. That gives us our reconstruction procedure. Here is a recipe:

(1) First, find all the connected hyperplanes $H_i$ of $M(G)$. Since any matroid is determined by its hyperplanes and you can tell if a matroid is connected (by looking at the circuits it contains, for example), this can be done.
(2) For each connected hyperplane $H_i$, write down its complement $E - H_i$. Call these sets $\{v_1, v_2, \ldots, v_k\}$.
(3) Since each $v_i$ corresponds to a vertex cut-set, each $x \in E$ is in precisely two of the $v_i$.
(4) Now draw $G$: label the $k$ vertices $\{v_1, v_2, \ldots, v_k\}$, and draw and label the edge $x$ joining vertices $v_i$ and $v_j$ if and only if $x \in v_i \cap v_j$.

□

One final comment is in order. Matroid glasses work pretty well when we look at 3-connected graphs, i.e., we can see the vertices. Much of the difficulty in generalizing theorems from graphs to matroids is that there is no real analog of a vertex for a general matroid.

## Exercises

### Section 4.1 – Graphs are matroids

(1) Let $G$ be the graph in Figure 4.14.
    (a) Find the rank of the cycle matroid $r(M(G))$.
    (b) Find all flats containing both edges $a$ and $c$.
    (c) Find the closure: $\overline{aeg}$.

(2) Suppose $M$ is a rank 3 matroid on five points that is not graphic. Does this information uniquely determine $M$? If so, find the unique matroid; otherwise, find two different matroids $M$ satisfying the given conditions.

(3) We know the Fano matroid $F_7$ is not graphic – see Exercise 15 in Chapter 1. Show that the matroids $F_7/x$ and $F_7 - x$ are both graphic for any $x$.

(4) Here are a few more proofs that graphs give matroids.

 (a) Show that the cycles of a graph $G$ satisfy the three circuit axioms (C1), (C2) and (C3). (Note: (C3) follows from Theorem 4.5.) Conclude $M(G)$ is a matroid.

 (b) Let $G$ be a connected graph. Show directly that the spanning trees of $G$ satisfy the three basis axioms (B1), (B2) and (B3).

 (c) Let $A_G$ be the vertex–edge incidence matrix for a graph $G$ as in Example 4.17. Show that a subset of columns of $A_G$ is linearly independent over $\mathbb{F}_2$, the field with two elements, if and only if the corresponding edges are acyclic. Conclude $M(G)$ is a representation of $G$. (Note: we haven't proven matrices give matroids yet – see Theorem 6.1 of Chapter 6. It turns out graphic matroids are representable over all fields – see Theorem 8.28.)

(5) Show that Theorem 4.5 is false for the non-graphic uniform matroid $U_{2,4}$.

(6) Find the number of flats for the cycle matroid $M(K_5)$, where $K_5$ is the complete graph on five vertices. How many circuits does $M(K_5)$ have? Bases? (See Exercise 7 for more on the number of flats in $M(K_n)$.)

(7) Let $K_n$ be the complete graph on $n \geq 1$ vertices. Let's count the number of rank $r$ flats of $M(K_n)$.

 (a) Show that the number of rank 1 flats of $M(K_n)$ is $\binom{n}{2}$, i.e., the number of edges of $K_n$.

 (b) Show that the number of hyperplanes equals the number of ways to partition an $n$ element set into exactly two pieces. Conclude there are $2^{n-1} - 1$ hyperplanes. (By the way, this tells you $M(K_n)$ also has $2^{n-1} - 1$ cocircuits.)

 (c) *Stirling numbers of the second kind.* The number of ways to partition the set $\{1, 2, \ldots, n\}$ into exactly $k$ non-empty parts is $S[n, k]$, the *Stirling number of the second kind.* Show there are precisely $S[n, r]$ flats of rank $r$ in $M(K_n)$. (These numbers are important in combinatorics. For example, they come up when you count the number of onto functions from $\{1, 2, \ldots, n\}$ to $\{1, 2, \ldots, k\}$. There are no "simple" formulas for $S[n, k]$.)

(8) We saw that every flat in the cycle matroid $M(G)$ arises from a vertex partition (Table 4.1). Show that each flat corresponds to a *unique* partition if and only if $G$ is a complete graph.

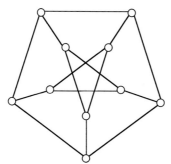

Figure 4.15. The Petersen graph for Exercise 9.

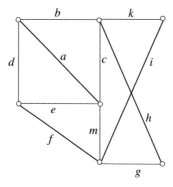

Figure 4.16. The graph for Exercise 11.

(9) Show that the dual of the Petersen graph (Figure 4.15) is not graphic. (Hint: try to modify the procedure outlined in Example 3.25 in Chapter 3. Alternatively, use Exercise 10 to first show the Petersen graph is not planar, then apply Theorem 3.27.)

(10) Suppose $G$ is a connected planar graph with $v$ vertices and $e$ edges. Suppose you know that every cycle in $G$ has length at least $k$.
 (a) Use Euler's formula (Theorem 3.24 from Chapter 3) and the discussion following Example 3.25 to show $(k - 2)e \leq k(v - 2)$.
 (b) Use part (a) to show $K_{3,3}$ and the Petersen graph (Figure 4.15) are not planar. (This gives a proof that the duals $M(K_{3,3})^*$ and $M(G)^*$ are not graphic, where $G$ is the Petersen graph. An approach that uses Kuratowski's Theorem can also be used to show the Petersen graph is not planar – see Example 8.23.)

## Section 4.2 – Circuits, cocircuits and bases

(11) Let $G$ be the graph in Figure 4.16.
 (a) First, verify strong circuit elimination for the two circuits $C_1 = \{b, d, f, m, c\}$ and $C_2 = \{h, b, d, f, g\}$. You can pick whatever you like for the $x \in C_1 \cap C_2$ and $y \in C_1 - C_2$.
 (b) Now verify strong basis exchange for the two bases $B_1 = \{b, d, e, f, g, i\}$ and $B_2 = \{a, c, e, m, k, h\}$. Again, choose

whatever you like for $x$ (you won't have that freedom for $y$, though.)

(c) Let's verify that circuits and cocircuits have even intersection in graphic matroids. Let $C^* = \{a, b, e, f\}$ be a cocircuit. Find circuits $C_1$, $C_2$ and $C_3$ with $|C_1 \cap C^*| = 0$, $|C_2 \cap C^*| = 2$ and $|C_3 \cap C^*| = 4$.

(12) Show that Proposition 4.11 is false for the non-graphic uniform matroid $U_{2,4}$.

(13) Prove strong basis exchange (B3′) directly for graphs: if $T_1$ and $T_2$ are spanning trees in a graph $G$, then there are edges $x \in T_1$ and $y \in T_2$ so that *both* $T_1 - x \cup y$ and $T_2 - y \cup x$ are spanning trees of $G$.

(14) This exercise extends the definition of the symmetric difference $A \triangle B$ to more than two sets.

(a) Let $A, B, C \subseteq S$. Show $(A \triangle B) \triangle C = A \triangle (B \triangle C)$. This justifies the notation $A \triangle B \triangle C$. Draw a Venn diagram and prove that $A \triangle B \triangle C$ is the set of all elements in exactly one of $A$, $B$ or $C$, or in $A \cap B \cap C$.

(b) Show that $A_1 \triangle A_2 \triangle \cdots \triangle A_n$ is well defined, and give a set theoretic description of $A_1 \triangle A_2 \triangle \cdots \triangle A_n$.

(15) Prove Theorem 4.13: Let $G$ be a graph with no cut-vertices.

(a) The symmetric difference of any two minimal cut-sets is a disjoint union of minimal cut-sets.

(b) Every minimal cutset is a symmetric difference of vertex cutsets, i.e., every cocircuit of the cycle matroid $M(G)$ is a symmetric difference of the vertex cocircuits.

(c) Let $S \subseteq V$ and let $C_S^*$ be the symmetric difference of the vertex cut-sets over the vertices of $S$. Then $C_S^* = C_{V-S}^*$.

(Direct, graph theoretic proofs are possible for all three parts. We remark that part (1) follows from Theorem 4.5 and duality for planar graphs, but this can be extended to prove part (1) for all graphs using Theorem 8.15 from Chapter 8 and the fact that graphic matroids are binary.)

(16) Here's a pretty problem. Draw $n$ chords of a circle, as in Figure 4.17. Prove that it is always possible to 2-color the regions formed inside the circle so that adjacent regions receive different colors. (Regions that only share a vertex are not considered neighbors.)

It's easy to do this by induction. Try a matroid-duality inspired proof:

(i) Assume no two chords meet along the bounding circle. Use the picture to create a finite planar graph by placing a vertex at all the points of intersection. This gives an even number of edges along the bounding circle (why?).

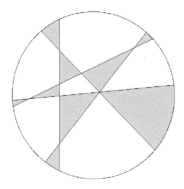 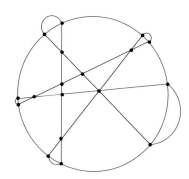

Figure 4.17. Left: a 2-coloring of the regions determined by $n$ chords of a circle. Right: a graph associated to the drawing. See Exercise 16.

(ii) Then "double" every other edge along the bounding circle (see Figure 4.17). Show that all the vertices in this graph have even degree.

(iii) Now show each vertex cut-set is even, so every cut-set is even.

(iv) Show this implies that every circuit in the dual is even, so the dual graph is bipartite. Show that this gives a legal 2-coloring of the regions in the original chord configuration.

(v) Finally, if two (or more) chords do meet along the bounding circle, show how to modify the above procedure.

## Section 4.3 – Duality and incidence matrices

(17) Let $G$ be the graph in Figure 4.14.

(a) Show that $adh$ is a cocircuit of the cycle matroid $M(G)$, and express this cocircuit as a symmetric difference of vertex cocircuits.

(b) Show that the largest cocircuit in $M(G)$ as four edges, and find one such cocircuit.

(c) Draw the dual graph $G^*$ and then draw the dual matroid $M(G)^*$. Show that $adh$ is a circuit in the dual.

(d) Show that the largest circuit in the dual has four elements (this is a check on your answer to part (b)).

(18) For the complete graph $K_4$:

(a) Find the vertex–edge incidence matrix $A$ and show that the cycles of $K_4$ correspond to linearly dependent columns of $A$ (using mod 2 arithmetic: $1 + 1 = 0$).

(b) Perform row operations on $A$ (and reorder the columns, if necessary) to get a matrix $A' = [I_3 \mid D]$, where $I_3$ is a $3 \times 3$ identity matrix and $D$ is a matrix of 0's and 1's. (Note: you are still working mod 2. If you do this correctly, you should get a row of 0's at the bottom of the matrix. Feel free to remove this row.)

Figure 4.18. Two, three and
four-dimensional hypercubes.
See Exercise 19.

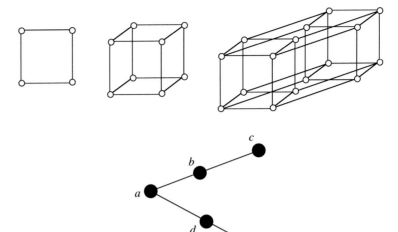

Figure 4.19. The matroid for
Exercise 21.

(c)   Now apply Theorem 6.6 (presented in Chapter 3) to the matrix
you obtained in part (b) to get a new matrix $A^* = [D^T \mid I_3]$,
and show this matrix represents $M(K_4)$, too. ("Represents"
should be understood to mean the same thing it means in
part (a).)

(d)   Interpret your result in part (c) via duality. (Note: $K_4$ is a planar
graph, so $M(K_4)^*$ is graphic.)

(19)   The *hypercube* $Q_n$ is a graph whose $2^n$ vertices are the $n$-tuples
$(\pm 1, \pm 1, \ldots, \pm 1)$. Two vertices are joined by an edge if they differ
in precisely one coordinate. Thus, $Q_2$ is a square, $Q_3$ a cube, $Q_4$
a four-dimensional cube, and so on. See Figure 4.18.

(a)   Find the incidence matrix for $Q_1$, $Q_2$ and $Q_3$.

(b)   Show $Q_n$ has $n \cdot 2^{n-1}$ edges.

(c)   Let $M(Q_n)$ be the cycle matroid defined on the graph $Q_n$. Show
$M(Q_n)^*$ is not graphic if $n \geq 4$. (Note: this would follow from
the fact that $Q_n$ is not planar for $n \geq 4$, but try to use an
argument like the one given in Example 3.25.)

## Section 4.4 – 2-isomorphism of graphs

(20)   Show that the relation $G$ is 2-isomorphic to $G'$ is an equivalence
relation.

(21)   Here's a slightly different connection between matroids and
graphs. Given a matroid $M$, form a graph $G_M$ called the *basis
graph* as follows. The vertices of $G_M$ are the bases of $M$, with an
edge joining $B_1$ and $B_2$ if $|B_1 \triangle B_2| = 2$, i.e., $B_2 = B_1 - x \cup y$ and
$B_1 = B_2 - y \cup x$ for some $x \in B_1$ and $y \in B_2$.

Basis graph

(a)   Draw $G_M$ for the matroid of Figure 4.19.

(b)   Use deletion and contraction to construct an inductive proof
that $M_G$ always has a Hamilton cycle. (A *Hamilton cycle* in

a graph $G$ is just a cycle through all of the vertices of $G$.)
Conclude that $G_M$ is a connected graph for all matroids $M$.

(c) Suppose $M = M_1 \oplus M_2$ is a direct sum of two smaller matroids. Describe how to obtain $G_M$ from $G_{M_1}$ and $G_{M_2}$.

(22) Making more connections. Suppose $G$ is a graph with cycle matroid $M(G)$. We are interested in determining when $M(G)$ is a connected matroid. (Recall $M$ is connected if every pair of elements is in some circuit.) This exercise develops a proof of Theorem 3.50 in Chapter 3, and is an important step in the proof of Theorem 4.23.

(a) Show that if $G$ is disconnected or if $G$ has a cut-vertex, then $M(G)$ is not a connected matroid.

(b) If $G$ has loops, then $M(G)$ is disconnected (you don't need to show this – it's immediate). Assume $G$ has no loops and is a block with at least three vertices. We want to show that $M(G)$ is connected. This will take two steps.

  (i) Show that, given any two vertices $u$ and $v$ in $G$, there are two *internally disjoint* paths joining $u$ and $v$. (*Internally disjoint* means the paths share no vertices except $u$ and $v$.) (Hint: try this by induction on the distance from $u$ to $v$.)

  (ii) Now show that, given any two *edges* $x$ and $y$, there is a cycle containing $x$ and $y$. (Hint: to use part (i), place vertices $u$ and $v$ at the midpoints of $x$ and $y$ and note this new graph is still a block with at least three vertices. Then piece the two internally disjoint paths together to get your cycle.)

This theorem is due to Whitney (1932).

# 5

## Finite geometry

Matroids can be thought of in many different ways; we tried to make that point in Chapter 2. But the common thread running through all of our different approaches to the subject is the underlying connection to geometry. When we "draw a picture of a matroid," we are thinking of the elements of the matroid as points and the dependences as lines, planes, and so on.

Geometry in the plane motivates our treatment of *affine geometry*. Although the word "affine" may be unfamiliar, affine geometry based on coordinates covers very familiar material; points in the plane correspond to ordered pairs $(x, y)$, points in three-dimensions correspond to ordered triples $(x, y, z)$, and so on.[1] Lines in the plane are given by linear equations of the form $ax + by = c$ for constants $a, b$ and $c$, planes in 3-space are described by equations of the form $ax + by + cz = d$, and this also generalizes to higher dimensions.

From the geometric point of view, here's what you learned long ago about points and lines in the plane:

**A:** Every pair of points determines a unique line, and
**B:** Given a point $P$ and a line $l$ not containing $P$, there is a unique line through $P$ *parallel* to $l$.

Our matroid interpretation for property A is direct;

- If $a$ and $b$ are non-parallel points in a matroid, then they determine a unique rank 2 flat of the matroid – see Figure 5.1.

Restating this in terms of the matroid rank function, if $r(abc) = 2$ (so $a, b$ and $c$ are collinear) and $r(abd) = 2$ (so $a, b$ and $d$ are also collinear), then $r(abcd) = 2$. (Quick proof: if $r(abcd) \geq 3$, we violate the independent augmentation axiom (I3): set $I = \{a, b\}$ and $J = \{a, c, d\}$.)

Property B above is the famous *parallel postulate*. There is a long and somewhat complicated history associated with this property. Briefly,

---

[1] This is standard Euclidean geometry.

180

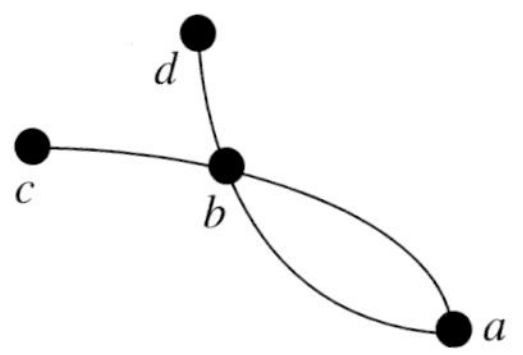

Figure 5.1. This is *not* a matroid. It's a fish.

Euclid stated this as an *axiom* (his fifth axiom of geometry), but attempts were made subsequently to *prove* this property as a theorem, based on Euclid's four previous axioms. All of these attempts failed. The failed proofs led to the discovery of strange, but perfectly valid geometries, geometries in which the parallel postulate does not hold. We'll see one of these "strange" geometries – the *projective plane* – in Section 5.3. Projective planes are very important for us because they are matroids.

Our motivation in this chapter is two-fold. First, affine and projective geometry are beautiful, well-developed areas of geometry that are of interest on their own. But, more importantly for us, they give us the proper setting for all representable matroids, i.e., matroids defined on the columns of a matrix. We'll study representable matroids in more detail in Chapter 6.

## 5.1 Affine geometry and affine dependence

When you study analytic geometry as a student (for instance, as preparation for studying calculus), you are introduced to Cartesian coordinates. There are several good reasons for this.[2] Coordinates can make many of the standard geometry proofs easier, and this approach also gives you good practice with algebra. So, with this motivation in the back of your mind, let's look at a *finite* version of analytic geometry.

### 5.1.1 Affine planes and cartesian coordinates

**Example 5.1.** Our first example is based on arithmetic modulo 3. Let's figure out the geometric structure for the collection of all ordered pairs $(a, b)$, where $a, b \in \{0, 1, 2\}$. We list the nine points as columns in a matrix.

$$
\begin{array}{ccccccccc}
P_1 & P_2 & P_3 & P_4 & P_5 & P_6 & P_7 & P_8 & P_9
\end{array}
$$
$$
\begin{bmatrix}
0 & 0 & 0 & 1 & 1 & 1 & 2 & 2 & 2 \\
0 & 1 & 2 & 0 & 1 & 2 & 0 & 1 & 2
\end{bmatrix}.
$$

---

[2] And a few bad ones. Analytic proofs of geometric theorems can sometimes hide some beautiful geometry. For example, consider the following fact you might have proven in a geometry course: the line joining the midpoints of two sides of a triangle is parallel to the third side, and half the length of that side. An analytic proof uses coordinates, slopes, and so on. A *synthetic* proof is coordinate-free, and it uses properties of similar triangles. Generally, in the context of geometric proofs, the term *synthetic* refers to axiomatic approaches.

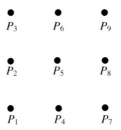

In Figure 5.2, we draw the nine points in a $3 \times 3$ grid, as we do in the
Euclidean plane. What are the lines in this geometry? Imitating ordinary
analytic geometry, a line is the collection of points $(x, y)$ satisfying $ax +
by = c$, where $a, b, c \in \{0, 1, 2\}$. So, for instance, the line $x + 2y = 1$
includes three points: $(1, 0)$, $(0, 2)$ and $(2, 1)$.

How many lines are there? Assume our line has equation $ax + by =
c$, and let's count the number of distinct equations in a systematic way.

- *Case 1: $a \neq 0$.* Then we can divide both sides of $ax + by = c$ by
  $a$, yielding an equation that looks like $x + b'y = c'$. There are three
  choices for $b'$ and three for $c'$, so there are nine lines in this case.
- *Case 2: $a = 0$.* Then, dividing by $b$ (which must be non-zero), we get
  an equation of the form $y = c'$, so we get three more lines in this case.

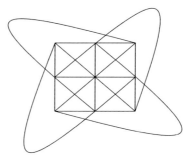

This gives us a total of 12 lines. A picture of the entire collection of
points and lines is shown in Figure 5.3. You can check this geometry
satisfies both of our properties above:

- Every pair of distinct points determines a unique line, and
- The parallel postulate holds: given any line $l$ and point $P$ not on $l$,
  there is a unique line $l'$ containing $P$ that does not intersect $l$.

The collection of points and lines in Example 5.1 is an *affine geometry*. It is usually denoted $AG(2, 3)$, where the $AG$ stands for *affine
geometry*, the 2 indicates the dimension of the space (it's a plane), and
the 3 indicates all of our computations are done mod 3.

In general, we will want the points of our affine geometries to correspond to ordered $n$-tuples $(a_1, a_2, \ldots, a_n)$, where the $a_i$ are elements of
a *field* $\mathbb{F}$. These ordered $n$-tuples form a *vector space* over the field $\mathbb{F}$.

We present a brief summary of what we will need to know about fields and vector spaces before proceeding with our development of affine geometry. (A standard reference is Gallian's *Contemporary Abstract Algebra* text.)

## 5.1.2 Fields and vector spaces

### Fields

A *field* is a set closed under the commutative operations of addition and multiplication in which every element (except 0) has a multiplicative inverse. Further, the operations obey the distributive law: $(a + b)c = ab + ac$.

Familiar fields include the rational numbers $\mathbb{Q}$, the real numbers $\mathbb{R}$ and the complex numbers $\mathbb{C}$. The integers $\mathbb{Z}$ do not form a field; they are missing multiplicative inverses.

*Finite fields* are fields with a finite number of elements. When $p$ is a prime number, the set $\mathbb{F}_p = \{0, 1, \ldots, p - 1\}$ forms a finite field, where addition and multiplication are defined modulo $p$. If $n$ is not prime, then $\{0, 1, 2, \ldots, n - 1\}$ does not form a field: For instance, in $\mathbb{Z}_4 = \{0, 1, 2, 3\}$, we have $2 \times 1 = 2, 2 \times 2 = 0$ and $2 \times 3 = 2$, so 2 has no multiplicative inverse.[3]

Finite field

The key thing you need to remember about finite fields is this:

- If $\mathbb{F}$ is a finite field, then $|\mathbb{F}| = p^n$ for some prime $p$ and positive integer $n$, and there is exactly one field (up to isomorphism) with $p^n$ elements.

We write $\mathbb{F}_q$ for the unique finite field with $|\mathbb{F}_q| = q = p^n$. When $n > 1$ and $q = p^n$ with $p$ prime, we point out that $\mathbb{F}_q$ is *not the same as* $\mathbb{Z}_q$. The construction of $\mathbb{F}_q$ is more involved.[4]

In any field, either $1 + 1 + \cdots + 1$ eventually equals 0, or it never equals 0. For example, in $\mathbb{F}_3$, we have $1 + 1 + 1 = 0$. The *characteristic* of a field $\mathbb{F}$ is the smallest positive integer $m$ such that $\underbrace{1 + 1 + \cdots + 1}_{m} = 0$ in $\mathbb{F}$. If such an integer $m$ exists, $m$ must be prime. Then if $q = p^k$ for some prime $p$ and positive integer $k$, the characteristic of the field $\mathbb{F}_q$ is $p$.

Field characteristic

If $\underbrace{1 + 1 + \cdots + 1}_{m}$ is never equal to 0 in $\mathbb{F}$ for any $m$, we say the characteristic of $\mathbb{F}$ is 0. Thus, the fields $\mathbb{Q}, \mathbb{R}$ and $\mathbb{C}$ all have characteristic 0.

---

[3] The notation $\mathbb{Z}_n$ is often used when you are thinking about $\{0, 1, \ldots, n - 1\}$ as a group under addition, or as a ring.

[4] Given the finite field $\mathbb{F}_p$ for $p$ prime and $n > 1$, we construct $\mathbb{F}_q$ for $q = p^n$ as follows. First form the polynomial ring $\mathbb{F}_p[x]$. Then find an irreducible polynomial $f(x)$ in $\mathbb{F}_p[x]$ of degree $n$ and form the quotient ring $\mathbb{F}_p[x]/\langle f(x)\rangle$. Then this is a field (as the ideal $\langle f(x)\rangle$ is a maximal ideal), and $\mathbb{F}_p[x]/\langle f(x)\rangle = \mathbb{F}_q$ for $q = p^n$. Note this also implies $\mathbb{F}_q$ is an $n$-dimensional vector space over $\mathbb{F}_p$.

The set of all possible field characteristics includes all primes, together with 0: $\{0, 2, 3, 5, 7, 11 \ldots\}$. Field characteristics will be important when we study matroid representation in more detail in Chapter 6.

## Vector spaces

A *vector space* $V$ over a field $\mathbb{F}$ (the *scalars*) is a collection of objects, called *vectors*, satisfying the following properties:

- $V$ is closed under addition: for all $u, v \in V$, the sum $u + v \in V$;
- $V$ is closed under scalar multiplication: for all $a \in \mathbb{F}$ and $v \in V$, we have $av \in V$;
- addition of vectors is commutative and associative;
- there is a zero vector $\bar{0}$ satisfying $u + \bar{0} = u$ for all $u \in V$;
- for every vector $u \in V$, there is an additive inverse $v$ such that $u + v = \bar{0}$;
- for every $v \in V$, $1v = v$, i.e., scalar multiplication by 1 doesn't change $v$;
- scalar multiplication is associative: $(ab)v = a(bv)$;
- for all $a \in \mathbb{F}$ and $u, v \in V$, we have $a(u + v) = au + av$; and
- for all $a, b \in \mathbb{F}$ and $v \in V$, we have $(a + b)v = av + bv$.

The most familiar vector space is $\mathbb{R}^n$, the ordered $n$-tuples of real numbers. But we can replace $\mathbb{R}$ with any other field $\mathbb{F}$, including finite fields, and still have a vector space $\mathbb{F}^n$. We will need the following facts about finite-dimensional vector spaces.

(1) The dimension of a vector space $V$ is the number of vectors in a basis of $V$. All of our vector spaces will be finite dimensional.

(2) A *subspace* $W \subseteq V$ is a subset of $V$ that is also a vector space. Given any subset of vectors $S = \{v_1, v_2, \ldots, v_k\}$, we can form a subspace $W$ by forming all linear combinations of these vectors: $\{\sum_{i=1}^{k} a_i v_i \mid a_i \in \mathbb{F}\}$ is a subspace. We say $W$ is *generated* by the set $S$. Since $W$ is also a vector space, every subspace contains $\bar{0}$.

(3) In general, if $W \subseteq V$ for a vector space $V$, then $W$ is a subspace of $V$ if and only if $W \neq \emptyset$ and $W$ is closed under addition and scalar multiplication.

(4) If $W_1$ and $W_2$ are subspaces, then so are $W_1 \cap W_2$ and $W_1 + W_2$, where $W_1 + W_2 = \{u + v \mid u \in W_1 \text{ and } v \in W_2\}$.

(5) If $W_1$ and $W_2$ are subspaces, then $\dim(W_1) + \dim(W_2) = \dim(W_1 + W_2) + \dim(W_1 \cap W_2)$.

(6) A set of vectors $\{v_1, v_2, \ldots, v_n\}$ is *linearly independent* if $\sum a_i v_i = \bar{0}$ implies $a_i = 0$ for all $i$. A set $S$ of vectors *spans* a subspace $W$ if every vector in $W$ can be expressed as a linear combination of vectors in $S$. A *basis* is a linearly independent spanning set, and all bases have the same number of vectors.

(7) If $W$ is a subspace of $V$, then $W^{\perp} = \{v \in \mathbb{F} \mid w \cdot v = 0 \text{ for all } w \in W\}$ is also a subspace of $V$. (The inner product $u \cdot v$ is the usual dot product: $(u_1, u_2, \ldots, u_n) \cdot (v_1, v_2, \ldots, v_n) = u_1 v_1 + u_2 v_2 + \cdots + u_n v_n$.)

(8) $\dim(W) + \dim(W^{\perp}) = \dim(V)$, and $(W^{\perp})^{\perp} = W$.

(9) If $A$ is an $m \times n$ matrix with entries in the field $\mathbb{F}$, then there are three subspaces we associate with $A$:

   (a) The *row space* of $A$: this is the subspace $R(A)$ of $\mathbb{F}^n$ generated by the rows of $A$.

   (b) The *column space* of $A$: this is the subspace $C(A)$ of $\mathbb{F}^m$ generated by the columns of $A$.

   (c) The *null space* or *kernel* of $A$: this is denoted $N(A)$, and is defined by

$$N(A) = \{v \in \mathbb{F}^n \mid Av = 0\}.$$

   Then $N(A)$ is a subspace of $\mathbb{F}^n$.

(10) For any $m \times n$ matrix $A$, we have $N(A) = R(A)^{\perp}$.

(11) Row rank = column rank: for any matrix $A$, the row and column spaces have the same dimension: $\dim(R(A)) = \dim(C(A))$.

(12) Rank–Nullity Theorem: For any $m \times n$ matrix $A$,

$$\dim(R(A)) + \dim(N(A)) = n.$$

### 5.1.3 Affine geometry and matroids

Back to geometry. The construction from Example 5.1 based on ordered pairs $(x, y)$ works over any field $\mathbb{F}$. (If our coordinates do not come from a field, we can get into some trouble – see Exercise 3.)

**Definition 5.2.** Let $q = p^k$ for some prime $p$ and positive integer $k$. The points of the *affine plane* $AG(2, q)$ are the $q^2$ ordered pairs $(x, y)$, where $x, y \in \mathbb{F}_q$. The lines are collections of points $(x, y)$ satisfying equations of the form $ax + by = c$, where $a, b, c \in \mathbb{F}_q$.

Affine plane $AG(2, q)$

You can think of the points of $AG(2, q)$ as a collection of $q^2$ points in the Euclidean plane, where certain subsets of points are (not necessarily straight) lines. The total number of lines is $q^2 + q$ and every line has exactly $q$ points – see Exercises 1 and 5.

Our next goal is to generalize our construction of the affine plane $AG(2, q)$ to $n$-dimensional affine space $AG(n, q)$. To do this, just take the points to be all ordered $n$-tuples $(a_1, a_2, \ldots, a_n)$, where each $a_i \in \mathbb{F}_q$. This is the finite vector space $\mathbb{F}_q^n$ of dimension $n$ over the field $\mathbb{F}_q$.

We need to define and understand the points, lines, planes and hyperplanes of all dimensions in $AG(n, q)$, just as we do in $\mathbb{R}^n$, ordinary Euclidean space. We call these objects the *flats* of the affine geometry;

Figure 5.4. The eight points of the affine geometry $AG(3, 2)$.

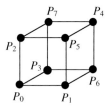

this terminology will be especially apt when we get around to discussing matroids.[5]

For a simple example, consider the affine geometry $AG(2, 3)$ whose points are all ordered triples $(x, y, z)$, where $x, y, z \in \{0, 1\} = \mathbb{F}_2$. Then we have a three-dimensional geometry on eight points (Figure 5.4). The eight points correspond to the eight columns in the matrix

$$
\begin{array}{cccccccc}
P_0 & P_1 & P_2 & P_3 & P_4 & P_5 & P_6 & P_7
\end{array}
$$
$$
\begin{bmatrix}
0 & 1 & 0 & 0 & 1 & 1 & 1 & 0 \\
0 & 0 & 1 & 0 & 1 & 1 & 0 & 1 \\
0 & 0 & 0 & 1 & 1 & 0 & 1 & 1
\end{bmatrix}.
$$

A typical plane has equation $ax + by + cz = d$, for $a, b, c, d \in \{0, 1\}$. So, for instance, the four points $P_0$, $P_5$, $P_6$ and $P_7$ form a plane because these four points all satisfy the equation $x + y + z = 0$. You can check there are a total of 14 (rank 3) planes, each of which is a four-point matroid circuit. See Exercise 2.

How should we define *flats* in an affine geometry? Subspaces will play the central role. In $\mathbb{R}^n$, the one-dimensional subspaces of $\mathbb{R}^n$ are lines through the origin, the two-dimensional subspaces are planes through the origin, and so on. Then our flats will just be *translates* of these subspaces.

**Definition 5.3.** Let $F \subseteq \mathbb{F}_q^n$. Then $F$ is a *flat* if $F = u + W := \{u + w \mid w \in W\}$ for some subspace $W \subseteq \mathbb{F}_q^n$ and some vector $u \in \mathbb{F}_q^n$. We also call $F = \emptyset$ a flat.

In Example 5.1, $\emptyset$ is a flat, the nine points are flats of dimension 0, the 12 lines are the flats of dimension 1, the entire plane is the only flat of dimension 2. (In general, the flats are cosets of the quotient group $V/W$, but we won't need this level of abstraction.) We will use the flats to describe affine dependence in $AG(n, q)$ algebraically.

How are the flats in $AG(n, q)$ related to flats of a matroid? The next result tells us they coincide.[6] Thus, $AG(n, q)$ is a matroid.

**Theorem 5.4.** *Let M be defined on the ground set E, where E consists of the $q^n$ points of $AG(n, q)$. Let $\mathcal{F}$ be the family of all flats of $\mathbb{F}_q^n$ along with $\emptyset$. Then $\mathcal{F}$ are the flats of a matroid.*

---

[5] This will happen very soon, we promise.
[6] If they didn't, we made a huge mistake in terminology.

*Proof Theorem 5.4.* Let $\mathcal{F}$ be the collection of flats from Definition 5.3. We need to show $\mathcal{F}$ satisfies the three flat axioms (F1), (F2) and (F3) given in Theorem 2.52:

(F1) $E \in \mathcal{F}$.
(F2) If $F_1, F_2 \in \mathcal{F}$, then $F_1 \cap F_2 \in \mathcal{F}$.
(F3) If $F \in \mathcal{F}$ and $\{F_1, F_2, \ldots, F_k\}$ is the set of flats that cover $F$, then the collection $\{F_1 - F, F_2 - F, \ldots, F_k - F\}$ partitions $E - F$.

For (F1), $E = \mathbb{F}_q^n$ is a vector space, and $\mathbb{F}_q^n$ a subspace of itself; thus $E \in \mathcal{F}$. For (F2), suppose $F_1, F_2 \in \mathcal{F}$. Now if $F_1 \cap F_2 = \emptyset$, then (F2) is satisfied, so assume $u \in F_1 \cap F_2$. Then $F_1 = u + W_1$ and $F_2 = u + W_2$ for some subspaces $W_1$ and $W_2$, so $F_1 \cap F_2 = u + (W_1 \cap W_2) \in \mathcal{F}$ since $W_1 \cap W_2$ is a subspace.

To prove (F3), we suppose $F \in \mathcal{F}$ and $F_1, F_2, \ldots, F_k$ are the flats that cover $F$. The proof strategy is straightforward. Translate everything back to the origin, then use the fact that all the flats have been translated to subspaces.

Now $F \in \mathcal{F}$ implies $F = u + W$ for some $u \in \mathbb{F}_q^n$ and some subspace $W$. Then, for all $1 \le i \le k$, we have $u \in F_i$ and $F_i = u + W_i$, where the $W_i$ are subspaces. Further, since $F_i$ covers $F$, we know $\dim(W_i) = \dim(W) + 1$. Then $\{W_1, W_2, \ldots, W_k\}$ are the subspaces that cover $W$, and $\{W_1 - W, W_2 - W, \ldots, W_k - W\}$ partitions $\mathbb{F}_q^n - W$. Translating back to the $F_i$'s finishes the argument. □

We'll see another proof that $AG(n, q)$ is a matroid that uses *affine dependence* – see Proposition 5.8. We conclude this section with a connection between flats in $AG(n, q)$ and matrices. Since flats are translates of subspaces, and since we can use matrices to define subspaces, we can find a direct correspondence between matrices and flats of $AG(n, q)$. Here's the connection:

---

### Flat–matrix correspondence

- Given a flat $F$ in $AG(n, q)$, find a vector $u$ and a subspace $W$ so that $F = u + W$.
- Next, suppose $\dim(W) = r$. Then $\dim(W^{\perp}) = n - r$, where $W^{\perp} = \{v \in \mathbb{F}_q^n \mid w \cdot v = 0 \text{ for all } w \in W\}$.
- Now find a basis for $W^{\perp}$ and form an $(n - r) \times n$ matrix $A$ whose rows are these basis vectors for $W^{\perp}$.
- Then the null space $N(A) = W$ (since $(W^{\perp})^{\perp} = W$). This means $W = \{x \in \mathbb{F}_q^n \mid Ax = 0\}$.
- Now add $u$: this gives $F = \{y \in \mathbb{F}_q^n \mid y = x + u \text{ and } Ax = 0\}$. Equivalently, $F = \{y \in \mathbb{F}_q^n \mid Ay = Au\}$.

---

This process is completely reversible – given *any* rank $r$ flat in $AG(n, q)$, we can find an $(n - r) \times n$ matrix $A$ and a vector $u$ so that $F$ is the set of $n$-tuples $y$ satisfying $Ay = Au$.

Executive summary:

**Proposition 5.5.** *For every rank $r$ flat $F$ in $AG(n, q)$, there is an $(n - r) \times n$ matrix $A$ and a vector $u$ so that $F$ is the set of all $n$-tuples $y \in \mathbb{F}_q^n$ satisfying $Ay = Au$.*

Skip ahead to Example 5.10 to see Proposition 5.5 in action.

## 5.2 Affine dependence

Since the affine geometry $AG(n, q)$ is a matroid, it makes sense to ask about dependent and independent sets in $AG(n, q)$. From our experience with geometry in general (and matroid flats in particular), we expect the following to be true in $AG(n, q)$.

- Every point is an independent set.
- Every pair of points is independent.
- A set of three points is dependent if and only if those points are collinear.
- A set of four points is dependent if and only if they are coplanar.
- Five points are dependent when they all lie on the same three-dimensional flat, and so on.

Affine dependence

This is precisely how we've been drawing "pictures" of matroids[7] throughout the text. We call a subset $S$ of points in an affine geometry $AG(n, q)$ *affinely dependent* if $S$ is a dependent set in the matroid on $E = \mathbb{F}_q^n$. Our immediate goal is to find an algebraic way to determine when a set is affinely dependent. The answer is closely related to *linear dependence* in a vector space.

**Proposition 5.6.** *Let $S = \{P_0, P_1, \ldots, P_k\}$ be a collection of points in the affine geometry $AG(n, q)$. Then $S$ is affinely dependent if and only if there are scalars $a_0, a_1, \ldots, a_k \in \mathbb{F}_q$, not all 0, such that both of the following conditions hold:*

(1) $\displaystyle\sum_{i=0}^{k} a_i = 0$, *and*

(2) $\displaystyle\sum_{i=0}^{k} a_i P_i = \overline{0}$ *(where $\overline{0}$ is the zero vector in $\mathbb{F}_q^n$).*

---

[7] Some authors call these pictures "affine drawings" of a matroid.

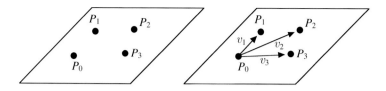

Figure 5.5. The four points $P_0$, $P_1$, $P_2$ and $P_3$ are affinely dependent because they are coplanar. These four points give rise to three linearly dependent vectors $v_1$, $v_2$ and $v_3$.

*Proof Proposition 5.6.* Let $S = \{P_0, P_1, \ldots, P_k\}$ be a collection of $k + 1$ points, and form $k$ vectors as follows: $v_1 = P_1 - P_0$, $v_2 = P_2 - P_0, \ldots, v_k = P_k - P_0$. Then the key observation we require is this:

> A set $\{P_0, P_1, \ldots, P_k\}$ of $k + 1$ points is affinely dependent if and only if the corresponding $k$ vectors $\{v_1, v_2, \ldots, v_k\}$ are linearly dependent.

This follows immediately from the definition of linear dependence of vectors, and it makes sense geometrically (see Figure 5.5). Then $S$ is affinely dependent if and only if there are scalars $b_i \in \mathbb{F}_q$ for $1 \le i \le k$, not all 0, such that $\sum_{i=1}^{k} b_i v_i = \bar{0}$. Since $v_i = P_i - P_0$, this is equivalent to $\sum_{i=1}^{k} b_i (P_i - P_0) = \bar{0}$. Then $\sum_{i=1}^{k} b_i (P_i - P_0)$

$$= (-b_1 - b_2 - \cdots - b_k)P_0 + b_1 P_1 + b_2 P_2 + \cdots + b_k P_k = \bar{0}.$$

Setting $a_0 = -(b_1 + b_2 + \cdots + b_k)$ and $a_i = b_i$ for $1 \le i \le k$ gives $\sum_{i=0}^{k} a_i = 0$ and $\sum_{i=0}^{k} a_i P_i = \bar{0}$, where not all of the $a_i$ are 0. This completes one direction of our proof.

For the converse, suppose there are scalars $a_i$ for $0 \le i \le k$, not all 0, such that $\sum_{i=0}^{k} a_i = 0$ and $\sum_{i=0}^{k} a_i P_i = \bar{0}$. Then $a_0 = -(a_1 + a_2 + \cdots + a_k)$, so we can reverse the above procedure:

$$\sum_{i=0}^{k} a_i P_i = -(a_1 + a_2 + \cdots + a_k)P_0 + a_1 P_1 + a_2 P_2 + \cdots + a_k P_k$$

$$= a_1(P_1 - P_0) + a_2(P_2 - P_0) + \cdots + a_k(P_k - P_0)$$

$$= a_1 v_1 + a_2 v_2 + \cdots + a_k v_k.$$

Since $\sum_{i=0}^{k} a_i P_i = \bar{0}$, we now have $a_1 v_1 + a_2 v_2 + \cdots + a_k v_k = \bar{0}$, where not all of the $a_i$ are 0. Thus, the set of vectors $\{v_1, v_2, \ldots, v_k\}$ is linearly dependent, so the points $P_0, P_1, \ldots, P_k$ are affinely dependent. $\square$

As an example of affine dependence in $AG(2, 3)$, let $P_0 = (0, 0)$, $P_1 = (1, 2)$ and $P_2 = (2, 1)$. Then these three points are affinely dependent because $1 \cdot P_0 + 1 \cdot P_1 + 1 \cdot P_2 = (0, 0)$ and $1 + 1 + 1 = 0$ in $\mathbb{F}_3$. (These three points form the line with equation $x + y = 0$.)

When is a collection of points $S$ affinely *independent*? $S$ is affinely independent precisely when it's not affinely dependent.[8] For example, the points $Q_0 = (0, 0)$, $Q_1 = (1, 0)$ and $Q_2 = (1, 2)$ are affinely independent because, if $aQ_0 + bQ_1 + cQ_2 = (0, 0)$, then we get:

$$a + b + c = 0$$
$$b + c = 0$$
$$2c = 0.$$

This forces $a = b = c = 0$, so the points are not affinely dependent.

There is a useful bookkeeping[9] device we can use that will help determine affine dependence and affine independence. As is the case in almost all things mathematical, it's easiest to explain the technique via an example.

**Example 5.7.** The affine geometry $AG(2, 5)$ consists of $5^2 = 25$ points which are ordered pairs $(x, y)$, where $x, y \in \{0, 1, 2, 3, 4\} = \mathbb{F}_5$. Consider the three points $P = (1, 0)$, $Q = (0, 2)$ and $R = (3, 1)$ (and remember, we're doing our arithmetic modulo 5). These points are affinely dependent (by Proposition 5.6) because

$$3 \cdot P + 3 \cdot Q + 4 \cdot R = (0, 0) \text{ and } 3 + 3 + 4 \equiv 0 \pmod 5.$$

Now change your three points in $\mathbb{F}_5^2$ to three vectors in $\mathbb{F}_5^3$ by placing a 1 in the first coordinate of each point:

$$
\begin{array}{ccc} P & Q & R \end{array}
\begin{bmatrix} 1 & 0 & 3 \\ 0 & 2 & 1 \end{bmatrix}
\mapsto
\begin{array}{ccc} P' & Q' & R' \end{array}
\begin{bmatrix} 1 & 1 & 1 \\ 1 & 0 & 3 \\ 0 & 2 & 1 \end{bmatrix}.
$$

Then the three vectors $P'$, $Q'$ and $R'$ in $\mathbb{F}_5^3$ are *linearly* dependent:

$$3 \cdot P' + 3 \cdot Q' + 4 \cdot R' = (0, 0, 0).$$

This combines the two requirements of affine dependence in Proposition 5.6 into one nice, neat, linear dependence package: the last two coordinates of $3 \cdot P' + 3 \cdot Q' + 4 \cdot R'$ are zero because $3 \cdot P + 3 \cdot Q + 4 \cdot R = (0, 0)$ in $\mathbb{F}_5^2$, while the first coordinate of $3 \cdot P' + 3 \cdot Q' + 4 \cdot R'$

---

[8]  This is our candidate for "The most useless answer in this text." As a research question, test this assertion against other answers in the text.

[9]  Did you know "bookkeeping" is the only word in the English language with three consecutive sets of double letters?

is zero because of our clever use of the 1's in the first coordinate of each of the vectors: $3 \cdot 1 + 3 \cdot 1 + 4 \cdot 1 \equiv 0 \pmod 5$.

We state this bookkeeping trick as a proposition:

**Proposition 5.8.** *Let* $x = (x_1, x_2, \ldots, x_n) \in \mathbb{F}_q^n$ *be an* $n$*-tuple, and define an* $(n+1)$*-tuple* $x' \in \mathbb{F}_q^{n+1}$ *by placing a 1 in the first coordinate of* $x$: $x' = (1, x_1, x_2, \ldots, x_n)$. *Then the vectors* $v_1, v_2, \ldots, v_m$ *are affinely dependent in* $\mathbb{F}_q^n$ *if and only if the corresponding vectors* $v_1', v_2', \ldots, v_m'$ *are linearly dependent in* $\mathbb{F}_q^{n+1}$.

Thus, for instance, *affine* dependence (and independence) of the columns of the nine points of $AG(2, 3)$

$$\begin{array}{ccccccccc} v_1 & v_2 & v_3 & v_4 & v_5 & v_6 & v_7 & v_8 & v_9 \end{array}$$
$$\begin{bmatrix} 0 & 0 & 0 & 1 & 1 & 1 & 2 & 2 & 2 \\ 0 & 1 & 2 & 0 & 1 & 2 & 0 & 1 & 2 \end{bmatrix}$$

matches precisely *linear* dependence (and independence) of the corresponding columns of

$$\begin{array}{ccccccccc} v_1' & v_2' & v_3' & v_4' & v_5' & v_6' & v_7' & v_8' & v_9' \end{array}$$
$$\begin{bmatrix} 1 & 1 & 1 & 1 & 1 & 1 & 1 & 1 & 1 \\ 0 & 0 & 0 & 1 & 1 & 1 & 2 & 2 & 2 \\ 0 & 1 & 2 & 0 & 1 & 2 & 0 & 1 & 2 \end{bmatrix}.$$

We omit the proof of Proposition 5.8, but it should be immediately clear why this works. The affine dependence of $v_1, v_2, \ldots, v_m$ in $\mathbb{F}_q^n$ guarantees $n$ coefficients $a_1, a_2, \ldots, a_n$ with $\sum_{i=0}^{k} a_i = 0$, and $\sum_{i=0}^{k} a_i P_i = \bar{0}$.

Then these same coefficients give a linear dependence for $v_1', v_2', \ldots, v_m'$, and this argument is clearly reversible.

We don't want to oversell this idea, but Proposition 5.8 gives us more. For example, since it shows affine dependence matches linear dependence (of a corresponding set of vectors in a vector space of dimension 1 greater), it will give us another proof of Theorem 5.4 that the affine geometry $AG(n, q)$ forms a matroid.[10] It will also be important when we embed affine geometries in projective geometries in Section 5.4.

## 5.2.1 Rank, closure and hyperplanes for affine geometries

It's worth interpreting three more matroid concepts in the context of affine geometry: rank, closure and hyperplanes. For rank, note that if $F$

---

[10] That is, once we give our proof that linear independence satisfies the independent set axioms. See Theorem 6.1.

is a flat, then the dimension of $F$ is a well-defined concept: since $F = u + W$ for some vector $u$ and subspace $W$, we have $\dim(F) = \dim(W)$.

**Proposition 5.9.** *Let $S = \{P_1, P_2, \ldots, P_n\}$ be a collection of points in the affine geometry $AG(n, q)$.*

(1) **Rank** $r(S) = \dim(F) + 1$, *where $F$ is the smallest flat containing $S$.*

(2) **Closure** $P \in \overline{S}$ *if and only if there are scalars $a_i \in \mathbb{F}_q$ with*

(a) $\displaystyle\sum_{i=1}^{n} a_i = 1$, *and*

(b) $P = \displaystyle\sum_{i=1}^{n} a_i P_i$.

(3) **Hyperplanes** *$S$ is a hyperplane if and only if $S$ is the set of all $n$-tuples $(x_1, x_2, \ldots, x_n) \in \mathbb{F}_q^n$ satisfying $a_1 x_1 + a_2 x_2 + \cdots + a_n x_n = b$ for some scalars $a_1, a_2, \ldots, a_n, b \in \mathbb{F}$.*

We leave the proof of Proposition 5.9 to you in Exercises 4, 11 and 12. But, as a check of these descriptions of closure and hyperplanes, let's do an example.

**Example 5.10.** Let $P = (4, 3, 3)$, $Q = (1, 2, 4)$ and $R = (3, 1, 0)$ be points in $AG(3, 5)$. Then $P = 2 \cdot Q + 4 \cdot R$ and $2 + 4 \equiv 1 \pmod 5$, so $P \in \overline{QR}$ by part (2) of Proposition 5.9. So these three points are collinear in $AG(3, 5)$.

These three points are on a line, and a line is a rank 2 flat in $AG(3, 5)$, which is a rank 4 matroid. How can we describe this flat algebraically? We use Proposition 5.5: if $F$ is a flat in $AG(n, q)$, then $F = u + W$ for a vector $u$ and a subspace $W$. Then if $A$ is a matrix whose rows form a basis for $W^\perp$, we have $v \in F$ if and only if $Av = Au$.

To apply Proposition 5.5, we take $u = (4, 3, 3)$, the vector $\overline{OP}$ from the origin to the point $P$. For the subspace $W$, we can use $W = t \cdot \overline{PQ} = t \cdot ((1, 2, 4) - (4, 3, 3)) = t \cdot (2, 4, 1)$, for $t \in \mathbb{F}_5$. (Think of $(2, 4, 1)$ as the *direction vector* for the line.)

How do we get the matrix $A$? In this case, $\dim(W) = 1$, so $\dim(W^\perp) = 2$. A basis for $W^\perp$ is the solution set to the equation $2x + 4y + z = 0$. Working mod 5, we get a basis for $W^\perp$ of $\{(3, 1, 0), (2, 0, 1)\}$. This gives

$$A = \begin{pmatrix} 3 & 1 & 0 \\ 2 & 0 & 1 \end{pmatrix}.$$

Then $v = (x, y, z)$ is on the flat $F$ if and only if $Av = Au$, i.e.,

$$\begin{pmatrix} 3 & 1 & 0 \\ 2 & 0 & 1 \end{pmatrix} \begin{pmatrix} x \\ y \\ z \end{pmatrix} = \begin{pmatrix} 3 & 1 & 0 \\ 2 & 0 & 1 \end{pmatrix} \cdot \begin{pmatrix} 4 \\ 3 \\ 3 \end{pmatrix} = \begin{pmatrix} 0 \\ 1 \end{pmatrix}.$$

This reduces to two equations: $3x + y = 0$ and $2x + z = 1$. You can check that each of the points $P$, $Q$ and $R$ satisfy both of these equations. Note this also expresses the flat $F$ as the intersection of two hyperplanes; one hyperplane is given by the solutions of $3x + y = 0$, and the other is given by $2x + z = 1$.

We can use Proposition 5.8 to check the closure characterization in Proposition 5.9(2). Let's define an *affine combination* as follows.   Affine combination

**Definition 5.11.** $P \in \mathbb{F}_q^n$ is an *affine combination* of $S = \{P_1, P_2, \ldots, P_n\}$ if $\sum_{i=1}^{n} a_i = 1$ and $P = \sum_{i=1}^{n} a_i P_i$, i.e., $P \in \overline{S}$.

Then our bookkeeping trick in Proposition 5.8 allows us to interpret *affine* combinations in $\mathbb{F}_q^n$ as *linear* combinations in $\mathbb{F}_q^{n+1}$.

For instance, the affine combination in Example 5.10 that gave us $2 \cdot Q + 4 \cdot R = P$ in $\mathbb{F}_5^3$ is transformed into the linear combination $2 \cdot Q' + 4 \cdot R' = P'$ in $\mathbb{F}_5^4$:

$$
\left(
2 \overset{Q}{\begin{bmatrix} 1 \\ 2 \\ 4 \end{bmatrix}} + 4 \overset{R}{\begin{bmatrix} 3 \\ 1 \\ 0 \end{bmatrix}} = \overset{P}{\begin{bmatrix} 4 \\ 3 \\ 3 \end{bmatrix}}
\right)
\longrightarrow
\left(
2 \overset{Q'}{\begin{bmatrix} 1 \\ 1 \\ 2 \\ 4 \end{bmatrix}} + 4 \overset{R'}{\begin{bmatrix} 1 \\ 3 \\ 1 \\ 0 \end{bmatrix}} = \overset{P'}{\begin{bmatrix} 1 \\ 4 \\ 3 \\ 3 \end{bmatrix}}
\right).
$$

As before, the extra row of 1's in an extra dimension allows us to encode the affine condition on the sum of the coefficients (Proposition 5.9(2a)): $\sum_{i=1}^{n} a_i = 1$. In general, here's what we get:

**Proposition 5.12.** *Let $x = (x_1, x_2, \ldots, x_n) \in \mathbb{F}_q^n$ be an n-tuple, and, as before, define an $(n + 1)$-tuple $x' \in \mathbb{F}_q^{n+1}$ by placing a 1 in the first coordinate of $x$: $x' = (1, x_1, x_2, \ldots, x_n)$. Let $u, v_1, v_2, \ldots, v_m \in \mathbb{F}_q^n$. Then u is an affine combination of $v_1, v_2, \ldots, v_m$ if and only if the corresponding vector $u'$ is a linear combination of $v_1', v_2', \ldots, v_m'$ in $\mathbb{F}_q^{n+1}$.*

See Exercise 13 for an outline of a proof.

For example, consider $AG(3, 3)$, which is a rank 4 matroid. The 27 points of this geometry are the ordered triples $(x, y, z) \in \mathbb{F}_3^3$. Then the equation $x + 2y + z = 1$ is a hyperplane, and you can check there are exactly nine points that satisfy this equation:

$$(1, 0, 0) \quad (0, 2, 0) \quad (0, 0, 1) \quad (2, 1, 0) \quad (2, 0, 2)$$
$$(0, 1, 2) \quad (1, 1, 1) \quad (2, 2, 1) \quad (1, 2, 2).$$

These nine points form a copy of $AG(2, 3)$, and each point is in the affine span of any three non-collinear points in the set. So, for instance, if we

Table 5.1. *Matroid interpretations for affine geometry.*

| Matroid concept | Interpretation in $AG(n, q)$ |
|---|---|
| Points | Ordered $n$-tuples in $\mathbb{F}_q^n$. |
| Independent sets | $\{P_1, P_2, \ldots, P_n\}$ is affinely independent if there are no non-trivial scalars $a_i$ such that $\sum_{i=1}^{k} a_i = 0$ and $\sum_{i=1}^{k} a_i P_i = \bar{0}$. |
| Closure | Affine closure: for $S = \{P_1, P_2, \ldots, P_n\}$, $Q \in \bar{S}$ if there are non-trivial scalars $a_i$ such that $\sum_{i=1}^{n} a_i = 1$ and $Q = \sum_{i=1}^{n} a_i P_i$. |
| Flats | Affine flats: translates of subspaces of $\mathbb{F}_q^n$. Equivalently, all $x \in \mathbb{F}_q^n$ satisfying $Ax = Au$ for some $(n - r) \times n$ matrix $A$ and $u \in \mathbb{F}_q^n$. |
| Hyperplanes | Maximal flats: $\left\{ (x_1, x_2, \ldots, x_n) \mid \sum_{i=1}^{n} a_i x_i = b \text{ for } a_i, b \in \mathbb{F}_q \right\}$. |
| Rank | $r(AG(n, q)) = n + 1$, and $r(S) = \dim(\bar{S}) + 1$. |

take $B = \{(1, 2, 2), (2, 1, 0), (1, 1, 1)\}$, then $(2, 0, 2)$ is in the affine span of $B$ because

$$(2, 0, 2) = 2 \cdot (1, 2, 2) + 1 \cdot (2, 1, 0) + 1 \cdot (1, 1, 1),$$

where the sum of the coefficients $2 + 1 + 1 \equiv 1 \pmod{3}$, as required by Proposition 5.9(2).

We celebrate the end of this section by summarizing the connections between matroids and affine geometry in Table 5.1.

## 5.3 The projective plane $PG(2, q)$

In an affine plane, the parallel postulate holds. Projective planes have no parallel lines. It will turn out that the projective viewpoint is more general than the affine approach: every affine space naturally resides in a larger projective space. But that's getting a little ahead of ourselves.[11]

From the axiomatic point of view, a *projective plane* will satisfy the following two properties:

**P1:** Every pair of points determines a unique line, and
**P2:** Every pair of lines intersect in a unique point.

To rule out some trivial examples, it is traditional to also assume there are at least four points, no three on a line. Such a configuration is called a *four-point*.

---

[11] This is a very common expression, but, like many other common expressions, it doesn't really make sense.

**Definition 5.13.** A *projective plane* is a triple $(\mathcal{P}, \mathcal{L}, \mathcal{I})$ of points $\mathcal{P}$, lines $\mathcal{L}$ and incidence $\mathcal{I}$ between the points and lines satisfying **P1** and **P2**, and containing a four-point.

Projective plane

Before returning to coordinates, column vectors and linear dependence, we make three quick comments:

(1) There is a beautiful symmetry between **P1** and **P2**. Just interchange the words "point" and "line" to interchange **P1** and **P2**. One important consequence: any theorem you prove for projective planes gives you a *dual* theorem for free by interchanging "point" and "line."
(2) The parallel postulate is completely false here: there are *no* pairs of parallel lines, since every pair of lines intersect.
(3) You can use this abstract definition to prove that every line has the same number of points. If each line has $q + 1$ points, we say the projective plane has *order q*.[12]

### 5.3.1 The projective plane $PG(2, q)$ from affine space $AG(3, q)$

How can we construct projective planes? In order to ensure every pair of lines intersect, do the following:

- First, consider the three-dimensional vector space $V = \mathbb{F}_q^3$ over the field $\mathbb{F}_q$.
- Identify each one-dimensional subspace (i.e., line through $(0, 0, 0)$) as a *point* of your geometry.
- Identify each two-dimensional subspace (i.e., plane through $(0, 0, 0)$) as a *line* of your geometry.

Let's look at an example, shall we?

**Example 5.14.** The vector space $\mathbb{F}_2^3$ consists of the eight ordered triples $(x, y, z)$, where each coordinate is 0 or 1. This is the three-dimensional affine geometry $AG(3, 2)$ of Figure 5.4, and our construction of the associated projective plane is based on this geometry. Every line in $AG(3, 2)$ (there are $\binom{8}{2} = 28$ lines) has two points, and seven of these lines contain the point $(0, 0, 0)$. Thus, there are seven one-dimensional subspaces; we represent each such line by the one non-zero vector it contains. We present these seven vectors as the columns in a matrix:

$$\begin{array}{ccccccc} a & b & c & d & e & f & g \end{array}$$
$$\begin{bmatrix} 1 & 0 & 0 & 1 & 1 & 1 & 0 \\ 0 & 1 & 0 & 1 & 1 & 0 & 1 \\ 0 & 0 & 1 & 1 & 0 & 1 & 1 \end{bmatrix}.$$

---

[12] This is a little unfortunate, but will make sense when the finite field $\mathbb{F}_q$ enters the story.

Figure 5.6. Left: the seven lines through $(0, 0, 0)$ in $AG(2, 3)$. Right: the associated projective plane $PG(2, 2)$. (Note: $PG(2, 2)$ is the Fano plane.)

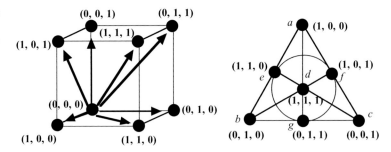

So, for instance, the projective *point* $f$ (corresponding to the column vector $(1, 0, 1)$) really represents the *line* through the two points $(0, 0, 0)$ and $(1, 0, 1)$ in $AG(3, 2)$. We denote the associated projective plane $PG(2, 2)$, where the "$PG$" stands for *projective geometry*, the first "2" gives the dimension of the object (it's a plane, so two-dimensional) and the second "2" indicates the field we are working over (it's $\mathbb{F}_2 = \{0, 1\}$). (This is completely analogous to our notation $AG(n, q)$ for an affine geometry.)

Now the lines of $PG(2, 2)$ correspond to the planes through $(0, 0, 0)$ in $AG(3, 2)$. For example, the four points $S = \{(0, 0, 0), (1, 0, 0), (0, 1, 0), (1, 1, 0)\}$ are a plane containing $(0, 0, 0)$ in $AG(3, 2)$, so the three points $\{(1, 0, 0), (0, 1, 0), (1, 1, 0)\}$ form a three-point line in $PG(2, 2)$.

How many lines are there in $PG(2, 2)$? We know the planes in $AG(3, 2)$ are generated by two linearly independent vectors: for instance, the vectors $(1, 0, 0)$ and $(0, 1, 0)$ generate the plane $S$. There are $\binom{7}{2} = 21$ pairs of lines through the origin in $AG(3, 2)$, but this counts each plane three times since each of the $\binom{3}{2} = 3$ pairs of non-zero vectors in the plane gives the same plane. Thus, there are seven lines in $PG(2, 2)$, so the number of points equals the number of lines.[13]

So $PG(2, 2)$ has seven points and seven lines. We can draw this projective plane, remembering each point corresponds to a line through the origin in $\mathbb{F}_2^3$ and each line corresponds to a plane through the origin. See Figure 5.6 for a drawing of the connection between $AG(3, 2)$ and $PG(2, 2)$.

The "circular" line $\{(1, 1, 0), (1, 0, 1), (0, 1, 1)\}$ in $PG(2, 2)$ corresponds to the plane $\{(0, 0, 0), (1, 1, 0), (1, 0, 1), (0, 1, 1)\}$ in $AG(3, 2)$. These four points form a plane in $AG(3, 2)$ because the three vectors $\{(1, 1, 0), (1, 0, 1), (0, 1, 1)\}$ are linearly dependent over $\mathbb{F}_2$.

We've seen this matroid before! The projective plane $PG(2, 2)$ is just the Fano plane we met way back in Example 1.10 in Chapter 1. More importantly, our scheme for drawing the matroid whose points

---

[13] This is what we expected from the point–line symmetry of projective planes.

correspond to the column vectors matches the procedure we've been
using throughout the text. Our main point is this:

- Three points are collinear in $PG(2, 2)$ if and only if the corresponding
  three vectors in $\mathbb{F}_2^3$ are linearly dependent.

It is also worth pointing out that $AG(3, 2)/x = PG(2, 2)$, where $x$
is any point in $AG(3, 2)$ and $AG(3, 2)/x$ is ordinary matroid contraction. This works nicely because $\mathbb{F}_2$ has only one non-zero element. A
generalization appears in Exercise 19.

We can generalize this procedure by changing $\mathbb{F}_2$ to $\mathbb{F}_q$ for any
prime power $q = p^k$. This will produce the projective plane $PG(2, q)$.
Each one-dimensional subspace of the three-dimensional affine geometry $AG(3, q)$ will correspond to a point in the projective plane $PG(2, q)$,
and each two-dimensional subspace of $AG(3, q)$ will correspond to a
line in $PG(2, q)$.

Let's look at a slightly larger example.

**Example 5.15.** Let's repeat the procedure from Example 5.14 for $\mathbb{F}_3 =
\{0, 1, 2\}$. To count the points of $PG(2, 3)$, first note there are 27 points
in the affine geometry $AG(3, 3)$. The points of the projective plane
$PG(2, 3)$ correspond to the lines through $(0, 0, 0)$ in $AG(3, 3)$, and all
lines in $AG(3, 3)$ have three points. For instance, the line consisting of
the three points $\{(0, 0, 0), (1, 0, 0), (2, 0, 0)\}$ will correspond to a point
in the projective plane $PG(2, 3)$. We list all the points as column vectors
in the matrix $A_3$.

$$
A_3 =
\begin{array}{c}
\begin{array}{ccccccccccccc}
a & b & c & d & e & f & g & h & i & j & k & l & m
\end{array} \\
\left[
\begin{array}{ccccccccccccc}
1 & 1 & 1 & 1 & 1 & 1 & 1 & 1 & 1 & 0 & 0 & 0 & 0 \\
0 & 0 & 0 & 1 & 1 & 1 & 2 & 2 & 2 & 1 & 1 & 1 & 0 \\
0 & 1 & 2 & 0 & 1 & 2 & 0 & 1 & 2 & 0 & 1 & 2 & 1
\end{array}
\right].
\end{array}
$$

Thus, there are exactly 13 points in $PG(2, 3)$ because the lines
through $(0, 0, 0)$ in $AG(3, 3)$ partition the 26 points of $AG(3, 3) -
(0, 0, 0)$ into 13 sets of size two.[14] So, by point–line symmetry, we
also have 13 lines in $PG(2, q)$.

How many points are on a line? Note that a plane through $(0, 0, 0)$ in
the affine geometry $AG(3, 3)$ has nine points, so it contains four lines
through $(0, 0, 0)$. This means every line in $PG(2, 3)$ will contain four
points. See Figure 5.7 for a picture of the 13 points and seven of the
13 lines of $PG(2, 3)$.

There are a few things you might notice when you stare at the 13
columns of the matrix $A_3$ in Example 5.15. We'll prove some of this
soon, but let's try to build some intuition first.

---

[14] Also because $m$ is the thirteenth letter of the alphabet.

Figure 5.7. The projective
plane $PG(2, 3)$. For clarity,
only seven of the 13 lines have
been drawn!

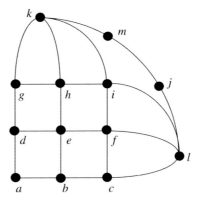

Figure 5.7. The projective plane $PG(2, 3)$. For clarity, only seven of the 13 lines have been drawn!

(1) First, notice that every pair of column vectors is linearly independent. That's because we only took one vector from each line through
the origin in $AG(3, 3)$. (For example, we chose the vector $(1, 0, 0)$,
excluding $(2, 0, 0)$.) Note also that the first non-zero entry in each
column equals 1.

(2) The first nine columns of $A_3$ have first coordinate equal to 1. Furthermore, the other two coordinates of these nine columns are the
nine ordered pairs $(x, y)$ (with $x, y \in \mathbb{F}_3$) we saw in $AG(2, 3)$, the
affine plane over $\mathbb{F}_3$. This strongly suggests that the affine plane
$AG(2, 3)$ is a subset of the projective plane $PG(2, 3)$.[15]

(3) The remaining columns of $A_3$, columns $j, k, l$ and $m$, all have 0 as
their first coordinate. Any subset of three of these columns will be
linearly dependent, hence the corresponding points in $PG(2, 3)$ will
be collinear. So $\overline{jklm}$ is a four-point line disjoint from the affine
plane formed by the first nine columns.

Here's our general definition of the projective plane $PG(2, q)$.

Projective plane

**Definition 5.16.**   Let $q = p^k$ for some prime $p$. The points of the
*projective plane $PG(2, q)$* are the one-dimensional subspaces of the
three-dimensional vector space $\mathbb{F}_q^3$, and the lines are the two-dimesnional
subspaces.

More formally, we introduce an equivalence relation $\sim$ on the non-
zero vectors of $\mathbb{F}_q^3$ as follows:

• $u \sim v$ if and only if $u = k \cdot v$ for some non-zero scalar $k \in \mathbb{F}_q$.

Then the *points* of $PG(2, q)$ are defined to be the equivalence classes
$[v]$ of $\sim$. Three such points are then defined to be collinear if we
can choose three linearly dependent vectors from the corresponding

---

[15]  This doesn't show $AG(2, 3)$ is a sub-object of $PG(2, 3)$ because we also need to
check that lines in $AG(2, 3)$ are correctly represented in $PG(2, 3)$. But don't worry –
this all works out quite nicely.

three equivalence classes. So, in this setting, we write $PG(2, q) = (AG(3, q) - (0, 0, 0))/ \sim$.

It's a bit awkward to constantly use "equivalence classes of vectors" whenever we want to discuss a point in $PG(2, q)$. Rather than worry excessively[16] about equivalence classes, we will use the words "point" and "vector" interchangeably in this context. So, for instance, the point $P$ in $PG(2, 3)$ corresponding to the one-dimensional subspace $t(1, 2, 2)$ of $AG(3, 3)$ will simply be the vector $(1, 2, 2)$. Of course, the vector $2 \cdot (1, 2, 2) = (2, 1, 1)$ represents the same point of $PG(2, 3)$. For that reason, we will usually use a canonical form for our representing vector, one whose first non-zero coordinate equals 1.

We now have two notions of a projective plane: on the one hand, we have the synthetic point–line incidence structure of Definition 5.13. On the other, we used the vector space $\mathbb{F}_q^3$ for a coordinate based, analytic definition of $PG(2, q)$ in Definition 5.16. What's the connection?

**Proposition 5.17.** *$PG(2, q)$ is a projective plane, i.e., $PG(2, q)$ satisfies the two axioms* **P1** *and* **P2***, and it always contains a four-point.*

The proof is left for you in Exercise 14. Although $PG(2, q)$ always satisfies Definition 5.13, not every projective plane arises this way. We return to this issue in Section 5.6. For now, we concentrate on $PG(2, q)$.

Why do we care about projective planes?[17] You should not be surprised by our answer:

> Projective planes are matroids.

The points of $PG(2, q)$ (which we can think of as vectors in $\mathbb{F}_q^3$) form the ground set $E$ of our matroid, and a collection of points is independent in the matroid if and only if the corresponding vectors in $\mathbb{F}_q^3$ are linearly independent.

Here is a pile of facts about the projective planes $PG(2, q)$.

**Proposition 5.18.** *Let $PG(2, q)$ be the projective plane over the field $\mathbb{F}_q$.*

(1) *$PG(2, q)$ is a matroid whose ground set is the set of points of the geometry.*
(2) *Every pair of points determines a unique line, and every pair of lines intersects in exactly one point.*
(3) *$PG(2, q)$ has $q^2 + q + 1$ points and $q^2 + q + 1$ lines.*
(4) *Three points are collinear if and only if the corresponding vectors are linearly dependent.*

---

[16] We leave the reader to find the proper amount of worrying this approach warrants.
[17] This is a (very) dangerous question to ask in a math book.

(5) *For scalars $a, b, c \in \mathbb{F}_q$, not all zero, the equation $ax + by + cz = 0$ defines a line.*

(6) *Removing any line from $PG(2, q)$ leaves the affine plane $AG(2, q)$.*

*Proof Proposition 5.18.*

(1) This follows from checking the independent set axioms (I1), (I2) and (I3) for matrices – see Theorem 6.1.

(2) This is (most of) Proposition 5.17, and is just a way of restating the fact that $PG(2, q)$ satisfies the projective plane axioms **P1** and **P2**.

(3) There are $q^3 - 1$ non-zero vectors in $\mathbb{F}_q^3$, and there are $q - 1$ non-zero points on each line through $(0, 0, 0)$ in $AG(3, q)$. This gives the total number of one-dimensional subspaces:

$$\frac{q^3 - 1}{q - 1} = q^2 + q + 1.$$

The number of lines follows by symmetry.

(4) Three points are collinear if and only if the three corresponding one-dimensional subspaces of $AG(3, q)$ are coplanar, i.e., the three vectors corresponding to the three points of $PG(2, q)$ are linearly dependent.

(5) This follows from a more general result concerning matrix representations of flats – see Proposition 5.25.

(6) This follows from the construction of $PG(2, q)$ from the affine plane $AG(2, q)$ – see the discussion in Section 5.3.2 below.  □

## 5.3.2 The projective plane $PG(2, q)$ from the affine plane $AG(2, q)$

We can also create the projective plane $PG(2, q)$ by building up from the affine plane $AG(2, q)$. This is explicit in our matrix in Example 5.15, and it's worth examining this method in a bit more detail.

Let's re-examine $PG(2, 2)$, this time concentrating on the connections between the affine plane $AG(2, 2)$ and $PG(2, 2)$.

**Example 5.19.** First, consider the affine plane $AG(2, 2)$. This plane has four points:

$$\begin{array}{cccc} a & b & c & d \end{array}$$
$$\begin{bmatrix} 0 & 1 & 0 & 1 \\ 0 & 0 & 1 & 1 \end{bmatrix}.$$

There are six lines in $AG(2, 2)$, and they can be partitioned into three *parallel classes*. For instance, the line through the two points $a$ and $b$ is parallel to the line through $c$ and $d$: this means $\{\overline{ab}, \overline{cd}\}$ is a parallel class. Note that each parallel class contains two lines.

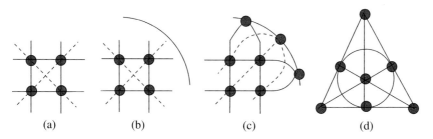

Figure 5.8. From $AG(2, 2)$ to $PG(2, 2)$.

We create the projective plane $PG(2, 2)$ by *adding a line at infinity*.[18] This new line will have three points, and each point on that line will correspond to one of our three parallel classes. Then *every* pair of lines will intersect, since lines in $AG(2, 2)$ that were parallel now meet along the line at infinity.

Figure 5.8 shows this process in stages. Figure 5.8(a) shows $AG(2, 2)$ with three sets of parallel lines, (b) shows the addition of a line at infinity, (c) shows the parallel lines intersecting that line, and (d) gives a more familiar picture of $PG(2, 2)$, the Fano plane. (There is nothing special about the line at infinity in our final picture of $PG(2, 2)$: removing *any* line from $PG(2, 2)$ leaves $AG(2, 2)$.)

Algebraically, we embed the four points $a, b, c, d$ of $AG(2, 2)$ into $PG(2, 2)$ as follows:

(1) First, add a third coordinate to each ordered pair, and set this new coordinate equal to 1. (This is our bookkeeping trick from Proposition 5.8.) Here we get

$$
\begin{array}{cccc}
a & b & c & d \\
\end{array}
$$
$$
\begin{bmatrix}
\boxed{1} & \boxed{1} & \boxed{1} & \boxed{1} \\
0 & 1 & 0 & 1 \\
0 & 0 & 1 & 1
\end{bmatrix}.
$$

(2) Now add the line at infinity. This line now has equation $x = 0$, i.e., it consists of all projective points whose first coordinate is 0. There are three such points, $e, f$ and $g$.

$$
\begin{array}{ccccccc}
a & b & c & d & e & f & g \\
\end{array}
$$
$$
\begin{bmatrix}
1 & 1 & 1 & 1 & 0 & 0 & 0 \\
0 & 1 & 0 & 1 & 1 & 0 & 1 \\
0 & 0 & 1 & 1 & 0 & 1 & 1
\end{bmatrix}.
$$

For example, the parallel class $\{\overline{ab}, \overline{cd}\}$ intersects the line at infinity at the point $e$ because the sets of vectors $\{a, b, e\}$ and $\{c, d, e\}$ are both linearly dependent over the field $\mathbb{F}_2 = \{0, 1\}$.

---

[18] In perspective painting the line where all parallels meet is called the line at the horizon.

Figure 5.9. The nine points $a, b, \ldots, i$ form the affine plane $AG(2, 3)$. Adding the line $\overline{jklm}$ gives the projective plane $PG(2, 3)$. Note: each parallel class of lines in the affine plane intersects the line $\overline{jklm}$ at a unique point.

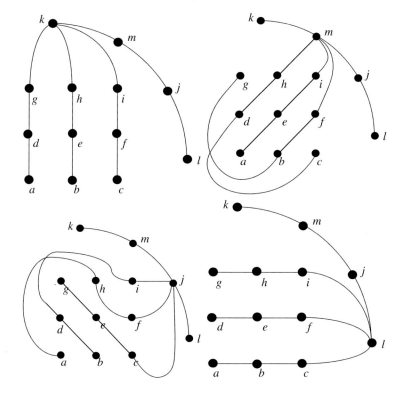

Figure 5.8 shows how to augment the affine plane $AG(2, 2)$ geometrically to create the Fano plane $PG(2, 2)$. As noted above, the reverse process of deletion of a line in the projective plane $PG(2, q)$ leaves the affine plane $AG(2, q)$. See Figure 5.9 for a picture of how the four parallel classes in $AG(2, 3)$ can be used to create $PG(2, 3)$.

When we represent the points of $PG(2, q)$ as the columns of a matrix, here's a quick summary of what we get:

- Our matrix will have three rows and $q^2 + q + 1$ columns.
- We may assume the first non-zero entry of each column equals 1.
- We can order the columns so that the first $q^2$ columns all have first coordinate 1, and the remaining $q + 1$ columns all have first coordinate 0.
- Three points will be collinear if and only if the corresponding columns are linearly dependent.

For instance, our matrix for $PG(2, 5)$ is $3 \times 31$:

$$
\begin{bmatrix}
1 & 1 & 1 & 1 & 1 & 1 & 1 & 1 & \cdots & 1 & 1 & 0 & 0 & 0 & 0 & 0 & 0 \\
0 & 0 & 0 & 0 & 0 & 1 & 1 & 1 & \cdots & 4 & 4 & 1 & 1 & 1 & 1 & 1 & 0 \\
0 & 1 & 2 & 3 & 4 & 0 & 1 & 2 & \cdots & 3 & 4 & 0 & 1 & 2 & 3 & 4 & 1
\end{bmatrix}.
$$

We close this section with an interesting topological non-sequitar. If we replace $\mathbb{F}_q$ by $\mathbb{R}$, then we get the *real projective plane*. It's a fun exercise in visualization to try to understand what this object looks like. Starting with ordinary three-dimensional Euclidean space $\mathbb{R}^3$, we first remove the origin. Now, for each ray through the origin, map all the points of that ray to the point on the unit sphere where the ray meets the sphere. (This just replaces a ray $t \cdot v$, where $t > 0$, by the unit vector $v/||v||$.) Finally, to get the real projective plane, identify each point on the sphere with its antipodal point. This gives the real projective plane, a topologically non-orientable surface, often called a *cross-cap* by topologists.

## 5.4 Projective geometry

In Section 5.3, we gave two (secretly) equivalent[19] ways to construct the projective plane $PG(2, q)$ over the field $\mathbb{F}_q$. Both methods are based on affine geometry.

### From affine 3-space

We created $PG(2, q)$ from the three-dimensional affine geometry $AG(3, q)$ by defining *points* of $PG(2, q)$ to be lines through the origin of $AG(3, q)$. Then the points of our projective plane $PG(2, q)$ are *equivalence classes* of non-zero vectors in $AG(3, q)$.

### From the affine plane

This time, we started with the affine plane $AG(2, q)$, then added a *line at infinity*. Then, for each of the $q + 1$ *parallel classes* of lines in the affine plane, we place one point on this line at infinity. The result is the projective plane $PG(2, q)$.

The two constructions are equivalent, and each one has its merits. For instance, the first method makes it easy to see why the number of points (and lines) in $PG(2, q)$ is the quotient $\dfrac{q^3 - 1}{q - 1}$. On the other hand, the second method makes it obvious that the affine plane $AG(2, q)$ is a subset of the projective plane $PG(2, q)$. We will want to keep both constructions handy, and you should, too. Both constructions generalize to higher dimensions, and that's the point of this section.

---

[19] One might be tempted to say "cryptomorphic" ways, but we would never fall into that trap. Never.

**Definition 5.20.** The *projective geometry* $PG(n, q)$ can be defined as follows:

- Points: the points of $PG(n, q)$ are the lines through the origin in the affine space $AG(n + 1, q)$ (i.e., the one-dimensional subspaces of the $(n + 1)$-dimensional vector space $\mathbb{F}_q^{n+1}$).
- Lines: the lines of $PG(n, q)$ are the (two-dimensional) planes through the origin in the affine space $AG(n + 1, q)$.
- $k$-dimensional planes: the $k$-dimensional planes of $PG(n, q)$ are the $(k + 1)$-dimensional planes through the origin in the affine space $AG(n + 1, q)$.

So, the number of points in $PG(n, q)$ is $\frac{q^{n+1}-1}{q-1} = q^n + q^{n-1} + \cdots + q + 1$. $PG(n, q)$ is a matroid, and the flats are simply the subspaces of $\mathbb{F}_q^{n+1}$ of various dimensions from Definition 5.20. This is important enough for us to state it as a proposition.

**Proposition 5.21.** *Let $E$ be the set of points of $PG(n, q)$, and let $\mathcal{F}$ be the collection of all planes of all dimensions of $PG(n, q)$, along with $\emptyset$ and $E$. Then $\mathcal{F}$ is the family of flats of a matroid.*

The proof of Proposition 5.21 follows immediately from Theorem 6.1, the theorem that tells us linearly independent subsets of a vector space satisfy (I1), (I2) and (I3). Once we have proven that theorem, we just need to show the flats correspond to the subsets of vectors closed under the operation of taking linear combinations. But these are just the subspaces of $\mathbb{F}_q^{n+1}$.

When $n = 1$, the projective space $PG(1, q)$ is a *line*. We look at this simple case in the next example.

**Example 5.22.** The points of $PG(1, q)$ are the lines through $(0, 0)$ in the affine plane $AG(2, q)$. There are $q + 1$ lines through $(0, 0)$ in $AG(2, q)$, and we list these points as the column vectors in a matrix:

$$\begin{bmatrix} 1 & 1 & 1 & 1 & \cdots & 1 & 0 \\ 0 & 1 & 2 & 3 & \cdots & q-1 & 1 \end{bmatrix}.$$

The matroid is just the $(q + 1)$-point line $U_{2,q+1}$. Alternatively, you can create $PG(1, q)$ by building up from $AG(1, q)$ (which is a $q$-point line) by adding a "point at infinity."

In Figure 5.7, you can see how the projective plane $PG(2, 3)$ is partitioned into the affine plane $AG(2, 3)$ and the projective line $PG(1, 3)$. This always works.

**Proposition 5.23.** *Let H be a hyperplane of $PG(n, q)$.*

(1) *H is isomorphic to $PG(n - 1, q)$.*
(2) *$PG(n, q) - H$ is isomorphic to $AG(n, q)$.*

*(Here, the isomorphism preserves all point–line–plane– $\cdots$ incidences. Alternatively, these can be viewed as matroid isomorphisms.)*

We omit the straightforward proof of Proposition 5.23. In general, we can build $PG(n, q)$ up from the affine geometry $AG(n, q)$ by adding a "hyperplane at infinity." This is entirely analogous to the "line at infinity" we added to $AG(2, q)$ when we created the projective plane $PG(2, q)$. Parallel lines in $AG(n, q)$ will intersect in that hyperplane, which is isomorphic to $PG(n - 1, q)$. We don't need to exploit this relationship in any detail, but it's helpful to remember this when considering hyperplanes and their complements in $PG(n, q)$.

More generally, every rank $k$ flat of $PG(n, q)$ is isomorphic to $PG(k - 1, q)$. For example, a rank 3 flat in $PG(n, q)$ is isomorphic to the projective plane $PG(2, q)$, and this arises from a three-dimensional subspace of $\mathbb{F}_q^{n+1}$. Since the rank $k$ flats of $PG(n, q)$ are simply the dimension $k$ subspaces of $\mathbb{F}_q^{n+1}$, one consequence of Proposition 5.21 is the following.

**Theorem 5.24.** *The lattice of subspaces of $\mathbb{F}_q^{n+1}$ ordered by inclusion is a geometric lattice, i.e., the lattice of flats of the matroid $PG(n, q)$.*

As we did for flats in $AG(n, q)$ in Proposition 5.5, we can find a matrix-based description of the projective flats. The proof, which is similar to the proof of Proposition 5.5, is left for you in Exercise 20.

**Proposition 5.25.** *Let F be a flat in $PG(n, q)$ with $r(F) = k$. Then there is an $(n - k) \times n$ matrix A such that $F = \{v \mid Av = \bar{0}\}$.*

**Example 5.26.** Consider the four-dimensional (rank 5) projective space $PG(4, 3)$. Suppose we would like to describe the line determined by the two points $P = (1, 2, 0, 0, 1)$ and $Q = (0, 1, 1, 2, 0)$. Then one way to proceed is to simply use linear span: $R$ is on the line $\overline{PQ}$ if $R = a \cdot P + b \cdot Q$, where $a, b \in \mathbb{F}_3$.

Alternatively, we can use Proposition 5.25 by producing an appropriate matrix $A$. Setting $B = \begin{pmatrix} 1 & 2 & 0 & 0 & 1 \\ 0 & 1 & 1 & 2 & 0 \end{pmatrix}$, we find a basis for the three-dimensional null space of $B$, and arrange those three basis vectors as the rows of our matrix $A$:

$$A = \begin{pmatrix} 1 & 0 & 0 & 0 & 2 \\ 1 & 1 & 0 & 1 & 0 \\ 1 & 1 & 2 & 0 & 0 \end{pmatrix}.$$

Table 5.2. *Matroid interpretations for the projective geometry $PG(n, q)$.*

| Matroid concept | Interpretation in $PG(n, q)$ |
| --- | --- |
| Points | Lines through $(0, 0, \ldots, 0)$ in $AG(n + 1, q)$. Equivalently, equivalence classes $[v]$ for $\bar{0} \neq v \in \mathbb{F}_q^{n+1}$, where $v \sim v'$ if $v = k \cdot v'$ for a non-zero scalar $k$. |
| Independent sets | Linear independence: $\{P_1, P_2, \ldots, P_n\}$ is independent if there are no non-trivial scalars $a_i$ such that $\sum_{i=1}^{k} a_i P_i = \bar{0}$. |
| Closure | Linear closure: for $S = \{P_1, P_2, \ldots, P_n\}$, $Q \in \bar{S}$ if there are non-trivial scalars $a_i$ such that $Q = \sum_{i=1}^{n} a_i P_i$. |
| Flats | The points of $F$ correspond to a subspace of $\mathbb{F}_q^{n+1}$. Equivalently, a $k$-flat consists of all vectors $v$ satisfying $Av = \bar{0}$ for some $(n - k) \times n$ matrix $A$. |
| Hyperplanes | Maximal flats: $\left\{ (x_0, x_1, \ldots, x_n) \mid \sum_{i=0}^{n} a_i x_i = 0 \text{ for } a_i \in \mathbb{F}_q \right\}$. |
| Rank | $r(S)$ is the matrix rank of the matrix whose column vectors are the points of $S$. $r(PG(n, q)) = n + 1$. |

Then you can quickly check that the matrix $A$ kills[20] the vectors $P$ and $Q$: $AP = \bar{0}$ and $AQ = \bar{0}$. Since the rank of $A$ is 3, the rank of the null space of $A$ is 2, which is what we need, since a line is a rank 2 flat. So our line is $\{v \mid Av = \bar{0}\}$. In terms of coordinates, the point $(x_1, x_2, x_3, x_4, x_5)$ is on the line $\overline{PQ}$ if and only if $(x_1, x_2, x_3, x_4, x_5)$ satisfies all three of these equations:

$$x_1 + 2x_5 = 0$$
$$x_1 + x_2 + x_4 = 0$$
$$x_1 + x_2 + 2x_3 = 0.$$

Note that this expresses our (rank 2) line as the intersection of three (rank 4) hyperplanes.

Studying the structure of flats in projective geometries is then reduced to a problem of (finite) linear algebra. We'll count the number of subspaces of various dimensions in the next section.

As we did in Table 5.1 for the affine geometry $AG(n, q)$, we interpret several matroid concepts for projective geometry in Table 5.2. As usual, in the table, we identify the point $P$ with any vector in the equivalence class of vectors that correspond to a line through the origin in $(k + 1)$-dimensional affine space.

---

[20] We apologize for the violent metaphor, but we have found "$A$ picks a flower for $P$ and $Q$" is less useful as a pedagogical technique for describing the null space.

It's useful to compare Tables 5.1 and 5.2. For example, the points of a hyperplane in $AG(n, q)$ satisfy a linear equation $\sum a_i x_i = b$, while a hyperplane in $PG(n, q)$ is determined by a *homogeneous* linear equation $\sum a_i x_i = 0$.[21]

## 5.5 Counting k-flats and q-binomial coefficients

A big difference between the vector spaces $\mathbb{F}_q^n$ and $\mathbb{R}^n$ is that the former only has a finite number of vectors. This (blindingly obvious) observation will lead us to some pretty neat counting arguments. For example, we can count the number of subspaces of $\mathbb{F}_q^n$ of each dimension, and this will allow us to count the number of flats of rank $k$ in both $PG(n, q)$ and $AG(n, q)$.

**Definition 5.27.** Denote by $\begin{bmatrix} n \\ k \end{bmatrix}_q$ the number of $k$-dimensional subspaces of the $n$-dimensional vector space $\mathbb{F}_q^n$. These numbers are called *q-binomial* (or *Gaussian*) coefficients.[22]

*q-binomial coefficients*

### 5.5.1 Computing $\begin{bmatrix} n \\ k \end{bmatrix}_q$

We count the number of $k$-dimensional subspaces of the $\mathbb{F}_q^n$ by counting the number of ordered $k$-tuples of linearly independent vectors chosen from $\mathbb{F}_q^n$ in two ways. Counting something in two different ways is a standard (and very valuable) combinatorial tool.

Our immediate goal is to find a reasonable formula for $\begin{bmatrix} n \\ k \end{bmatrix}_q$. To that end, we'll first do some counting, then give you the formula. Don't peek![23]

First, we let $N(n, k)$ be the number of ways to select an *ordered* collection of $k$ linearly independent vectors from $\mathbb{F}_q^n$. So, for example, if $\{v_1, v_2, \ldots, v_k\}$ is a linearly independent set, this collection would contribute $k!$ *ordered* collections counted by $N(n, k)$. (Using order here may seem a little strange, but it will make the counting easier.)

Let's get started:

- First, choose any non-zero vector $v_1$ to be the first vector in our ordered list. There are $q^n - 1$ ways to choose a non-zero vector.

---

[21] One way to see Tables 5.1 and 5.2 side-by-side is to tear out one page of the text. If you then decide to replace your damaged text with a new copy, the authors would not object.

[22] We use the term *q-binomial* instead of *Gaussian* coefficients for two reasons: First, "q-binomial" emphasizes a close connection with ordinary binomial coefficients, and, second, Gauss already has enough theorems, algorithms and concepts named for him. In our opinion.

[23] Go ahead and peek – the formula isn't all that surprising.

- Next, choose a vector $v_2$ so that $v_1$ and $v_2$ are linearly independent. This eliminates all the scalar multiples of $v_1$, of which there are precisely $q$ (including the zero vector). Thus, we can choose $v_2$ in $q^n - q$ ways.
- For $v_3$, we eliminate any linear combination of $v_1$ and $v_2$ (corresponding to any vector in the plane determined by $v_1$ and $v_2$). Thus, there are precisely $q^2$ vectors that are in the span of $v_1$ and $v_2$ who are not available to be chosen for $v_3$. So the number of choices for $v_3$ is $q^n - q^2$.

Continuing in this way, we obtain:

$$N(n, k) = (q^n - 1)(q^n - q)(q^n - q^2) \cdots (q^n - q^{k-1}).$$

Now let's count $N(n, k)$ in a second way. This time, we first choose a subspace $W$ of dimension $k$ in $\begin{bmatrix} n \\ k \end{bmatrix}_q$ ways, and then we pick an ordered basis for this subspace. Then we can repeat our counting argument above for the $q^k$ vectors of $W$ to get:

$$N(n, k) = \begin{bmatrix} n \\ k \end{bmatrix}_q (q^k - 1)(q^k - q)(q^k - q^2) \cdots (q^k - q^{k-1}).$$

Setting our two expressions for $N(n, k)$ equal to each other gives us a formula for $\begin{bmatrix} n \\ k \end{bmatrix}_q$:

$$\begin{bmatrix} n \\ k \end{bmatrix}_q = \frac{(q^n - 1)(q^n - q)(q^n - q^2) \cdots (q^n - q^{k-1})}{(q^k - 1)(q^k - q)(q^k - q^2) \cdots (q^k - q^{k-1})}.$$

Some elementary algebra then gives us the following theorem.[24]

**Theorem 5.28.** *The number of k-dimensional subspaces of the n-dimensional vector space* $\mathbb{F}_q^n$ *is*

$$\begin{bmatrix} n \\ k \end{bmatrix}_q = \frac{(q^n - 1)(q^{n-1} - 1)(q^{n-2} - 1) \cdots (q^{n-k+1} - 1)}{(q^k - 1)(q^{k-1} - 1)(q^{k-2} - 1) \cdots (q - 1)}.$$

When $k = 0$, we note $\begin{bmatrix} n \\ 0 \end{bmatrix}_q = 1$. This corresponds to the single subspace $W$ of dimension 0, namely $W = \{0\}$.

You can check the formula for $\begin{bmatrix} n \\ k \end{bmatrix}_q$ for some specific values of $n, k$ and $q$. For instance, $\begin{bmatrix} 3 \\ 1 \end{bmatrix}_2 = 7$ and $\begin{bmatrix} 3 \\ 2 \end{bmatrix}_2 = 7$. The first formula tells us there are seven points in the Fano plane $PG(2, 2)$, and the second tells us the Fano plane also has seven lines.

By Proposition 5.21, the dimension $k$ subspaces of $\mathbb{F}_q^{n+1}$ are the rank $k$ flats of the matroid $PG(n, q)$. (For instance, the two-dimensional subspaces of $\mathbb{F}_q^{n+1}$ correspond to lines in $PG(n, q)$, which are rank 2 flats.) Putting this together with what we now know about the $q$-binomial coefficients gives us the number of rank $k$ flats in $PG(n, q)$.

**Theorem 5.29.** *The number of flats of rank k in* $PG(n, q)$ *is* $\begin{bmatrix} n+1 \\ k \end{bmatrix}_q$.

---

[24] This formula appears on comprehensive exams in graduate school with surprising frequency.

What about flats in the affine geometry $AG(n, q)$? Can we use $q$-binomials somehow to count these, too? The answer is yes.[25] The key to computing the number of flats in $AG(n, q)$ is another combinatorial argument in which we count something in two different ways.

**Theorem 5.30.** *Let $1 \leq k \leq n$. Then the number of flats of rank $k$ in $AG(n, q)$ is $q^{n-k+1} \left[ {n \atop k-1} \right]_q$.*

*Proof Theorem 5.30.* Let $f_k$ be the number of rank $k$ flats. We count the *point–flat incidences* in two different ways. First, note that each point is in the same number of rank $k$ flats (this is obvious from the symmetry of $AG(n, q)$). Then we might as well count the number of rank $k$ flats which contain $\bar{0}$. But these are just the subspaces of dimension $k - 1$ in $\mathbb{F}_q^n$, and there are $\left[ {n \atop k-1} \right]_q$ such subspaces. Since there are $q^n$ points, this gives an incidence count of $q^n \left[ {n \atop k-1} \right]_q$.

On the other hand, we can fix a flat[26] $F$ and count the number of points $F$ contains. But every rank $k$ flat is a translate of a $(k - 1)$-dimensional subspace, and each of these subspaces has $q^{k-1}$ points. Thus, this incidence count gives $f_k q^{k-1}$.

Equating the two incidence counts gives us our formula. □

Let's check the formula for $k = 1$ and $k = 2$. When $k = 1$, the formula gives us $q^n$ rank 1 flats, i.e., $q^n$ points in $AG(n, q)$. For $k = 2$, we get $q^{n-1} \left[ {n \atop 1} \right] = q^{n-1} \frac{q^n-1}{q-1}$ lines in $AG(n, q)$.

Incidence counts are extremely important in combinatorics. Any incidence count like the one we just gave can be expressed via a bipartite graph. In this case, the vertices of the graph are the points of $AG(n, q)$ (on one side of the bipartite graph) and the rank $k$ flats (on the other side). We join a point $P$ to a flat $F$ with an edge in the bipartite graph precisely when $P$ is on $F$. Then the incidence count is simply the number of edges in the bipartite graph.

Alternatively, you might prefer to think of these counts as follows. You are omniscient, and can see all the points and rank $k$ flats. Then you shout every time you observe a point $P$ on a rank $k$ flat $F$. Now simply count the number of times you shouted in two different ways: once when you look at $AG(n, q)$ point by point, and the second time when you look at it $k$-flat by $k$-flat.

Incidence counts

## 5.5.2 Connection to binomial coefficients

It's fun[27] to simplify the formula for $\left[ {n \atop k} \right]_q$ for specific values of $k$, with $n$ and $q$ unspecified. For example, you can check the following:

---

[25] By now, you should know we wouldn't ask this question if the answer were "no."
[26] Does "fix a flat" suggest a joke? Feel free to patch one in.
[27] Here you should interpret the word "fun" as "an exercise in factoring polynomials."

(1) $\begin{bmatrix} n \\ 0 \end{bmatrix}_q = \begin{bmatrix} n \\ n \end{bmatrix}_q = 1.$

(2) $\begin{bmatrix} n \\ 1 \end{bmatrix}_q = \dfrac{q^n - 1}{q - 1} = q^n + q^{n-1} + \cdots + q + 1.$

(3) $\begin{bmatrix} n \\ n-1 \end{bmatrix}_q = \dfrac{(q^n - 1)(q^{n-1} - 1) \cdots (q^3 - 1)(q^2 - 1)}{(q^{n-1} - 1)(q^{n-2} - 1) \cdots (q^2 - 1)(q - 1)}$

$$= \dfrac{q^n - 1}{q - 1} = \begin{bmatrix} n \\ 1 \end{bmatrix}_q.$$

Then (2) counts the number of points in the projective geometry $PG(n, q)$, and (3) counts the number of hyperplanes. Note the symmetry in (1) and between (2) and (3). This suggests the following.

**Proposition 5.31.** *For all non-negative integers $k, n$, with $k \leq n$, we have* $\begin{bmatrix} n \\ k \end{bmatrix}_q = \begin{bmatrix} n \\ n - k \end{bmatrix}_q.$

*Proof Proposition 5.31.* Specifying the subspace $W$ also uniquely determines the orthogonal complement $W^\perp$. Further, $\dim(W) + \dim(W^\perp) = n$. Since $(W^\perp)^\perp = W$, we get a bijection between the subspaces of dimension $k$ and those of dimension $n - k$.   □

In the vector space $\mathbb{R}^n$, the pair of subspaces $W$ and $W^\perp$ have zero-dimensional intersection: $W \cap W^\perp = \{0\}$. This may no longer be true when we replace the reals with a finite field. For example, consider the vector space $\mathbb{F}_2^4$, with $W = \{(0, 0, 0, 0), (1, 1, 0, 0)\}$. Then, since $(1, 1, 0, 0) \cdot (1, 1, 0, 0) = 0$, we get $W \subseteq W^\perp$. But it is still true that $\dim(W^\perp) = 3 = n - \dim(W)$.

The notation for $q$-binomial coefficients $\begin{bmatrix} n \\ k \end{bmatrix}_q$ (and the name "$q$-binomial coefficients") should suggest a strong connection with the ordinary binomial coefficients $\binom{n}{k}$. For example, Proposition 5.31 is a generalization of a symmetry property for ordinary binomial coefficients:

$$\binom{n}{k} = \binom{n}{n - k}.$$

Additionally, binomial coefficients satisfy Pascal's identity:

$$\binom{n + 1}{k} = \binom{n}{k - 1} + \binom{n}{k}$$

while $q$-binomial coefficients satisfy:

**Proposition 5.32.**

$$\begin{bmatrix} n + 1 \\ k \end{bmatrix}_q = \begin{bmatrix} n \\ k - 1 \end{bmatrix}_q + q^k \begin{bmatrix} n \\ k \end{bmatrix}_q.$$

You are asked to prove this identity in Exercise 21(a).

If you like patterns, you should find plenty in our list of $\begin{bmatrix} 8 \\ k \end{bmatrix}_q$, for $0 \le k \le 8$:

---

$\begin{bmatrix} 8 \\ 0 \end{bmatrix}_q$    1

$\begin{bmatrix} 8 \\ 1 \end{bmatrix}_q$    $q^7 + q^6 + q^5 + q^4 + q^3 + q^2 + q + 1$

$\begin{bmatrix} 8 \\ 2 \end{bmatrix}_q$    $q^{12} + q^{11} + 2q^{10} + 2q^9 + 3q^8 + 3q^7 + 4q^6 + 3q^5 + 3q^4$
$+ 2q^3 + 2q^2 + q + 1$

$\begin{bmatrix} 8 \\ 3 \end{bmatrix}_q$    $q^{15} + q^{14} + 2q^{13} + 3q^{12} + 4q^{11} + 5q^{10} + 6q^9 + 6q^8 + 6q^7$
$+ 6q^6 + 5q^5 + 4q^4 + 3q^3 + 2q^2 + q + 1$

$\begin{bmatrix} 8 \\ 4 \end{bmatrix}_q$    $q^{16} + q^{15} + 2q^{14} + 3q^{13} + 5q^{12} + 5q^{11} + 7q^{10} + 7q^9 + 8q^8$
$+ 7q^7 + 7q^6 + 5q^5 + 5q^4 + 3q^3 + 2q^2 + q + 1$

$\begin{bmatrix} 8 \\ 5 \end{bmatrix}_q$    $q^{15} + q^{14} + 2q^{13} + 3q^{12} + 4q^{11} + 5q^{10} + 6q^9 + 6q^8 + 6q^7$
$+ 6q^6 + 5q^5 + 4q^4 + 3q^3 + 2q^2 + q + 1$

$\begin{bmatrix} 8 \\ 6 \end{bmatrix}_q$    $q^{12} + q^{11} + 2q^{10} + 2q^9 + 3q^8 + 3q^7 + 4q^6 + 3q^5 + 3q^4$
$+ 2q^3 + 2q^2 + q + 1$

$\begin{bmatrix} 8 \\ 7 \end{bmatrix}_q$    $q^7 + q^6 + q^5 + q^4 + q^3 + q^2 + q + 1$

$\begin{bmatrix} 8 \\ 8 \end{bmatrix}_q$    1

---

In particular, note that $\begin{bmatrix} n \\ k \end{bmatrix}_q$ is always a *polynomial* in $q$. Furthermore, evaluating $\begin{bmatrix} n \\ k \end{bmatrix}_q$ (expressed as a polynomial) at $q = 1$ gives the ordinary binomial coefficient $\binom{n}{k}$. See Exercises 21(c) and 22.

For more counting using the $q$-binomial coefficients, here is a quick guide to the exercises:

---

| Exercise # | Topic |
|---|---|
| 23 | Counting lines in $PG(n, q)$. |
| 24 | Counting three-point circuits in $AG(n, q)$ and $PG(n, q)$. |
| 25 | Counting independent sets in $AG(n, q)$ and $PG(n, q)$. |
| 26 | Counting $k$-point circuits in $AG(n, q)$. |
| 27 | Counting $k$-point circuits in $PG(n, q)$. |

If this fails to satisfy your need to count things, Project P.4 studies three probability problems, all of which are solved using $q$-binomials:

(1) Suppose you are given an $n \times n$ matrix, with entries in the finite field $\mathbb{F}_q$. What is the probability your matrix is invertible?
(2) Suppose you randomly choose a set $B$ of $n + 1$ points from the rank $n + 1$ affine geometry $AG(n, q)$. What is the probability $B$ is a basis?
(3) Same question as (2), but for the rank $n + 1$ projective space $PG(n, q)$.

## 5.6 Abstract projective planes

The definition of a projective plane (Definition 5.13) did not depend on coordinates, equations or finite fields. Did we really need vectors to represent the points?

This question occupied nineteenth-century geometers as they attempted to prove that the axioms for projective planes forced the existence of coordinates, finite fields, and all that. It was not until late in that century that some pathological, *non-Desarguesian* planes were discovered. These projective planes satisfy the axioms of Definition 5.13, but do not arise from fields or division rings.

It is a standard exercise to count the number of points, lines, points per line and lines through a point.

**Theorem 5.33.** *Let $P$ be a projective plane, and assume some line of $P$ contains exactly $m + 1$ points. Then*

(1) *Every line has $m + 1$ points.*
(2) *Every point is on exactly $m + 1$ lines.*
(3) *There are a total of $m^2 + m + 1$ points.*
(4) *There are a total of $m^2 + m + 1$ lines.*

*We say $P$ is a projective plane of order $m$.*

See [17] for a proof based solely on the abstract Definition 5.13.[28] One of the most important open problems in finite geometry concerns the existence of projective planes.

For what integers $m$ does there exist a projective plane of order $m$?

If $m = p^k$ is a prime power, then our construction of $PG(2, m)$ gives us a projective plane of order $m$. In fact, all known projective planes have orders that are prime powers: no other orders have been discovered. Furthermore, when the exponent $k = 1$ (so $m = p$ is prime),

---

[28]  Or try it yourself – it's fun, and not too hard.

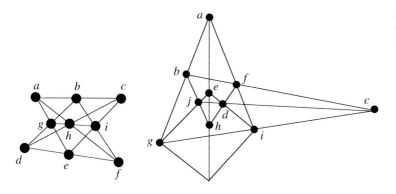

Figure 5.10. Left: Pappus'
configuration. Right:
Desargues' configuration.

the projective plane $PG(2, m)$ is unique. This is false for prime powers: there are four projective planes of order 9: $PG(2, 9)$ and three "others."

The most famous theorem concerning possible orders of projective planes is the following number-theoretic result.

**Theorem 5.34.** *[Bruck–Ryser Theorem] Suppose P is a projective plane of order m, where $m \equiv 1$ (mod 4) or $m \equiv 2$ (mod 4). Then m can be written as the sum of two squares.*

A proof of this theorem can be found in [17], for example. The Bruck–Ryser Theorem is usually used to rule out possible orders of projective planes. For example, if $m = 6$, then $m \equiv 2$ (mod 4), but 6 cannot be written as the sum of two squares. Thus there is no projective plane of order 6, i.e., there is no projective plane with seven points on each line.

The smallest integer which is not a prime power and does not fall under the hypothesis of Bruck–Ryser is 10. It was proven that there is no projective plane of order 10 in 1989 [21]; the proof made extensive use of a computer. (The proof examined the point–line incidence for a purported projective plane of order 10. Such an incidence structure would be a $111 \times 111$ matrix of 0's and 1's, where each row and each column would have exactly 11 1's and 100 0's. Searching through all such matrices is not a job for a human being, but a computer program eventually showed no such incidence matrices exist.)

What's up with affine planes? Nothing new:

**Theorem 5.35.** *A projective plane of order n exists if and only if an affine plane of order n exists.*

This is easy: removing any line from *any* projective plane will leave an affine plane (assuming we've axiomatized what it means to be an affine plane).

One major difference between $PG(2, q)$ and the projective planes that do not arise from fields concerns two famous configurations, the configurations of *Pappus* and *Desargues* (Figure 5.10).

**Theorem 5.36.** *[Pappus' Theorem] Let $a, b, c, d, e, f$ be points in a projective plane $PG(2, q)$ such that $a, b, c$ and $d, e, f$ are collinear. Let $g = \overline{ae} \cap \overline{bd}, h = \overline{af} \cap \overline{cd}$ and $i = \overline{bf} \cap \overline{ce}$. Then, $g, h$ and $i$ are collinear.*

**Theorem 5.37.** *[Desargues' Theorem] Let $\overline{bg}, \overline{eh},$ and $\overline{fi}$ be three lines in $PG(2, q)$ all meeting at the point $a$. Let $j = \overline{bh} \cap \overline{eg}, d = \overline{fh} \cap \overline{ei}$ and $c = \overline{bf} \cap \overline{gi}$. Then, $c, d$ and $j$ are collinear.*

We/you will prove these theorems are valid in the coordinate-based projective plane $PG(2, q)$ in Chapter 6 (Theorem 6.21 and Exercise 26). But the conclusions of these theorems do not hold in the "other" projective planes, i.e., those that do not arise from a field.

**Theorem 5.38.** *Suppose $P$ is a finite projective plane. Then the following are equivalent:*

(1)  *$P = PG(2, q)$ for some prime power $q$.*
(2)  *Desargues' Theorem holds.*
(3)  *Pappus' Theorem holds.*

You can find a proof of Theorem 5.38 in [34], for example.

It's possible to extend this discussion to include the higher-dimensional projective spaces. To do this, you need to axiomatize $n$-dimensional projective space. We omit the details, but mention the main result.

**Theorem 5.39.** *[Veblen–Young Theorem] Every finite projective space of dimension at least three is isomorphic to the projective space $PG(n, q)$ for some $n \geq 3$ and $q$ a prime power.*

This comforting result tells us all projective geometries of dimension 3 or more are the familiar $PG(n, q)$, so the pathologies that arise in dimension 2 do not trouble us in higher dimensions.

## Exercises

### Section 5.1 – Affine planes

(1) Let $AG(2, q)$ be an affine plane. Show there are $q$ points on every line by showing there are exactly $q$ solutions to the equation $ax + by = c$, where $a, b$ and $c$ are fixed elements of $\mathbb{F}_q$.

(2) Show there are precisely 14 hyperplanes in $AG(2, 3)$ by counting all possible equations of the form $ax + by + cz = d$, where $a, b, c, d \in \{0, 1\} = \mathbb{F}_2$. Describe the geometric structure of each of the 14 planes in terms of faces of the cube in Figure 5.4. (You should find 12 "Euclidean" planes and two "twisted" planes.)

(3) Let's try to define an affine plane based on $\mathbb{Z}_4 = \{0, 1, 2, 3\}$, with our arithmetic modulo 4. (Recall $\mathbb{Z}_4$ is not a field since 2 has no multiplicative inverse. However, there is a field of order 4.) Let $\mathcal{P}$ be the collection of all ordered pairs $(x, y)$, with $x, y \in \mathbb{Z}_4$. Then, assuming the "lines" are defined as the solutions of equations of the form $ax + by = c$ for $a, b, c \in \mathbb{Z}_4$, find two "lines" that intersect in more than one point. Conclude that this structure is not a matroid.

(4) [Proposition 5.9(1)] Let $S = \{P_1, P_2, \ldots, P_n\}$ be a collection of points in the affine geometry $AG(n, q)$. Show that the rank of $S$ equals $\dim(F) + 1$, where $F$ is the smallest flat containing $S$.

(5) Let $AG(2, q)$ be an affine plane.
  (a) Show that there are exactly $q^2 + q$ distinct lines in $AG(2, q)$ by finding the number of distinct equations of the form $ax + by = c$ for $a, b, c \in \mathbb{F}_q$.
  (b) Show the $q^2 + q$ lines can be partitioned into $q + 1$ parallel classes, with each class containing $q$ lines. (Suggestion: show there are exactly $q - 1$ lines parallel to a given line with equation $ax + by = c$.)

(6) Suppose $q > 2$. Show that it is not possible to represent the affine plane $AG(2, q)$ in the Euclidean plane so that the lines of $AG(2, q)$ are all given by straight-line segments. (See Exercise 25 of Chapter 1.)

### Section 5.2 – Affine dependence

(7) Consider the three points $P = (1, 0)$, $Q = (0, 2)$ and $R = (3, 1)$ in $AG(2, 5)$. These three points are collinear (see Example 5.7). Find scalars $a, b, c \in \mathbb{F}_5$ so that $P$, $Q$ and $R$ satisfy the equation $ax + by = c$.

(8) Let $S = \{(x, y, z) \mid x = 0 \text{ or } y = 0\}$ for $x, y, z \in \mathbb{F}_3$. Show $S$ is not a flat of $AG(3, 3)$ and find $\overline{S}$.

(9) Let $v$ be a non-zero vector. Show $v$ and $2v$ are affinely independent over $\mathbb{F}_q$, where $q$ is not a power of 2.

(10) Every flat is the intersection of hyperplanes: let $P = (1, 2, 1)$ and $Q = (4, 0, 1)$ be two points in the affine geometry $AG(3, 5)$. Find scalars $a_i, b_i, c_i$ and $d_i$ (for $i = 1, 2$) so that the line $\overline{PQ}$ is the intersection of the two hyperplanes $a_1 x + b_1 y + c_1 z = d_1$ and $a_2 x + b_2 y + c_2 z = d_2$. How many different pairs of hyperplanes work?

(11) [Proposition 5.9(2)] Let $S = \{P_1, P_2, \ldots, P_n\}$ be a collection of points in the affine geometry $AG(n, q)$. Show that $x \in \overline{S}$ if and only if there are scalars $a_i \in \mathbb{F}_q$ with $x = \sum_{i=1}^{n} a_i P_i$ and $\sum_{i=1}^{n} a_i = 1$. (Hint: use the observation from the proof of Proposition 5.6 that $\{x, P_1, P_2, \ldots, P_n\}$ is affinely dependent if and only if the set of vectors $\{P_1 - x, P_2 - x, \ldots, P_n - x\}$ is linearly dependent.)

(12) [Proposition 5.9(3)] Show that $H$ is a hyperplane in $AG(n, q)$ if and only if $H$ is the set of $n$-tuples $(x_1, x_2, \ldots, x_n)$ satisfying an equation of the form $a_1 x_1 + a_2 x_2 + \cdots + a_n x_n = b$ for scalars $a_i, b \in \mathbb{F}_q$. (Hint: use the procedure outlined in the discussion preceding Proposition 5.5.)

(13) [Proposition 5.12] For an $n$-tuple $x = (x_1, x_2, \ldots, x_n) \in \mathbb{F}_q^n$, let $x' \in \mathbb{F}_q^{n+1}$ be the corresponding $(n + 1)$-tuple formed by adding a 1 in the first coordinate of $x$: $x' = (1, x_1, x_2, \ldots, x_n)$. Now suppose $Q, P_1, P_2, \ldots, P_m \in \mathbb{F}_q^n$. Then prove $Q$ is an affine combination of $P_1, P_2, \ldots, P_m$ if and only if the corresponding vector $Q'$ is a linear combination of $P_1', P_2', \ldots, P_m'$ in $\mathbb{F}_q^{n+1}$. (Hint: if $Q = \sum_{i=1}^{m} a_i P_i$ and $\sum_{i=1}^{m} a_i = 1$, then use the same scalars to express $Q'$ as a linear combination of $P_1', P_2', \ldots, P_m'$.)

## Section 5.3 – Projective planes

(14) [Proposition 5.17] Show that $PG(2, q)$ is a projective plane, i.e., $PG(2, q)$ satisfies the two axioms **P1** and **P2**, and it always contains a four-point.

(15) Show that it is not possible to represent the projective plane $PG(2, q)$ in the Euclidean plane so that the lines of $PG(2, q)$ are all given by straight-line segments. (See Exercise 6 above and Exercise 25 of Chapter 1.)

(16) (a) Show that the $(q + 1)$-point line $U_{2,q+1}$ is not a submatroid of the affine plane $AG(2, q)$.
   (b) Show that the $(q + 2)$-point line $U_{2,q+2}$ is not a submatroid of the projective plane $PG(2, q)$.

(17) (a) Show that, for any point $x$ in the Fano plane, $PG(2, 2) - x$ is isomorphic to the graphic matroid $M(K_4)$.
   (b) Show that, for any prime power $q = p^k$, there is a set of points $S$ so that $PG(2, q) - S \equiv M(K_4)$. Conclude that the graphic matroid $M(K_4)$ is a subset of every projective plane $PG(2, q)$.
   (c) Show that $M(K_5)$ is contained in no projective plane $PG(2, q)$. (Hint: rank!)

## Section 5.4 – $AG(n, q)$ and $PG(n, q)$

(18) Semimodularity in affine and projective spaces. You know the semimodular law $r(A \cup B) + r(A \cap B) \le r(A) + r(B)$ holds for flats $A$ and $B$ in $AG(n, q)$ and $PG(n, q)$ because they are matroids.
   (a) Show the inequality may be strict in affine space: find flats $A$ and $B$ in $AG(n, q)$ with $r(A \cup B) + r(A \cap B) < r(A) + r(B)$. How large can you make the difference $r(A) + r(B) - (r(A \cup B) + r(A \cap B))$?
   (b) Show that, for all flats $A$ and $B$ in $PG(n, q)$, we get equality: $r(A \cup B) + r(A \cap B) = r(A) + r(B)$.

(c) Show that the result in (b) is "tight" in the following sense. If $M$ is a matroid whose ground set is a proper subset of $PG(n, q)$, then there are two flats $A$ and $B$ such that $r(A \cup B) + r(A \cap B) < r(A) + r(B)$.

(19) Let $a \in AG(n, q)$ and $b \in PG(n, q)$ be points in affine and projective space, respectively.

    (a) Show the matroid contraction $AG(n, q)/a \cong M$, where $M$ is the matroid obtained by replacing each point of $PG(n-1, q)$ with a multiple point of size $q - 1$. Conclude that when $q = 2$, we get $AG(n, 2)/a \cong PG(n-1, 2)$.

    (b) Show the matroid contraction $PG(n, q)/b \cong M'$, where $M'$ is the matroid obtained by replacing each point of $PG(n-1, q)$ with a multiple point of size $q$.

    (c) Conclude from parts (a) and (b) that the simplification $si(AG(n, q)/a)$ and $si(PG(n, q)/b)$ are both $PG(n-1, q)$.

(20) [Proposition 5.25] Let $F$ be a flat in $PG(n, q)$ with $r(F) = k$. Show that there is an $(n - k) \times n$ matrix $A$ such that $F = \{v \mid Av = \bar{0}\}$.

## Section 5.5 – $q$-binomial coefficients

(21) (a) Show the $q$-binomial coefficients satisfy

$$\begin{bmatrix} n + 1 \\ k \end{bmatrix}_q = \begin{bmatrix} n \\ k - 1 \end{bmatrix}_q + q^k \begin{bmatrix} n \\ k \end{bmatrix}_q$$

for $0 \le k \le n$.

    (b) Show $(q^k - 1)\begin{bmatrix} n \\ k \end{bmatrix}_q = (q^n - 1)\begin{bmatrix} n-1 \\ k-1 \end{bmatrix}_q$.

    (c) Show $\begin{bmatrix} n \\ k \end{bmatrix}_q$ is a polynomial in $q$ for fixed $k$ and $n$. (Hint: use part (a) and induction.) [Note: there is an interesting combinatorial interpretation for the coefficients of $q^m$ in $\begin{bmatrix} n \\ k \end{bmatrix}_q$. See [36] for more on this.]

(22) Show $\lim_{q \to 1} \begin{bmatrix} n \\ k \end{bmatrix}_q = \binom{n}{k}$.

(23) Let's count the number of lines $\begin{bmatrix} n+1 \\ 2 \end{bmatrix}_q$ in $PG(n, q)$ "directly." First, show that there are $\binom{\begin{bmatrix} n+1 \\ 1 \end{bmatrix}_q}{2}$ ways to select two points in $PG(n, q)$. Then, show

$$\begin{bmatrix} n + 1 \\ 2 \end{bmatrix}_q = \frac{\binom{\begin{bmatrix} n+1 \\ 1 \end{bmatrix}_q}{2}}{\binom{q+1}{2}}.$$

(24) Counting the circuits of size 3 in $AG(n, q)$ and $PG(n, q)$: you can count the number of three-point circuits in either $AG(n, q)$ of $PG(n, q)$ by first counting the number of lines, then selecting three points from that line. Use this idea to verify the following counts.

(a) Show the number of three-point circuits contained in $AG(n, q)$ is

$$q^{n-1} \begin{bmatrix} n \\ 1 \end{bmatrix}_q \binom{q}{3} = \frac{q^n(q^n - 1)(q - 2)}{6}.$$

(b) Show the number of three-point circuits contained in $PG(n, q)$ is

$$\begin{bmatrix} n+1 \\ 2 \end{bmatrix}_q \binom{q+1}{3} = \frac{(q^{n+1} - 1)(q^{n+1} - q)}{6(q - 1)}.$$

(25) How many independent sets are there in $AG(n, q)$ and $PG(n, q)$? Here is another chance to do some counting.

(a) Let $a_{n,k,q}$ be the number of independent sets of size $k$ in $AG(n, q)$.

(i) Show $a_{n,0,q} = 1$ and $a_{n,1,q} = \binom{q^n}{2} = \frac{q^n(q^n - 1)}{2}$.

(ii) For $2 \leq k \leq n + 1$, show

$$a_{n,k,q} = \frac{q^n(q^n - 1)(q^n - q)(q^n - q^2)\cdots(q^n - q^{k-2})}{k!}.$$

[Hint: modify the argument used to derive the formula for $\begin{bmatrix} n \\ k \end{bmatrix}_q$ (Theorem 5.28) by choosing an *ordered* independent set of size $k$.]

(b) Let $b_{n,k,q}$ be the number of independent sets of size $k$ in $PG(n, q)$. Show that $b_{n,0,q} = 1$ and, for $1 \leq k \leq n + 1$,

$$
\begin{aligned}
& b_{n,k,q} \\
&= \frac{\begin{bmatrix} n+1 \\ 1 \end{bmatrix}_q \left( \begin{bmatrix} n+1 \\ 1 \end{bmatrix}_q - \begin{bmatrix} 1 \\ 1 \end{bmatrix}_q \right) \left( \begin{bmatrix} n+1 \\ 1 \end{bmatrix}_q - \begin{bmatrix} 2 \\ 1 \end{bmatrix}_q \right) \cdots \left( \begin{bmatrix} n+1 \\ 1 \end{bmatrix}_q - \begin{bmatrix} k-1 \\ 1 \end{bmatrix}_q \right)}{k!} \\
&= \frac{(q^{n+1} - 1)(q^{n+1} - q)(q^{n+1} - q^2)\cdots(q^{n+1} - q^{k-1})}{k!(q - 1)^k}.
\end{aligned}
$$

It's worth checking both formulas for some small cases. We leave it to you to decide what small cases to check.

(26) Let's count the number of circuits of size $k$ for $3 \leq k \leq n + 1$ in $AG(n, q)$. We'll use an incidence count involving independent sets of size $k - 1$ and circuits of size $k$.

(a) First, let $g_{n,k,q}$ denote the number of $k$-element circuits in $AG(n, q)$, and suppose each independent set of size $k - 1$ is in precisely $z$ circuits. Use an incidence count to show

$$
\begin{aligned}
g_{n,k,q} &= \frac{a_{n,k-1,q} \cdot z}{k} \\
&= \frac{zq^n(q^n - 1)(q^n - q)(q^n - q^2)\cdots(q^n - q^{k-3})}{k!}
\end{aligned}
$$

where $a_{n,m,q}$ is the number of independent sets of size $m$ in $AG(n, q)$ from Exercise 25(a).

(b) It remains to compute $z$, the number of circuits of size $k$ that contain a given independent set of size $k - 1$. We'll do this using an inclusion–exclusion argument.[29] Let $A$ be an independent set of size $k - 1$, and suppose $A \cup \{x\}$ is a circuit of size $k$.

   (i) Show $x \in \overline{A}$, and $|\overline{A}| = q^{k-2}$ (in fact, $\overline{A}$ is a subset of $AG(n, q)$ isomorphic to $AG(k - 2, q)$).

   (ii) Show $A \cup \{x\}$ is a circuit of size $k$ if and only if $x \notin \overline{S}$ for any $S \subseteq A$ with $|S| = k - 2$.

   (iii) Use inclusion–exclusion to show

$$z = q^{k-2} - \binom{k-1}{k-2}q^{k-3} + \binom{k-1}{k-3}q^{k-4} - \cdots$$
$$+ (-1)^k(k-1).$$

   (iv) Use the binomial theorem to rewrite $z = (q-1)^{k-1} + (-1)^k$.

(c) Use parts (a) and (b) to get a formula for $g_{n,k,q}$:

$$g_{n,k,q}$$
$$= \frac{q^{n-1}(q^n-1)(q^n-q)(q^n-q^2)\cdots(q^n-q^{k-3})((q-1)^{k-1}+(-1)^k)}{k!}.$$

Check that this answer agrees with the formula given in Exercise 24(a) when $k = 3$.

(27) This time, we'll count the number of circuits of size $k$ for $3 \leq k \leq n + 1$ in the projective geometry $PG(n, q)$. As in Exercise 26, we use an incidence count involving independent sets of size $k - 1$ and circuits of size $k$.

(a) Let $h_{n,k,q}$ denote the number of $k$-element circuits in $PG(n, q)$, and, as in Exercise 26, suppose each independent set of size $k - 1$ is in precisely $z$ circuits. Show

$$h_{n,k,q} = \frac{b_{n,k-1,q} \cdot z}{k}$$
$$= \frac{z(q^{n+1}-1)(q^{n+1}-q)(q^{n+1}-q^2)\cdots(q^{n+1}-q^{k-2})}{k!(q-1)^{k-1}}$$

where $b_{n,m,q}$ is the number of independent sets of size $m$ in $PG(n, q)$ from Exercise 25(b).

(b) As before, it remains to compute $z$, the number of circuits of size $k$ that contain a given independent set of size $k - 1$. We use a vector representation (instead of inclusion–exclusion) this time, however. We may assume our independent set $A$ is represented by the vectors $e_1, e_2, \ldots, e_{k-1}$, where the vector $e_i$ has a 1 in the $i$th coordinate and 0's elsewhere. Then show

---

[29] If you've never seen this, it might be a good idea to look up "inclusion–exclusion" in a combinatorics book, or elsewhere (= the Internet).

$A \cup \{x\}$ is a circuit if and only if each of the first $k - 1$ coordinates of $x$ are non-zero, and the remaining coordinates are zero. Thus, projectively, we may assume the first coordinate of $x$ equals 1. Conclude that $z = (q - 1)^{k-2}$.

(c) Use parts (a) and (b) to get a formula for $h_{n,k,q}$:

$$h_{n,k,q} = \frac{(q^{n+1} - 1)(q^{n+1} - q)(q^{n+1} - q^2) \cdots (q^{n+1} - q^{k-2})}{(q - 1)k!}.$$

Check that this answer agrees with the formula given in Exercise 24(b) when $k = 3$.

# 6

# Representable matroids

## 6.1 Matrices are matroids

In Whitney's foundational paper [42], matroids were introduced as an abstraction of linear dependence. Throughout the history of the subject, connections between matrices and matroids have motivated an enormous amount of research. The first three questions we consider are fundamental:

Q1. Does every subset of vectors give rise to a matroid, i.e., do the subsets of linearly independent column vectors of a matrix always satisfy the independent set axioms? (Answer: Yes – Theorem 6.1.)

Q2. When do two different matrices give the same matroid? (Answer: Sometimes – Section 6.2.)

Q3. Does every matroid arise from the linear dependences of some collection of vectors? (Answer: No – Example 6.20.)

Concerning Q1, we've seen plenty of examples of matroids that come from specific matrices so far. In Chapter 1, matrices were the first examples we considered, and, in Chapter 3, we interpreted the matroid operations of deletion, contraction and duality for matrices. But, if you've been paying close attention, we never *proved* the subsets of linearly independent vectors satisfy the axioms (I1), (I2) and (I3). We fix that now by giving a proof that linear independence satisfies the matroid independence axioms.

### 6.1.1 Column dependences of a matrix

**Theorem 6.1.** *Let $E$ be the columns of a matrix $A$ with entries in a field $\mathbb{F}$, and let $\mathcal{I}$ be those subsets of $E$ that are linearly independent. Then $\mathcal{I}$ is the family of independent sets of a matroid.*

As a reminder, recall a set of vectors $\{v_1, v_2, \ldots, v_k\}$ in some vector space is *linearly independent* if $\sum_{i=1}^{k} c_i v_i = 0$ implies $c_i = 0$ for all $i$,

221

i.e., no $v_i$ can be written as a linear combination of the other vectors. Our proof will depend on one important result about the reduced row echelon form of a matrix. This is a standard fact from linear algebra – see Exercise 14 for an outline of a proof. (A slightly different version of this proposition appears as part (1) of Proposition 6.9.)

**Proposition 6.2.** *Let $A$ be an $r \times n$ matrix, and let $A'$ be the reduced row echelon form of the matrix $A$. Then a subset of columns of $A$ is linearly independent if and only if the corresponding columns of $A'$ are.*

*Proof Theorem 6.1.* We need to show that $\mathcal{I}$ satisfies (I1), (I2) and (I3). (I1) is trivial – $\emptyset$ is always a linearly independent set. (I2) is just as easy, following immediately from the definition of linear independence.

It remains to check (I3), which involves some matrix manipulation. Let $|I| = s$ and $|J| = t$, where $I$ and $J$ are linearly independent subsets with $s < t$, and let $A$ be the matrix whose columns correspond to $I \cup J$. Order the columns of $A$ so that the vectors in $I$ appear first. Then the matrix rank of $A$ is at least $t$, since $J$ is a linearly independent set.

Now row reduce $A$. Since $I$ is an independent set, we can transform the first $s$ columns into an $s \times s$ identity matrix, with rows of zeros below:

$$A \longrightarrow \left( \begin{array}{c|c} I_{s \times s} & B_1 \\ \hline 0 & B_2 \end{array} \right).$$

Since the matrix rank of $A$ is at least $t$, some element of the submatrix $B_2$ is non-zero. Then it is clear we can add some column from $J - I$ to $I$ and preserve linear independence. (Note: Proposition 6.2 is used here to ensure the column dependences of $A$ and those of the row reduced matrix are the same.) □

A longer, more painful proof that (mostly) avoids using matrices is outlined in Exercise 4.

Note that the proof of Theorem 6.1 does not depend on the field we are working over. From the viewpoint of linear algebra, this should not be a surprise; Proposition 6.2 does not depend on the particular field $\mathbb{F}$. Thus, subsets of vectors in the vector space $\mathbb{F}^m$ give the ground set of a matroid for any field $\mathbb{F}$. This rather mild comment is central to questions of matroid representation, and finite fields will play an extremely important role in answering those questions.

Recall that a matroid $M$ is *representable* over the field $\mathbb{F}$ if there is a matrix $A$ with entries in $\mathbb{F}$ whose column dependences match precisely the dependences in the matroid. Thus, the columns of the matrix form the ground set of a representable matroid, and the linearly independent sets (over $\mathbb{F}$) of columns of $A$ are the independent sets $\mathcal{I}$ of $M$. (See Definition 1.4.)

As we did in earlier chapters, we denote the column dependence matroid defined on a matrix $A$ by $M_{\mathbb{F}}[A]$ or simply $M[A]$ if we do not want to emphasize the field. Now that we have shown $M = M[A]$ is

Representable matroid

Table 6.1. *Matroid terms interpreted for matrices.*

| Matroid term | Symbol | Matrix description |
|---|---|---|
| Independent sets | $\mathcal{I}$ | Linearly independent sets of columns. |
| Bases | $\mathcal{B}$ | Maximal independent sets. |
| Circuits | $\mathcal{C}$ | Minimal linearly dependent sets. |
| Rank | $r$ | Rank of the corresponding submatrix. |
| Flats | $\mathcal{F}$ | Sets equal to their linear span. |
| Hyperplanes | $\mathcal{H}$ | Corank 1 flats. |
| Closure | – | Linear span. |
| Cocircuits | $\mathcal{C}^*$ | Complements of hyperplanes. |
| Spanning sets | $\mathcal{S}$ | Subsets of columns whose linear span contains all the columns of the matrix. |

Table 6.2. *Matroid axioms and the corresponding theorems from linear algebra.*

| Matroid axiom | Matroid object | Linear algebra theorem |
|---|---|---|
| (B2′) | Bases | Every basis of a finite-dimensional vector space has the same size. |
| (I3) | Independent sets | Every linearly independent set can be extended to a basis. |
| (F2) | Flats | The intersection of subspaces is a subspace. |
| (F3) | Flats | The subspaces that cover a given subspace $W$ partition $V - W$. |
| (r3) | Rank function | If $U$ and $W$ are subspaces, then $\dim(U) + \dim(W) = \dim(U \cap W) + \dim(U + W)$. |

a matroid, we can interpret other matroid notions in this setting. We summarize this in Table 6.1.

As we did for graphs in Chapter 4, we can interpret matroid axioms as theorems in linear algebra. A few of the standard theorems of linear algebra have direct interpretations, as listed in Table 6.2.

We have a confession to make: our translation of (F3) is not really a "standard" theorem[1] from linear algebra. But it's true – if $W$ is a $k$-dimensional subspace of a vector space, then the $(k + 1)$-dimensional subspaces that contain $W$ give rise to a partition of the vectors in $V - W$. In fact, more is true: if $W$ is a *translation* of a $k$-dimensional subspace, then the same thing is true – these are the $k$-dimensional flats in the affine geometry $AG(n, q)$, and we used this fact in our proof that the flats of $AG(n, q)$ satisfy (F3) in Theorem 5.4 in Chapter 5.

---

[1] See if you can: find a linear algebra book that states this theorem for translations of subspaces. Then tell your friends.

Figure 6.1. The matroid $M[A]$
for the matrix $A$ from
Example 6.3.

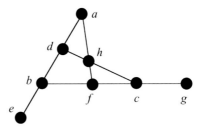

Let's test the connections between matroids and matrices in an example.

**Example 6.3.**  Let $A$ be the $3 \times 8$ matrix, and consider column dependences over the rational numbers.

$$A = \begin{array}{cccccccc} a & b & c & d & e & f & g & h \end{array}$$

$$A = \left[ \begin{array}{cccccccc} 1 & 0 & 0 & 1 & 1 & 0 & 0 & 1 \\ 0 & 1 & 0 & 1 & -1 & 1 & 1 & 1 \\ 0 & 0 & 1 & 0 & 0 & 1 & -1 & 1 \end{array} \right].$$

We then get a matroid $M = M[A]$ on ground set $E = \{a, b, c, d, e, f, g, h\}$. The matroid is shown in Figure 6.1. Let's look at the independent sets, bases, etc.

- Independent sets: as we know, these correspond to subsets of linearly independent columns of the matrix. All single columns are linearly independent since $\overline{0}$ does not appear as a column of $A$ (and so, $M[A]$ has no loops). We also see all pairs of columns are linearly independent; this is a simple matroid.

  Let's verify (I3): if $I = \{d, e\}$ and $J = \{a, b, c\}$, then we need to find an element of $J$ to add to $I$, preserving independence. In this case, we choose $c$, and the set $I \cup c = \{c, d, e\}$ is independent.
- Bases: since $r(M[A]) = 3$, a basis will have three elements. You can check every subset of size 3, for instance, by taking the determinant of a $3 \times 3$ submatrix. For example, the set $\{d, e, g\}$ has determinant

$$\begin{vmatrix} 1 & 1 & 0 \\ 1 & -1 & 1 \\ 0 & 0 & -1 \end{vmatrix} = 2,$$

  so $\{d, e, f\}$ is a basis.
- Rank: the rank of a subset of columns is just the matrix rank. For instance, the rank of any pair of columns is 2. Note $r(\{a, b, d, e\}) = 2$ since these four columns all have last coordinate 0. These four points are collinear in Figure 6.1.
- Closure: this is given by linear span. For instance, if $S = \{c, f\}$, then $\overline{S} = \{b, c, f, g\}$, since $b$ is a linear combination of columns $c$ and $f$, and so is $g$.

- Flats: flats are closed under the operation of linear span. We also verify the semimodular law (r3) for the flats $F_1 = \{c, d, h\}$ and $F_2 = \{a, g\}$. Then $r(F_1) = 2, r(F_2) = 2, r(F_1 \cup F_2) = 3$ and $r(F_1 \cap F_2) = 0$. In this example, we have $2 + 2 \geq 3 + 0$.
- Circuits: these are harder to recognize in the matrix – we know they correspond to subsets that are linearly dependent, but minimally so. For instance, the collection of vectors $\{a, d, f, g\}$ is a circuit because it's dependent with rank 3, and every proper subset of these column vectors is linearly independent. We remark that circuits never show up in standard linear algebra books, but there are linear algebra theorems involving them: it's not hard to verify the minimal linearly dependent sets satisfy the matroid circuit axioms – see Exercise 5.[2]

Just for fun, let's count the number of bases, independent sets and flats for $M[A]$. Bases have size 3, and there are $\binom{8}{3} = 56$ subsets of size 3 to check. But it's clear from Figure 6.1 there are 10 subsets of size 3 that are *not* bases. This gives us 46 bases. You can then add these 46 sets to the $\binom{8}{2} + \binom{8}{1} + \binom{8}{0} = 37$ independent sets of size less than 3. This gives us a total of 83 independent sets.

For the flats, note that each point is a flat, there are nine two-point lines, two three-point lines and two four-point lines. These 13 lines are the hyperplanes of $M[A]$. Adding $\emptyset$ and $E$ gives us all the flats; there are 23 flats in all.

Finally, we were working over the rational numbers in this example. Does this matter? We can determine the bases of $M[A]$ by computing 56 submatrix determinants. These determinants are all $0, \pm 1, \pm 2$ and 3. For instance, the determinant of the $3 \times 3$ submatrix formed by columns $e, g$ and $h$ equals 3. If we happen to be working over a field of characteristic 3, then these three columns will be dependent.

Similar problems occur if we are working over a field of characteristic 2, so the matroid depends on the specific characteristic. But if the field has characteristic $> 3$, then we always get the matroid in Figure 6.1. The issue of field characteristic is the focus of Section 6.3.

The matroid–linear algebra connections can run in both directions. For instance, theorems from linear algebra can also have matroid interpretations. The next result is another routine result from linear algebra.

**Theorem 6.4.** *[Linear Algebra Theorem] Let $I$ be a linearly independent set and $S$ a spanning set for a vector space $V$, with $I \subseteq S$. Then there is a basis $B$ with $I \subseteq B \subseteq S$.*

---

[2] Why are circuits missing from standard treatments of linear algebra? Actually, they do appear, but only in certain kinds of proofs involving minimal dependences. And don't look for the word "circuit" in your linear algebra text – you won't find it.

Is this true for all matroids? We can think of two ways to try to figure out if this is always true:

- Look in a linear algebra book and find this theorem. See if the proof only uses general properties of independence and spanning, properties that hold for all matroids.
- Don't look in a linear algebra book. Try to figure out a proof, or, failing that, find a counterexample.

It turns out Theorem 6.4 is true for all matroids;[3] see Exercise 6.

### 6.1.2 Isthmuses, loops, deletion and contraction

In Section 3.2.1 of Chapter 3, we described a procedure for deleting and contracting points in representable matroids. Let $M[A]$ be the matroid defined on the matrix $A$, and let $e$ be a point in $M[A]$. Then if $e$ is not an isthmus, we can represent the matroid $M[A] - e$ by simply removing the column corresponding to $e$. So, defining $A - e$ to be the matrix obtained from $A$ by removing $e$, we have $M[A] - e = M[A - e]$.

For contraction, we needed a more complicated procedure to construct a matrix that represents $M[A]/e$. Suppose $e$ is not a loop. Then the procedure we gave in Chapter 3 goes like this:

- First, row reduce the matrix $A$ so that there is exactly one non-zero entry in the column corresponding to $e$, and this non-zero entry appears in the first row. Further, assume this non-zero entry equals 1. Call this row reduced matrix $B$.
- Then remove *both* column $c$ and row $r$ (from the row reduced matrix). Call this derived matrix $A/e$.

We gave an example of how to use this procedure in Example 3.11. The (extra) restrictions in the first part of this procedure (making sure the first entry in the column $e$ is non-zero, and making sure this non-zero entry equals 1) are easily managed by elementary row operations. Let's justify this procedure now.

**Proposition 6.5.** *Let $A$ be a matrix and let $e$ be a column vector of $A$ that is neither a loop nor an isthmus in the matroid $M[A]$. Defining the matrices $A - e$ and $A/e$ as above, we have*

$$M[A] - e = M[A - e] \qquad M[A]/e = M[A/e].$$

*Proof* Deletion is straightforward, so we leave the proof as an exercise (Exercise 8). For contraction, we give a proof that you might see in a book.[4] We need to show that linearly independent subsets of columns

---

[3] So don't look too hard for counterexamples.
[4] Hey – this is a book!

of the matrix $A/e$ correspond to independent subsets of the matroid $M[A]/e$, and also that linearly dependent subsets of columns of $A/e$ correspond to dependent subsets of $M[A]/e$.

Now let $B$ be the matrix obtained by row reducing $A$ as in the procedure. We introduce some notation to help us with the proof. If $c$ is a column of the matrix $A$, we write $c'$ when we consider the same column in the matrix $B$ and we write $c''$ for that column in $A/e$. Thus $e' = (1, 0, 0, \ldots, 0)$ appears as a column in the matrix $B$.

Recall that $I$ is independent in $M/e$ if and only if $I \cup \{e\}$ is independent in $M$. So we let $I = \{c_1'', c_2'', \ldots, c_k''\}$ be a linearly independent subset of columns of $A/e$. We need to show $I \cup \{e\}$ is an independent subset of $M[A]$. So suppose

$$\sum_{i=1}^{k} a_i c_i + a_e e = \bar{0}.$$

We need to show $a_1 = \cdots = a_k = a_e = 0$.

First, by Proposition 6.2, $A$ and $B$ have the same column dependences, and we also have $\sum_{i=1}^{k} a_i c_i' + a_e e' = \bar{0}$. Now, since $e' = (1, 0, 0, \ldots, 0)$ in the matrix $B$, we get $\sum_{i=1}^{k} a_i c_i'' = \bar{0}$ in $A/e$ by the way we constructed the matrix $A/e$. But we assumed $\{c_1'', \ldots, c_k''\}$ is linearly independent in $A/e$, so $a_i = 0$ for all $i$. Then $\sum_{i=1}^{k} a_i c_i' + a_e e' = \bar{0}$ forces $a_e = 0$, too, so $I \cup \{e\}$ is independent in $M[A]$.

We also need to show that if $D$ is a linearly dependent set of columns in $A/e$, then $D \cup \{e\}$ is dependent in $M[A]$. Let $D = \{c_1'', c_2'', \ldots, c_k''\}$ be a linearly dependent subset of columns of $A/e$. Then $\sum_{i=1}^{k} a_i c_i'' = \bar{0}$ has a non-trivial solution in the $a_i$.

Now we focus on the first row of $B$, and let $r_i$ be the entry in row 1 and column $c_i'$ of $B$. Then we can solve the scalar equation $a_e + \sum_{i=1}^{k} a_i r_i = 0$ for the "unknown" coefficient $a_e$. Now setting $a_e = -\sum_{i=1}^{k} a_i r_i$ gives a non-trivial solution to $\sum_{i=1}^{k} a_i c_i' + a_e e' = 0$ in $B$. Thus, $D \cup \{e\}$ is a dependent set in $M[B]$, hence in $M[A]$. □

Let's revisit Example 3.11 quickly. We have

$$A = \begin{array}{c} \phantom{A =} \begin{matrix} a & b & c & d & e \end{matrix} \\ \begin{bmatrix} 1 & 0 & 0 & 1 & 1 \\ 0 & 1 & 0 & 2 & -1 \\ 0 & 0 & 1 & 0 & 1 \end{bmatrix} \end{array} \quad \text{and} \quad A/e = \begin{array}{c} \begin{matrix} a & b & c & d \end{matrix} \\ \begin{bmatrix} 1 & 1 & 0 & 3 \\ -1 & 0 & 1 & -1 \end{bmatrix} \end{array}.$$

Then, for instance, the linearly independent set $I = \{a, d\}$ in $A/e$ consisting of column vectors $(1, -1)$ and $(3, -1)$ in $\mathbb{R}^2$ lifts to the linearly independent set $\{a, d, e\}$ in $A$ since the corresponding vectors $(1, 0, 0), (1, 2, 0), (1, -1, 1)$ are linearly independent in $\mathbb{R}^3$. For the linearly dependent set $D = \{b, c, d\}$ in $A/e$, we note $D \cup e$ remains dependent in $A$. In fact, note the dependence

$$-3 \cdot (1, 0) + 1 \cdot (0, 1) + 1 \cdot (3, -1) = (0, 0)$$

lifts to the dependence

$$-3 \cdot (0, 1, 0) + 1 \cdot (0, 0, 1) + 1 \cdot (1, 2, 0) - 1 \cdot (1, -1, 1) = (0, 0, 0).$$

Loops and isthmuses for matrices were also described in Section 3.2.1 of Chapter 3. When $e$ is a loop, $e$ must correspond to a column of 0's. The column $e$ is an isthmus in $M[A]$ if, when row reducing $A$ so that $e$ has exactly one non-zero entry, that entry is also the only non-zero entry in its *row*. See Exercise 9.

### 6.1.3 Duality and the Rank–Nullity Theorem

Duals of representable matroids are also representable. We gave a preview of this fact in Section 3.4.2 of Chapter 3. In Example 3.30 of Chapter 3, we gave a procedure for finding a matrix $A^*$ representing the dual matroid $M[A]^*$, i.e., $M[A]^* = M[A^*]$.

We conclude this section by proving that procedure always works. Our proof uses the Rank-Nullity Theorem. See Section 5.1.2 from Chapter 5 for a review of this theorem.

**Theorem 6.6.** *The matrix* $A = [I_{r \times r} \mid D]$ *represents the matroid $M$ precisely when the matrix* $A^* = [-D^T \mid I_{(n-r) \times (n-r)}]$ *represents the dual matroid $M^*$.*

The entries of the matrices live in a field $\mathbb{F}$, and that field is the same for both matrices; if $M$ is representable over a given field, then so is $M^*$. The null space $N(A)$ of the matrix $A$ plays an important role here. Here is a quick summary of the linear algebra facts we will need.

- Suppose $A$ is an $r \times n$ matrix. Recall that if $x \in N(A)$, where $x = (x_1, x_2, \ldots, x_n)$ is an $n \times 1$ column vector, then $Ax = \overline{0}$. Then we can think of the matrix product $Ax$ as giving us a linear combination of the columns of $A$: $Ax = \sum_{i=1}^{n} x_i \overline{c_i}$, where the $\overline{c_i}$ are the $n$ columns of $A$.
- If $A$ is an $r \times n$ matrix with rank $r$, then the Rank–Nullity Theorem tells us the null space $N(A)$ is an $(n - r)$-dimensional subspace of $\mathbb{F}^n$.
- Suppose $B_u = \{u_1, u_2, \ldots, u_{n-r}\}$ and $B_v = \{v_1, v_2, \ldots, v_{n-r}\}$ are both bases for the null space $N(A)$, and form $(n - r) \times n$ matrices $D_u$ and $D_v$ whose rows are the $u_i$'s and $v_i$'s, respectively:

$$D_u = \begin{pmatrix} \text{---} & u_1 & \text{---} \\ \text{---} & u_2 & \text{---} \\ & \vdots & \\ \text{---} & u_{n-r} & \text{---} \end{pmatrix} \qquad D_v = \begin{pmatrix} \text{---} & v_1 & \text{---} \\ \text{---} & v_2 & \text{---} \\ & \vdots & \\ \text{---} & v_{n-r} & \text{---} \end{pmatrix}.$$

Then $D_u$ and $D_v$ are row equivalent, i.e., we can row reduce $D_u$ to get $D_v$. This follows from the fact that the row spaces of these two matrices are identical.

We will prove a slightly more general statement, and ask you to finish the proof of Theorem 6.6 in Exercise 10. Here's an example to illustrate the main ideas.

**Example 6.7.** Let

$$A = \begin{matrix} & a & b & c & d & e & f & g \\ & \begin{bmatrix} 1 & -1 & 1 & 0 & 2 & -3 & 0 \\ 1 & -1 & 0 & 1 & 1 & 1 & -1 \\ 2 & -2 & 2 & 1 & 1 & -1 & 0 \end{bmatrix} \end{matrix}.$$

From the viewpoint of linear algebra, $A$ represents a linear transformation $A : \mathbb{F}^7 \to \mathbb{F}^3$. Then the null space $N(A)$ is simply the solution space of the matrix equation $Ax = \bar{0}$, where $x \in \mathbb{F}^7$ is a 7-tuple. Solving $Ax = \bar{0}$ involves row reducing $A$:

$$A' = \begin{matrix} & a & b & c & d & e & f & g \\ & \begin{bmatrix} 1 & -1 & 0 & 0 & 4 & -4 & -1 \\ 0 & 0 & 1 & 0 & -2 & 1 & 1 \\ 0 & 0 & 0 & 1 & -3 & 5 & 0 \end{bmatrix} \end{matrix}.$$

Then the columns corresponding to $b, e, f$ and $g$ are *free* variables in the solution space for $Ax = \bar{0}$. A basis for $N(A)$ is then given by the four vectors

$$\{(1, 1, 0, 0, 0, 0, 0), (-4, 0, 2, 3, 1, 0, 0),$$
$$(4, 0, -1, -5, 0, 1, 0), (1, 0, -1, 0, 0, 0, 1)\}.$$

Let's assemble these four vectors in a $4 \times 7$ matrix (suggestively) called $(A')^*$:

$$(A')^* = \begin{matrix} & a & b & c & d & e & f & g \\ & \begin{bmatrix} 1 & 1 & 0 & 0 & 0 & 0 & 0 \\ -4 & 0 & 2 & 3 & 1 & 0 & 0 \\ 4 & 0 & -1 & -5 & 0 & 1 & 0 \\ 1 & 0 & -1 & 0 & 0 & 0 & 1 \end{bmatrix} \end{matrix}.$$

Now the connection with matroid duality should be clear: for the basis $\{a, c, d\}$ of $M[A]$, the complementary basis for $M[A]^*$ is $\{b, e, f, g\}$, which is obviously a basis in the matroid $M[(A')^*]$. The fact that the columns corresponding to $b, e, f$ and $g$ are free variables in the solution space $Ax = \bar{0}$ is obvious in the matrix $(A')^*$ – these four columns form a $4 \times 4$ identity matrix.

*Claim.* The matrix $(A')^*$ represents the dual matroid $M[A]^*$, i.e., $M[(A')^*] = M[A]^*$.

To verify this claim, we need to show this works for any pair of complementary subsets where $B$ is a basis for $M$ and $B^c$ is a basis for $M^*$. For example, if we choose the basis $B = \{b, d, g\}$ for $M$, then we can repeat the procedure we just used, first row reducing $A$ to transform columns $b$, $d$ and $g$ into a $3 \times 3$ identity matrix:

$$
A'' = \begin{array}{c} \begin{array}{ccccccc} a & \ b & \ c & \ d & \ e & \ f & \ g \end{array} \\ \left[ \begin{array}{ccccccc} -1 & 1 & -1 & 0 & -2 & 3 & 0 \\ 0 & 0 & 0 & 1 & -3 & 5 & 0 \\ 0 & 0 & 1 & 0 & -2 & 1 & 1 \end{array} \right] \end{array}.
$$

This time, we get a different basis for the solution space of $Ax = \bar{0}$. Proceeding as above, we create the matrix $(A'')^*$ whose rows form this new basis for the subspace $N[A]$:

$$
(A'')^* = \begin{array}{c} \begin{array}{ccccccc} a & \ b & \ c & \ d & \ e & \ f & \ g \end{array} \\ \left[ \begin{array}{ccccccc} 1 & 1 & 0 & 0 & 0 & 0 & 0 \\ 0 & 1 & 1 & 0 & 0 & 0 & -1 \\ 0 & 2 & 0 & 3 & 1 & 0 & 2 \\ 0 & -3 & 0 & -5 & 0 & 1 & -1 \end{array} \right] \end{array}.
$$

Then, as above, the matroid basis $\{b, d, g\}$ is clearly visible in the matrix $A''$, and the complementary basis $\{a, c, e, f\}$ is easy to see in $(A'')^*$. The fact that we used $A'$ and $A''$ interchangeably in describing the matroid $M[A]$ should not bother you; they are both row-equivalent to $A$, so Proposition 6.2 allows us to use either one as we see fit.

To complete this argument, we need to check the relationship between the two matrices $(A')^*$ and $(A'')^*$. The row vectors of both these matrices are bases for the null space $N[A]$. Since the row spaces are identical, the matrices are row-equivalent. That means (by Proposition 6.2) they represent the same matroid.

Finally, we are ready for our main lemma on representing duals.

**Lemma 6.8.**  *Let $A$ be an $r \times n$ matrix of rank $r$, and let $M = M[A]$. Let $\{v_1, v_2, \ldots, v_k\}$ be a basis for the null space $N(A)$, and let $A^*$ be the matrix with rows $v_1, v_2, \ldots, v_k$. Then $M[A^*] = M[A]^*$, i.e., a subset $S$ of columns in $A^*$ is a basis for $M[A]^*$ if and only if the complementary columns $E - S$ of $A$ are a basis for $M[A]$.*

*Proof Lemma 6.8.* Let $c_i$ be the $i$th column of $A$, and let $B$ be a basis of $M[A]$, i.e., a collection of $r$ linearly independent columns. For convenience, order the columns of $A$ so that the first $r$ correspond to the basis $B$, so $B = \{c_1, c_2, \ldots, c_r\}$. For each of the remaining columns $c_i$ for $r < i \leq n$, write $c_i$ as a linear combination of the vectors in $B$:

$$
c_{r+1} = a_{1,1}c_1 + a_{1,2}c_2 + \cdots + a_{1,r}c_r
$$
$$
c_{r+2} = a_{2,1}c_1 + a_{2,2}c_2 + \cdots + a_{2,r}c_r
$$
$$
\vdots
$$
$$
c_n \quad = a_{n-r,1}c_1 + a_{n-r,2}c_2 + \cdots + a_{n-r,r}c_r.
$$

These $n - r$ vector equations give us $n - r$ vectors:

$$v_1 = (a_{1,1}, a_{1,2}, \ldots, a_{1,r}, -1, 0, 0, \ldots, 0, 0)$$
$$v_2 = (a_{2,1}, a_{2,2}, \ldots, a_{1,r}, 0, -1, 0, \ldots, 0, 0)$$
$$\vdots$$
$$v_{n-r} = (a_{n-r,1}, a_{n-r,2}, \ldots, a_{n-r,r}, 0, 0, 0, \ldots, 0, -1).$$

*Claim.* $\{v_1, v_2, \ldots, v_{n-r}\}$ is a basis for the null space $N(A)$. To prove the claim, we first show $v_i \in N(A)$ for all $1 \leq i \leq n - r$. But $Av_i = \bar{0}$ then corresponds precisely to the linear combination $c_{r+i} = \sum_{j=1}^{r} a_{i,j} c_j$.

Thus, each $v_i \in N(A)$, and it's also clear that $\{v_1, v_2, \ldots, v_{n-r}\}$ is linearly independent (look at the last $n - r$ coordinates). But $N(A)$ is an $(n - r)$-dimensional subspace of $\mathbb{F}^n$ by the Rank–Nullity Theorem, so $\{v_1, v_2, \ldots, v_{n-r}\}$ is a basis for the null space $N(A)$ and the claim is established.

Now let $A^*$ be the matrix whose rows are the vectors $\{v_1, v_2, \ldots, v_{n-r}\}$. Then the $n - r$ columns $c_{r+1}, c_{r+2}, \ldots, c_n$ are linearly independent, so form a basis in the matroid $M[A^*]$. But this argument is independent of the basis $B$ chosen because choosing different bases produces row equivalent matrices. $\square$

If you have the time and the inclination, it is worthwhile to reread Example 3.30 from Chapter 3. In that example, the rows of the matrix $A^*$ we invented to represent the dual matroid $M[A]^*$ are also a basis for the null space $N[A]$. See Exercise 10.

## 6.2 Representing representable matroids

### 6.2.1 Finding a nice matrix

Suppose we know $M$ is a matroid representable over some field $\mathbb{F}$. It's not difficult to find different matrices $A$ and $B$ that represent the same matroid: $M[A] = M[B]$. This is the issue raised by Question Q2 in Section 6.1. This is closely related to the question we considered in Section 4.4 of Chapter 4, where Whitney's 2-isomorphism theorem (Theorem 4.21) tells you when two graphs give the same matroid.

The answer is not as nice this time, but we can still make some simplifying assumptions about our representing matrix. The key idea was given in Proposition 6.2: row reducing a matrix does not change the column dependences. But we can do a little more.

**Proposition 6.9.** *Let $A$ be an $r \times n$ matrix over a field $\mathbb{F}$, let $B$ be an $r \times r$ invertible matrix and let $D$ be an $n \times n$ invertible diagonal matrix.*

(1)  *The column dependences of A are the same as those of BA.*
(2)  *The column dependences of A are the same as those of AD.*

The proof is left for you: part (1) is Exercise 14 and part (2) is Exercise 15. (Note that part (1) is equivalent to Proposition 6.2.)

What does the proposition buy us? Lots. First, suppose we are given a rank $r$ matroid $M$ on $n$ points and we are asked to find a matrix whose column dependences match the matroid dependences. If an $m \times n$ matrix $A$ can be found with $M_A = M$, then we can use part (1) of Proposition 6.9 to row reduce $A$. Since the rank of the matroid (and the matrix $A$) is $r$, row reducing $A$ will produce a matrix $A'$ whose last $m - r$ rows are simply rows of zeros. Removing these rows from $A'$ clearly does not change column dependences, so we can assume $A'$ is an $r \times n$ matrix.

What else do we get? Let's assume the first $r$ columns of the (now) $r \times n$ matrix $A$ will correspond to a basis of $M$. (We can always reorder the columns to achieve this, if necessary.) Then part (1) of Proposition 6.9 says we may assume the first $r$ columns of $A$ form an identity matrix.

We can then use part (2) of Proposition 6.9 to make the first non-zero entry in each column of $A$ equal to 1. How do we accomplish this? The diagonal matrix $D$ can be chosen so that we multiply column $c_i$ by $a_i^{-1}$, where $a_i$ is the first non-zero term in column $c_i$ of $A$.

Let's see the proposition in action.

**Example 6.10.**  Let $M$ be the matroid of Figure 6.2. Then you can check[5] the matrix $A_1$ represents $M$, i.e., $M[A_1] = M$, where we are working over the rationals.

$$A_1 = \begin{bmatrix} \overset{a}{-1} & \overset{b}{1} & \overset{c}{1} & \overset{d}{8} & \overset{e}{2} & \overset{f}{6} & \overset{g}{4} \\ -1 & 3 & 2 & 22 & 10 & 15 & 14 \\ 0 & 1 & 1 & 10 & 4 & 9 & 5 \\ -1 & 1 & 2 & 14 & 2 & 15 & 4 \\ 0 & 1 & -1 & -2 & 4 & -9 & 5 \end{bmatrix}.$$

Figure 6.2. Matroid for
Example 6.10.

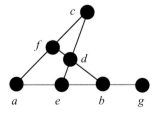

---

[5]  But we don't recommend this.

Let's apply Proposition 6.9 to simplify the matrix $A_1$. First, row reduce $A_1$ to get the matrix $A_2$:

$$A_2 = \begin{bmatrix} 1 & 0 & 0 & 2 & 2 & 3 & 1 \\ 0 & 1 & 0 & 4 & 4 & 0 & 5 \\ 0 & 0 & 1 & 6 & 0 & 9 & 0 \\ 0 & 0 & 0 & 0 & 0 & 0 & 0 \\ 0 & 0 & 0 & 0 & 0 & 0 & 0 \end{bmatrix}.$$

Next, drop the last two rows – our matroid has rank 3, so we should represent it by a $3 \times 8$ matrix. So we now have

$$A_3 = \begin{array}{ccccccc} a & b & c & d & e & f & g \\ \begin{bmatrix} 1 & 0 & 0 & 2 & 2 & 3 & 1 \\ 0 & 1 & 0 & 4 & 4 & 0 & 5 \\ 0 & 0 & 1 & 6 & 0 & 9 & 0 \end{bmatrix} \end{array}.$$

Finally, we multiply each column by the fraction that makes the first non-zero entry equal to 1. This gives us

$$A_4 = \begin{array}{ccccccc} a & b & c & d & e & f & g \\ \begin{bmatrix} 1 & 0 & 0 & 1 & 1 & 1 & 1 \\ 0 & 1 & 0 & 2 & 2 & 0 & 5 \\ 0 & 0 & 1 & 3 & 0 & 3 & 0 \end{bmatrix} \end{array}.$$

If you like, you can compute the invertible matrix $B$ (as a product of elementary matrices) and the invertible diagonal matrix $D$ that multiplies the columns to get the leading 1's. In this example, we find

$$B = \begin{bmatrix} -1 & 0 & 1 & 0 & 0 \\ -1 & 1 & -1 & 0 & 0 \\ 1 & -1 & 2 & 0 & 0 \\ -2 & 1 & -2 & 1 & 0 \\ 2 & -2 & 3 & 0 & 1 \end{bmatrix} \quad \text{and} \quad D = \begin{bmatrix} 1 & 0 & 0 & 0 & 0 & 0 & 0 \\ 0 & 1 & 0 & 0 & 0 & 0 & 0 \\ 0 & 0 & 1 & 0 & 0 & 0 & 0 \\ 0 & 0 & 0 & \frac{1}{2} & 0 & 0 & 0 \\ 0 & 0 & 0 & 0 & \frac{1}{2} & 0 & 0 \\ 0 & 0 & 0 & 0 & 0 & \frac{1}{3} & 0 \\ 0 & 0 & 0 & 0 & 0 & 0 & 1 \end{bmatrix}.$$

Although $A_4$ is much simpler than $A_1$, Proposition 6.9 gives us even more.[6] Consider the boxed 2 in the matrix $A_4$:

$$A_4 = \begin{array}{ccccccc} a & b & c & d & e & f & g \\ \begin{bmatrix} 1 & 0 & 0 & 1 & 1 & 1 & 1 \\ 0 & 1 & 0 & \boxed{2} & 2 & 0 & 5 \\ 0 & 0 & 1 & 3 & 0 & 3 & 0 \end{bmatrix} \end{array}.$$

---

[6] One might be tempted to say "It's the proposition that keeps on giving," but we resist that temptation.

We can multiply row 2 of this matrix by $1/2$ by choosing an appropriate matrix $B'$. Then

$$
\begin{bmatrix} 1 & 0 & 0 \\ 0 & 1/2 & 0 \\ 0 & 0 & 1 \end{bmatrix}
\begin{bmatrix} 1 & 0 & 0 & 1 & 1 & 1 & 1 \\ 0 & 1 & 0 & \boxed{2} & 2 & 0 & 5 \\ 0 & 0 & 1 & 3 & 0 & 3 & 0 \end{bmatrix}
$$

$$
= \begin{bmatrix} 1 & 0 & 0 & 1 & 1 & 1 & 1 \\ 0 & 1/2 & 0 & \boxed{1} & 1 & 0 & 5/2 \\ 0 & 0 & 1 & 3 & 0 & 3 & 0 \end{bmatrix}.
$$

This messes up the second column of our identity matrix, but we can fix this by multiplying column $b$ by 2 (using an appropriate diagonal matrix), so we get

$$
\begin{array}{ccccccc} a & b & c & d & e & f & g \end{array}
$$
$$
\begin{bmatrix} 1 & 0 & 0 & 1 & 1 & 1 & 1 \\ 0 & 1 & 0 & 1 & 1 & 0 & 5/2 \\ 0 & 0 & 1 & 3 & 0 & 3 & 0 \end{bmatrix}.
$$

Now repeat this process for the boxed 3:

$$
\begin{array}{ccccccc} a & b & c & d & e & f & g \end{array}
$$
$$
\begin{bmatrix} 1 & 0 & 0 & 1 & 1 & 1 & 1 \\ 0 & 1 & 0 & 1 & 1 & 0 & 5/2 \\ 0 & 0 & 1 & \boxed{3} & 0 & 3 & 0 \end{bmatrix}.
$$

Again, we can multiply row 3 of our matrix by $1/3$, then multiply the column $c$ by 3. The new matrix will be:

$$
\begin{array}{ccccccc} a & b & c & d & e & f & g \end{array}
$$
$$
A_5 = \begin{bmatrix} 1 & 0 & 0 & 1 & 1 & 1 & 1 \\ 0 & 1 & 0 & 1 & 1 & 0 & 5/2 \\ 0 & 0 & 1 & 1 & 0 & 1 & 0 \end{bmatrix}.
$$

Can we do any more? For instance, could we somehow replace the $5/2$ with another 1 by some clever choice of $B$ and $D$? The answer is no; doing so would now mess up something else that we could *not* fix. For instance, multiplying row 2 of the matrix by $2/5$ makes column $g$ look nice, but it messes up the columns $d$ and $e$ (it also messes up $b$, but that can be fixed).

By the way, the $5/2$ entry in column $g$ is a bit of a red herring. In fact, we could replace $5/2$ with any $x \neq 0, 1$, and our matrix (considered over $\mathbb{Q}$) would still represent the matroid $M$ from Figure 6.2.

But you can see this in Figure 6.2! The point $g$ in the matroid is not on two lines – it's "free" in that it can move freely along the line $ab$. This "degree of freedom" is reflected algebraically, since the value of $x$ is not uniquely determined. But this is a matroid observation – not a matrix one. We could *not* use Proposition 6.9 to change the $5/2$ entry

Geometric degree of freedom

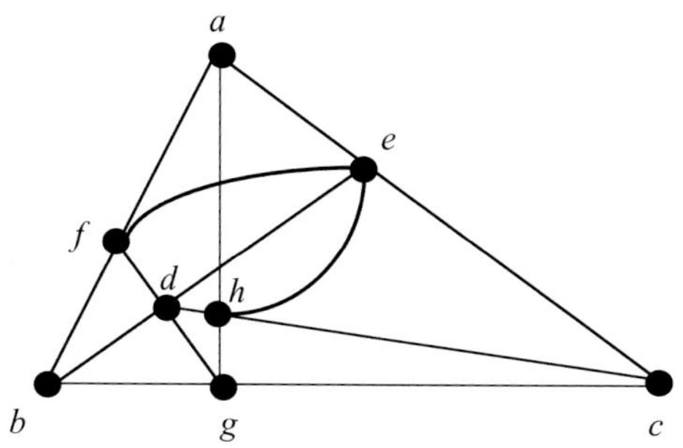

Figure 6.3. The Matroid $M_{\sqrt{-3}}$.

of $g$ to $x \neq 0, 1$ without changing other entries as well. The variable $x$ records fundamentally inequivalent ways to represent the matroid $M$. Thus, $A_5$ is as simple as this matrix will get.

The matrix operations that brought us from $A_4$ to $A_5$ depend on finding a *coordinatizing path*. We systematize this procedure in Section 6.2.3.

## 6.2.2 From a matroid to a matrix

If you have a matroid at hand, Example 6.10 shows how to find a nicer matrix to represent your matroid, assuming you were given a representing matrix to begin with. What if you aren't given a matrix? We can still use the ideas in Proposition 6.9 to attempt to *create* a matrix that represents our matroid. To see how this works, let's do another example.

**Example 6.11.** Let $M = M_{\sqrt{-3}}$ be the matroid in Figure 6.3.

Our goal is to create a matrix that represents $M$. We now know we may assume the following about our matrix $A$.

- The first non-zero entry in every column of $A$ can be taken to be 1.
- Since $r(M) = 3$, we may take $A$ to be a $3 \times 8$ matrix.
- Since $abc$ is a basis, we may take the first three columns of $A$ to be an identity matrix.
- Since $d$ is not on any of the lines determined by $a, b$ and $c$, we can also take $d$ to be the (column) vector $(1, 1, 1)$.

Here's what we have so far:

$$
A = \begin{array}{c c c c c} & a & b & c & d & \dots \\ & \begin{bmatrix} 1 & 0 & 0 & 1 \\ 0 & 1 & 0 & 1 \\ 0 & 0 & 1 & 1 \end{bmatrix} & & & & \dots \end{array}.
$$

We need to determine the coordinates for the remaining points in $M$. This will be easy and fun.[7]

- Since $e$ is on the line determined by $a$ and $c$, we know $e$ is a linear combination of $(1, 0, 0)$ and $(0, 0, 1)$, so the second coordinate is 0.
- Since $e$ is also on the line through $b$ and $d$, the first and last coordinates of $e$ must be equal.

Putting these together gives $e = (1, 0, 1)$. Continuing in this way, we determine the coordinates for the remaining points. At each stage, we use the coordinates of the points already assigned to determine restrictions on the coordinates of the point in question.

- $f$ is on the line through $a$ and $b$, so $f = (1, x, 0)$, where $x$ is temporarily undetermined.
- $g$ is on lines $bc$ and $df$. The first of these lines forces $g = (0, 1, y)$, where $y$ is undetermined. The line $dfg$ forces the determinant

$$\begin{vmatrix} 1 & 1 & 0 \\ 1 & x & 1 \\ 1 & 0 & y \end{vmatrix} = 0,$$

so $xy + 1 - y = 0$. This forces $y = \frac{1}{1-x}$.
- $h$ is the last point, and it's on three lines we can use to determine its coordinates: $agh$, $cdh$ and $efh$. The line $cdh$ forces the first two coordinates of $h$ to be equal, so we can assume $h = (1, 1, z)$:

$$
\begin{array}{cccccccc}
a & b & c & d & e & f & g & h
\end{array}
$$
$$
A = \begin{bmatrix} 1 & 0 & 0 & 1 & 1 & 1 & 0 & 1 \\ 0 & 1 & 0 & 1 & 0 & x & 1 & 1 \\ 0 & 0 & 1 & 1 & 1 & 0 & \frac{1}{1-x} & z \end{bmatrix}.
$$

We have two more dependences to consider: $agh$ and $efh$. One of these will determine the value of $z$, and the other will force $x$ to satisfy some relation. First, for $agh$, we get the determinant

$$\begin{vmatrix} 1 & 0 & 1 \\ 0 & 1 & 1 \\ 0 & \frac{1}{1-x} & z \end{vmatrix} = 0,$$

so $z = \frac{1}{1-x}$. Finally, for $efh$, we have the determinant

$$\begin{vmatrix} 1 & 1 & 1 \\ 0 & x & 1 \\ 1 & 0 & \frac{1}{1-x} \end{vmatrix} = 0,$$

which gives $x^2 - x + 1 = 0$.

---

[7] We promise. This should be better than $n$ Super Bowls, where the value of $n$ depends on several factors.

Let's collect our thoughts. At each stage of this process, we made the most general choice for coordinates possible. But this forced us to choose a value for $x$ that satisfies the quadratic equation $x^2 - x + 1 = 0$. Solving this equation forces $\sqrt{-3}$ into our field (use the quadratic formula). Thus, we get to assert the following:

**Proposition 6.12.** *The matroid $M_{\sqrt{-3}}$ is representable by the columns of a matrix over a field $\mathbb{F}$ if and only if $\mathbb{F}$ contains a root of the equation $x^2 - x + 1$.*

This example shows one way the representation can depend on the field $\mathbb{F}$. We saw another field dependence in Example 6.3, but Example 6.11 is different: the *matroid* imposes the restriction on the field $\mathbb{F}$, not the matrix. This is a fundamental distinction, and we'll return to it in Section 6.3.

### 6.2.3 Coordinatizing paths

We wish to investigate the procedure we used in the last steps of Example 6.10. This procedure produced a matrix with lots of 1's, and this simplification is useful in finding representations of a given matroid. We'll do this by finding a *coordinatizing path* in our matrix. This has a very nice algorithmic description, and it's just a consequence of Proposition 6.9.

**Example 6.13.** Let's return to the matroid $M_{\sqrt{-3}}$ in Figure 6.3. We've already found a matrix to represent $M_{\sqrt{-3}}$, and we presented some convincing arguments[8] that $M_{\sqrt{-3}}$ can be represented only over a field that has a solution to the equation $x^2 - x + 1 = 0$. We'll redo this, quickly, to illustrate how you can make your own coordinatizing path.   *Coordinatizing path*

---

**Coordinatizing path algorithm**

(1) First, pick a basis for $M$. We'll use $abc$, as before.
(2) Next, find the basic circuits that contain each of the remaining points. (Recall that if $B$ is a basis for a matroid $M$, then, for each $x \notin B$, the *basic circuit* determined by $B$ and $x$ is the unique circuit contained in $B \cup x$. See Exercise 13 in Chapter 2.)

| Point | $d$ | $e$ | $f$ | $g$ | $h$ |
|---|---|---|---|---|---|
| Basic circuit | $abcd$ | $ace$ | $abf$ | $bcg$ | $abch$ |

---

8  Well, we were convinced.

(3) Now record the incidence between the points $d, e, \ldots, h$ and their basic circuits in two ways: via a matrix $A'$ and a bipartite graph $G$. See Figure 6.4.

$$A' = \begin{array}{c} a \\ b \\ c \end{array} \begin{array}{cccccc} & d & e & f & g & h \\ \left[ \begin{array}{ccccc} * & * & * & 0 & * \\ * & 0 & * & * & * \\ * & * & 0 & * & * \end{array} \right] \end{array}.$$

The matrix $A'$ is the *basic-circuit incidence matrix*.

(4) Find a spanning tree for your bipartite graph $G$, if possible. See Figure 6.5. (This is *not* the same spanning tree we used in Example 6.11. See Exercise 16.)

(5) Highlight the entries of the matrix $A'$ that correspond to the edges of the spanning tree you just found. Replace each of those entries by a 1:

$$A' = \begin{array}{c} a \\ b \\ c \end{array} \begin{array}{cccccc} & d & e & f & g & h \\ \left[ \begin{array}{ccccc} * & * & \boxed{*} & 0 & \boxed{*} \\ \boxed{*} & 0 & * & \boxed{*} & * \\ * & \boxed{*} & 0 & \boxed{*} & \boxed{*} \end{array} \right] \end{array}$$

$$\longrightarrow \begin{array}{c} a \\ b \\ c \end{array} \begin{array}{cccccc} & d & e & f & g & h \\ \left[ \begin{array}{ccccc} * & * & \boxed{1} & 0 & \boxed{1} \\ \boxed{1} & 0 & * & \boxed{1} & * \\ * & \boxed{1} & 0 & \boxed{1} & \boxed{1} \end{array} \right] \end{array}.$$

(6) You can now safely assume your representing matrix $A$ has the form $[I \mid A']$.

$$A = \begin{array}{ccccccccc} & a & b & c & d & e & f & g & h \\ \left[ \begin{array}{cccccccc} 1 & 0 & 0 & * & * & \boxed{1} & 0 & \boxed{1} \\ 0 & 1 & 0 & \boxed{1} & 0 & * & \boxed{1} & * \\ 0 & 0 & 1 & * & \boxed{1} & 0 & \boxed{1} & \boxed{1} \end{array} \right] \end{array}.$$

Figure 6.4. Bipartite graph $G$ for basic-circuit incidence.

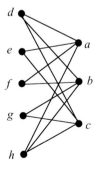

We'll outline a proof that this procedure is valid shortly. But a few comments are in order first.

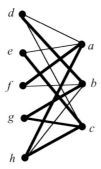

- The location of each 0 entry in $A'$ is completely determined by the basic circuits: if the initial $r$ columns are the basis $B = \{b_1, b_2, \ldots, b_r\}$, then entry $i$ of column $x$ will be 0 precisely when $b_i \notin C_x$, where $C_x$ is the basic circuit contained in $B \cup x$. Thus, the * entries cannot be 0.
- We can use the matrix operations of Proposition 6.9 to replace the boxed entries of $A$ with any non-zero field entries we desire. In our example, this would look like:

$$
\begin{array}{cccccccc}
a & b & c & d & e & f & g & h \\
\end{array}
$$
$$
\begin{bmatrix}
1 & 0 & 0 & * & * & \boxed{1} & 0 & \boxed{1} \\
0 & 1 & 0 & \boxed{1} & 0 & * & \boxed{1} & * \\
0 & 0 & 1 & * & \boxed{1} & 0 & \boxed{1} & \boxed{1}
\end{bmatrix}
$$
$$
\begin{array}{cccccccc}
& a & b & c & d & e & f & g & h \\
\end{array}
$$
$$
\longrightarrow
\begin{bmatrix}
1 & 0 & 0 & * & * & a_1 & 0 & a_2 \\
0 & 1 & 0 & a_3 & 0 & * & a_4 & * \\
0 & 0 & 1 & * & a_5 & 0 & a_6 & a_7
\end{bmatrix}.
$$

- The number of edges in a spanning tree of a graph is $v - 1$, where $v$ is the number of vertices (Theorem 1.15). Thus, we expect to find $r + (n - r) - 1 = n - 1$ non-zero entries determined by this process.
- What if the bipartite graph $G$ is not connected (as a graph)? Then we use a spanning forest. In fact, we could rephrase this last step via matroids as follows: find a basis of the graphic matroid associated to the graph $G$.

At this point, we have done everything we can (as far as Proposition 6.9 is concerned). We still have five *s, and these can be filled in temporarily by five indeterminates:

$$
\begin{array}{cccccccc}
a & b & c & d & e & f & g & h \\
\end{array}
$$
$$
\begin{bmatrix}
1 & 0 & 0 & x_1 & x_2 & 1 & 0 & 1 \\
0 & 1 & 0 & 1 & 0 & x_3 & 1 & x_4 \\
0 & 0 & 1 & x_5 & 1 & 0 & 1 & 1
\end{bmatrix}.
$$

Then the matroid dependences can be used to determine the relations these indeterminates must satisfy. Here's what we get:

| Dependence | $agh$ | $cdh$ | $bde$ | $efh$ | $dfg$ |
|---|---|---|---|---|---|
| Relation | $x_4 = 1$ | $x_1 x_4 = 1$ | $x_1 = x_2 x_5$ | $x_3(x_2 - 1) = -x_4$ | $x_1 x_3 + x_5 = 1$ |

Then we immediately get $x_1 = x_4 = 1$. We also find $x_2 = x_3$ and $x_3 = 1 - x_5$. Setting $x = x_2 = x_3$, we also have $x^2 - x + 1 = 0$, as in Example 6.11. Our matrix now looks like

$$
\begin{array}{ccccccccc}
 & a & b & c & d & e & f & g & h \\
\left[ \begin{array}{cccccccc}
1 & 0 & 0 & 1 & x & 1 & 0 & 1 \\
0 & 1 & 0 & 1 & 0 & x & 1 & 1 \\
0 & 0 & 1 & 1-x & 1 & 0 & 1 & 1
\end{array} \right]
\end{array}
$$

where $x$ satisfies $x^2 - x + 1 = 0$.

We summarize the entire "find-the-best-matrix-you-can" discussion of Examples 6.10 and 6.13 in the following theorem.

**Theorem 6.14.** *Let M be a rank r matroid on n points representable over a field $\mathbb{F}$. Then M is represented by an $r \times n$ matrix $A = \left[ I \mid A' \right]$, and the entries in A' that correspond to a coordinatizing path can all be taken to be 1.*

We omit the details of the proof of the theorem, but it's not too hard to do this inductively. First, find a spanning tree for the basic-circuit incidence bipartite graph $G$. That tree must have a vertex of degree 1 (a *leaf* of the tree), and that vertex of $G$ is the only boxed entry in its row or column of $A'$. Thus, you can multiply that row (using Proposition 6.9(1)) or column (using Proposition 6.9(2)) to make this single entry equal to 1. Now delete that row or column from $A'$ and the corresponding vertex of the bipartite graph $G$. Then you still have a spanning tree for $G - v$, so use induction to complete the proof.

## 6.3 How representations depend on the field

In Example 6.11, we saw how the representation of the matroid $M_{\sqrt{-3}}$ depends on an algebraic condition; $M_{\sqrt{-3}}$ is representable over a field $\mathbb{F}$ precisely when $\mathbb{F}$ contains the solution to a specific quadratic equation. This is an indirect restriction on the field; see Exercise 32. We can get a more direct dependence on the field pretty easily.

**Example 6.15.** Let $F_7$ be the Fano plane, shown on the left of Figure 6.6. We know $F_7$ is the projective plane $PG(2, 2)$. Let's try to

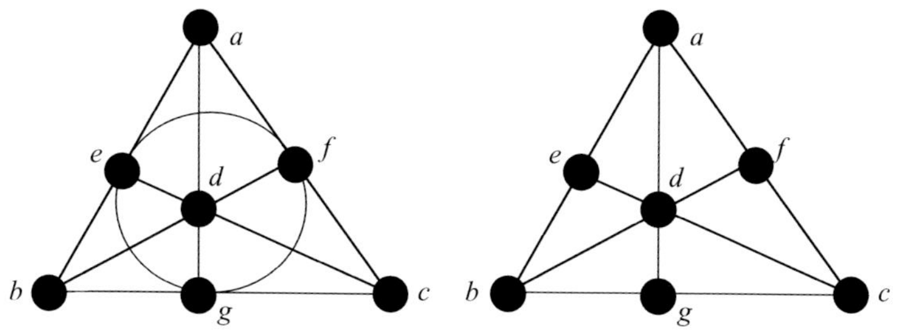

Figure 6.6. The Fano and non-Fano configurations.

represent $F_7$ by a matrix. As in the previous examples in this chapter, we'll assume our matrix is $3 \times 7$, where the first three columns of $A$ correspond to a basis of $F_7$. Then it's easy to check all of the following coordinates are forced (this follows from a standard coordinatizing path argument).

$$A = \begin{array}{c} \begin{array}{ccccccc} a & b & c & d & e & f & g \end{array} \\ \left[ \begin{array}{ccccccc} 1 & 0 & 0 & 1 & 1 & 1 & 0 \\ 0 & 1 & 0 & 1 & 1 & 0 & 1 \\ 0 & 0 & 1 & 1 & 0 & 1 & 1 \end{array} \right]. \end{array}$$

All the way back in Example 1.10 of Chapter 1, we showed the matrix $A$ represents $F_7$ over a field of characteristic 2. We can now see why this is the only way to represent $F_7$.

**Proposition 6.16.** *The Fano plane $F_7$ is representable over a field $\mathbb{F}$ if and only if the characteristic of $\mathbb{F}$ is 2.*

*Proof Proposition 6.16.* If a matrix represents $F_7$, we can use Proposition 6.9 to assume $A$ is our representing matrix, where $A$ is given above. Then $efg$ is dependent if and only if these three vectors are linearly dependent. This happens precisely when $2 = 0$. □

What's up with the non-Fano plane $F_7^-$, shown on the right in Figure 6.6? It is represented by the same matrix $A$ because $F_7^-$ and $F_7$ share all of the dependences used to determine the entries of $A$. The dependence $efg$ is the only difference, so the next result is immediate.

**Proposition 6.17.** *The non-Fano plane $F_7^-$ is representable over a field $\mathbb{F}$ if and only if the characteristic of $\mathbb{F}$ is not 2.*

If $M$ is representable over the field $\mathbb{F}_2$, we say $M$ is *binary*, so the Fano plane $F_7$ is a binary matroid. If $M$ is representable over $\mathbb{F}_3$, we say $M$ is *ternary*. Although the non-Fano plane $F_7^-$ is not binary, it is a ternary matroid.

Binary, ternary and regular matroids

Matroids representable over *all fields* are called *regular* or *unimodular*. We will study binary matroids in more detail in Section 8.2 in Chapter 8. It turns out that graphic matroids are regular; the vertex–edge

Figure 6.7. The "add 1"
matroid $M_5$.

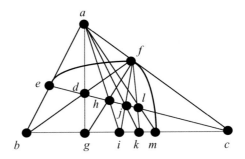

incidence matrix can be modified to give a representation that works
over any field. See Exercise 4(c) of Chapter 4 and Theorem 8.28.

We would like to generalize the Fano–non-Fano matroid representations.

**Example 6.18.** Just for some variety, we'll present the next example
as a sample homework problem.

Let $M_5$ be the matroid shown in Figure 6.7.

(1) Find a matrix that represents $M_5$.
(2) The dependence $\{e, f, m\}$ forces something interesting to happen.
What is it? Why is this called $M_5$?
(3) Generalize – find a matroid that is only representable over the field
$\mathbb{F}_p$ where $p$ is a prime. How many points does your matroid have?

Some answers:

(1) Here we go:

$$
A = \begin{array}{c} \\ \\ \end{array}
\begin{array}{ccccccccccccc}
a & b & c & d & e & f & g & h & i & j & k & l & m \\
\end{array}
$$

$$
A = \begin{bmatrix}
1 & 0 & 0 & 1 & 1 & 1 & 0 & 1 & 0 & 1 & 0 & 1 & 0 \\
0 & 1 & 0 & 1 & 1 & 0 & 1 & 1 & 1 & 1 & 1 & 1 & 1 \\
0 & 0 & 1 & 1 & 0 & 1 & 1 & 2 & 2 & 3 & 3 & 4 & 4
\end{bmatrix}.
$$

As in Example 6.15, each entry in the matrix is uniquely determined
(up to the use of Proposition 6.9). For instance, the column corresponding to $h$ must have its first two entries equal (since $h$ is on
the line $\overline{cd}$), and we can take this common value to be 1. But $h$ is
also on $\overline{fg}$, and this forces the last entry of $h$ to be 2. Note that all
of this makes sense geometrically: every point from $e$ on is on the
intersection of two lines, and so is geometrically uniquely placed.

Projectively unique representations       We say such a representation is *projectively unique*.

(2) The "final" dependence $efm$ forces the determinant

$$
\begin{vmatrix}
1 & 1 & 0 \\
1 & 0 & 1 \\
0 & 1 & 4
\end{vmatrix} = 0.
$$

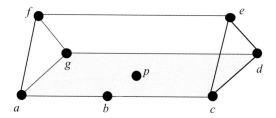

Figure 6.8. The point $p$ is freely added to the flat $F = \{a, b, c, d, g\}$ in $M$.

This forces $5 = 0$, so this matroid is only representable over fields of characteristic 5.

(3) See Exercise 22.

We conclude this section with an example of how you can sometimes *extend* a representation of a matroid $M$ to a larger matroid.

**Example 6.19.** Let $M$ be the rank 4 matroid in Figure 6.8. Then you can show $M - p$ is represented by the matrix $A$, where $z \neq 0, 1$. (If $z = 0$, then $a, e$ and $f$ are collinear. If $z = 1$, then $\{c, d, e, f\}$ is a dependent set.) Thus, this matrix represents $M - p$ over any field $\mathbb{F}$ *except* $\mathbb{F}_2$.

$$
A = \begin{array}{c} \\ \\ \\ \\ \end{array}
\begin{array}{cccccccc}
a & b & c & d & e & f & g \\
\end{array}
\left[
\begin{array}{ccccccc}
1 & 0 & 1 & 0 & 0 & 1 & 1 \\
0 & 1 & 1 & 0 & 0 & z & 1 \\
0 & 0 & 0 & 1 & 0 & 0 & 1 \\
0 & 0 & 0 & 0 & 1 & 1 & 0 \\
\end{array}
\right].
$$

Now note that the point $p$ is on the flat $F = \{a, b, c, d, g\}$ in $M$. How can we find a representation of the matroid $M$ from our representation of $M - p$?[9]

There is an easy way to answer this question, but it requires an *extension* field. First, since the flat $F$ is spanned by the points $a$, $b$ and $d$, the point $p$ is in the basic circuit $\{a, b, d, p\}$ (for the basis $\{a, b, d, e\}$). That means the last coordinate of column $p$ must equal 0. But, because $p$ satisfies no other non-trivial dependences, we make no additional assumptions on the first three coordinates. So, letting $x_1$, $x_2$ and $x_3$ be unspecified indeterminates for the moment, we get:

$$
\begin{array}{cccccccc}
a & b & c & d & e & f & g & p \\
\end{array}
\left[
\begin{array}{cccccccc}
1 & 0 & 1 & 0 & 0 & 1 & 1 & x_1 \\
0 & 1 & 1 & 0 & 0 & z & 1 & x_2 \\
0 & 0 & 0 & 1 & 0 & 0 & 1 & x_3 \\
0 & 0 & 0 & 0 & 1 & 1 & 0 & 0 \\
\end{array}
\right].
$$

We can simplify this matrix by using a coordinatizing path that includes the first entry of column $p$. This allows us to set that entry equal to 1.

---

[9] This is the reverse of what we've done before: given a matroid $M$ with a representing matrix $A$, it's easy to find a matrix that represents the deletion $M - x$.

Thus, column $p$ contains two indeterminates, $x$ and $y$, corresponding to the geometric property that "$p$ is freely placed on a plane."

$$
\begin{array}{cccccccc}
a & b & c & d & e & f & g & p \\
\end{array}
$$
$$
\begin{bmatrix}
1 & 0 & 1 & 0 & 0 & 1 & 1 & 1 \\
0 & 1 & 1 & 0 & 0 & z & 1 & x \\
0 & 0 & 0 & 1 & 0 & 0 & 1 & y \\
0 & 0 & 0 & 0 & 1 & 1 & 0 & 0 \\
\end{bmatrix}.
$$

Finally, we need to make sure $x$ and $y$ satisfy no "accidental" dependences over our field. We can find all these forbidden relations by taking the determinant of every $4 \times 4$ submatrix of our matrix.[10] Any such determinant involving $z$, $x$ or $y$ will give us an inequality. In this case, you can check the following inequalities are forced: $x \neq 0, 1, z$ and $y \neq 0, 1$, and $x \neq y$ (else $\{a, g, p\}$ is dependent). The last four columns force one additional inequality: $x - z - y + zy \neq 0$.

Thus, if $\mathbb{F}$ is large enough, we can choose elements $x$ and $y$ to satisfy all these requirements. In particular, the field $\mathbb{F}(x, y)$ will work. This is a *transcendental* extension of $\mathbb{F}$, and the transcendentals $x$ and $y$ satisfy no non-trivial polynomial dependences. (This is, admittedly, overkill. We only need to avoid a few values and relations.)

We conclude that $M$ is representable over every field characteristic, i.e., for any $p \in \{0, 2, 3, 5, 7, \dots\}$, there is a field $\mathbb{F}$ of characteristic $p$ such that $M$ is representable over $\mathbb{F}$.

Free extension    The operation of adding a point $p$ freely to a specified flat $F$ in a matroid $M$ is called *free extension*, and is usually expressed as $M +_F p$. This operation reverses deletion: $(M +_F p) - p = M$.

## 6.4 Non-representable matroids

Question Q3 at the beginning of this chapter asked if every matroid is representable:

> Does every matroid arise from the linear dependences of some collection of vectors?

The answer is no: we'll see several examples in this section. The question is of obvious importance; if every matroid were representable over some field, then the entire subject would be a branch of linear algebra. Searching for examples of matroids that are not representable over any field by column vectors occupied early researchers in matroid theory for a surprisingly long time. This search was complicated by the fact that every matroid on seven or fewer elements is representable over some field.

---

[10] Preferably via a computer.

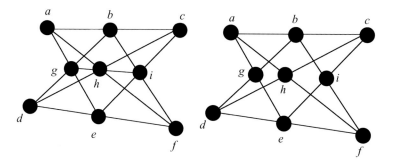

Figure 6.9. Left: the Pappus configuration. Right: the non-Pappus matroid. The points $g$, $h$ and $i$ must be collinear if $M$ is representable over a field.

In general, it's hard to prove something is impossible – a famous example is the P vs. NP problem.[11] It is (evidently) very difficult to prove no polynomial-time algorithm exists that solves one of the many problems known to be NP-complete.

It won't be that hard to produce non-representable matroids. Our first example of a non-representable matroid is based on a classic result in geometry.

**Example 6.20.** The configuration pictured on the left in Figure 6.9 is called the *Pappus* configuration. Its discovery dates to Pappus of Alexandria (c. 320 AD – see Exercise 3 in Chapter 2 and Theorem 5.36 in Chapter 5.) The configuration is constructed as follows:

- Start with two three-point lines $\overline{abc}$ and $\overline{def}$.
- Now form the point $g$ as the intersection of the two lines $\overline{ae}$ and $\overline{bd}$, i.e., $g = \overline{ae} \cap \overline{bd}$.
- Continue to form the points $h = \overline{af} \cap \overline{cd}$ and $i = \overline{bf} \cap \overline{ce}$.
- Then the three points $g$, $h$ and $i$ must be collinear.

Now let $M$ be the matroid depicted on the right of Figure 6.9. This matroid is called the *non-Pappus* matroid, and $g$, $h$ and $i$ are not collinear in $M$. This is the key to the next result.

**Theorem 6.21.** *The non-Pappus matroid is not representable over any field.*

*Proof Theorem 6.21.* As in Example 6.11, we attempt to find coordinates for the nine points of $M$. We use Proposition 6.9 to simplify our matrix as much as possible. This gives the matrix:

$$A = \begin{array}{c} \begin{array}{ccccccccc} a & c & d & f & h & b & e & g & i \end{array} \\ \left[ \begin{array}{ccccccccc} 1 & 0 & 0 & 1 & 0 & 1 & 1 & 1 & 1 \\ 0 & 1 & 0 & 1 & 1 & x & 1 & x & z \\ 0 & 0 & 1 & 1 & 1 & 0 & y & xy & y \end{array} \right] \end{array},$$

---

[11] The Clay Mathematics Institute lists this problem as one of its seven "Millennium Problems," and offers a $1,000,000 prize for its resolution. (The Poincaré conjecture is considered solved; this leaves six open problems for a potential $6,000,000 payoff.)

where $x$ and $y$ are indeterminates and $z$ will be determined. Note the order we list the columns; this ordering will make our argument a little easier computationally.

We comment first on the $xy$ term appearing in the last coordinate of column $g$. Since $g$ is on the line $bd$, we know $g$ must have the first two coordinates in the ratio $1 : x$, so $g = (1, x, w)$, and we still need to determine $w$. Since $g$ is also on the line $ae$, there is some linear combination of columns $a$ and $e$ that produce column $g$:

$$\alpha(1, 0, 0) + \beta(1, 1, y) = (1, x, w).$$

We get $\alpha = 1 - x$ and $\beta = x$; this gives $w = xy$.

For column $i$, a similar argument gives $i = (1, z, y)$, and $z$ is determined from the linear combination involving $bf$:

$$\alpha(1, 1, 1) + \beta(1, x, 0) = (1, z, y).$$

Choosing $\alpha = y$ and $\beta = 1 - y$ produces $z = x + y - yx$. Note that we are being careful about always multiplying on the left; this will be important in the next paragraph.

Now the points $g$, $h$ and $i$ will be linearly independent precisely when the determinant of the $3 \times 3$ submatrix formed by these three columns is not 0:

$$\begin{vmatrix} 0 & 1 & 1 \\ 1 & x & x + y - yx \\ 1 & xy & y \end{vmatrix} = 0.$$

But this determinant is $xy - yx$. Thus, the points $g, h$ and $i$ are independent precisely when $xy = yx$. But we are working over a field (where multiplication is commutative), so this completes the proof. □

The key to making the commutativity argument valid is that we always multiplied scalars on the left in determining the coordinates. That explains why we avoided using determinants until the last step, when the particular determinant does not involve multiplication of two indeterminates.

An interesting corollary of the proof[12] is that the non-Pappus matroid is representable over *division rings*. (A *division ring* or *skew-field* is an algebraic object with the two operations of addition (assumed to be commutative) and multiplication (not assumed to be commutative) where every non-zero element has a multiplicative inverse. There are no finite division rings.) Adding the dependence $bhe$ to the non-Pappus

---

[12] A result that follows from the *proof* of a result is called a *scholium*. Are you impressed?

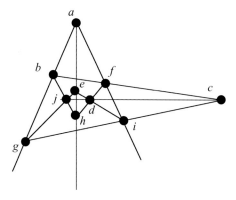

Figure 6.10. Desargues' configuration. Triangles *bfh* and *egi* are in *point perspective* from the point *a*. This forces the points *c*, *d* and *j* to be collinear in any $PG(n, q)$.

matroid produces a matroid not representable over any fields or division rings, either – see Exercise 27.

The non-Pappus matroid is a non-representable matroid that arises from a classic theorem in geometry. A similar construction is possible for another classical theorem – *Desargues' Theorem* – Theorem 5.37 of Chapter 5 (see Figure 6.10). See Exercise 26 for an interpretation of this theorem in terms of matroid representability.

Finding the non-Pappus matroid was lucky; we needed a theorem from geometry to get a non-representable matroid. We promised finding non-representable matroids wouldn't be that hard, so we need some systematic way to produce them. The next two results, a lemma and a proposition, will be very useful in that regard.

**Lemma 6.22.** *Suppose M is representable over the field $\mathbb{F}$ and N is obtained from M by some sequence of deletions and contractions. Then N is also representable over $\mathbb{F}$.*

The proof of Lemma 6.22 follows immediately from our description of how to represent the deletion $M - x$ and the contraction $M/x$ from a representation of $M$. See Proposition 6.5.

**Proposition 6.23.** *Let $\mathcal{P} = \{0, 2, 3, 5, 7, \ldots\}$ be the collection of all possible field characteristics. Suppose $M_1$ and $M_2$ are matroids on disjoint ground sets, with $M_1$ only representable over field characteristics $A_1 \subseteq \mathcal{P}$ and $M_2$ only representable over field characteristics $A_2 \subseteq \mathcal{P}$, with $A_1 \cap A_2 = \emptyset$. Then the direct sum $M_1 \oplus M_2$ is not representable over any field.*

*Proof Proposition 6.23.* Suppose $M_1 \oplus M_2$ is represented over some field $\mathbb{F}$. Then, by Lemma 6.22, $M_1$ and $M_2$ are both representable over $\mathbb{F}$, which contradicts the assumption that $A_1 \cap A_2 = \emptyset$. $\qquad\square$

The set $\mathcal{P}$ of all possible field characteristics comes up frequently enough for us to define the *characteristic set* of a matroid.

Characteristic set

Figure 6.11. Gluing $F_7$ and $F_7^-$
along a three-point line
produces a rank 4
non-representable matroid.

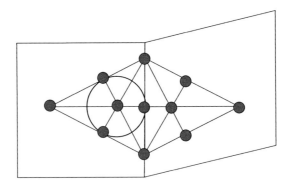

**Definition 6.24.** Let $M$ be a matroid and let $\mathcal{P} = \{0, 2, 3, 5, 7, \ldots\}$ be the collection of all possible field characteristics. Then the *characteristic set* $\chi(M)$ is the subset of $\mathcal{P}$ consisting of all characteristics over which $M$ is representable.

So, for instance, since the Fano plane is representable only over fields of characteristic 2, we know $\chi(F_7) = \{2\}$. Similarly, since the non-Fano plane is representable only over fields of characteristic $\neq 2$, we have $\chi(F_7^-) = \mathcal{P} - \{2\} = \{0, 3, 5, 7, \ldots\}$. Then as an application of Proposition 6.23, we immediately get that the matroid formed as the direct sum $F_7 \oplus F_7^-$ is a non-representable matroid (so $\chi(F_7 \oplus F_7^-) = \emptyset$). This gives us a rank 6 non-representable matroid.

While the direct sum $F_7 \oplus F_7^-$ is not representable, we can "glue" the Fano and non-Fano plane together in slightly different way to achieve the same result with a rank 4 matroid. Consider the matroid $M$ in Figure 6.11. Then $r(M) = 4$, and $M$ is non-representable over any field by Lemma 6.22.

How can we get more non-representable matroids? To use Proposition 6.23, we need to find pairs of matroids that generalize the Fano–non-Fano pair. A natural candidate pair is the "add 1" matroid $M_p$ from Example 6.18 and its companion $M_p'$ (see Figure 6.12). Then it is completely straightforward to show $M_p'$ is representable over a field $\mathbb{F}$ precisely when the characteristic of $\mathbb{F}$ is larger than $p$ (Exercise 23). Thus, Proposition 6.23 gives us an infinite collection of non-representable matroids.

**Proposition 6.25.** *Let* $M = M_p \oplus M_p'$. *Then $M$ is not representable over any field.*

As before, $r(M_p \oplus M_p') = 6$. But it's easy to see we can repeat the gluing operation shown in Figure 6.6 with $M_p$ and $M_p'$ to produce an infinite collection of rank 4 non-representable matroids.

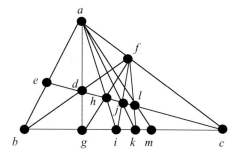

Figure 6.12. The matroid $M'_p$ for $p = 5$ is representable only over fields of characteristic $> 5$.

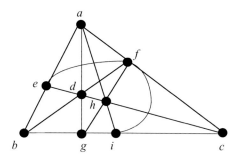

Figure 6.13. The matroid $M_3$ is representable over $\mathbb{F}_3$, so is a submatroid of $PG(2, 3)$.

## 6.5 Representations and geometry

In Chapter 5, we saw connections between matroid theory and geometry. We examine this connection in a bit more detail here. By analogy, if you are handed a simple graph, you can think of that graph as a subgraph of some complete graph. Thus, the complete graphs play the role of *universal embedding object*[13] for all simple graphs.

There is no corresponding universal object that works for all matroids, but projective geometries are a good substitute in the following respect.

**Theorem 6.26.** *Let M be a simple rank r matroid representable over the field* $\mathbb{F}$*. Then M is a submatroid of the projective geometry* $PG(r - 1, \mathbb{F})$*.*

The proof of Theorem 6.26 is immediate: if $M$ is a simple matroid representable over the field $\mathbb{F}$, then the points of $M$ correspond to vectors in $\mathbb{F}^n$. Since there are no loops or multiple points, each point in $M$ simply corresponds to a point in $PG(n, \mathbb{F})$.

For instance, the matroid $M_3$ in Figure 6.13 is representable over $\mathbb{F}_3$ by the matrix

$$A = \begin{array}{c} \begin{array}{ccccccccc} a & b & c & d & e & f & g & h & i \end{array} \\ \begin{bmatrix} 1 & 0 & 0 & 1 & 1 & 1 & 0 & 1 & 0 \\ 0 & 1 & 0 & 1 & 1 & 0 & 1 & 1 & 1 \\ 0 & 0 & 1 & 1 & 0 & 1 & 1 & 2 & 2 \end{bmatrix} \end{array}.$$

---

[13] This is mostly a made-up term, but aren't all terms made up?

Figure 6.14. Embedding the
matroid $M_3$ in the projective
plane $PG(2, 3)$. Note: for
clarity, not all of the lines in
$PG(2, 3)$ have been drawn.

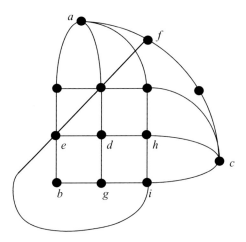

This matroid is an example of an "add 1" matroid with $p = 3$ from Example 6.18. Then $M_3$ is a submatroid of the projective plane $PG(2, 3)$. You can see this by labeling some of the points of $PG(2, 3)$ and checking that all of the lines in $M_3$ are represented by your labeling. See Figure 6.14 for one such embedding.

We can combine the observation that the matroid $M_p$ is only representable over fields of characteristic $p$ with Lemma 6.22 and Theorem 6.26 to get the following very attractive result.

**Theorem 6.27.** *Let $p$ be prime. Then the projective plane $PG(2, p)$ is only representable as a matroid over fields of characteristic $p$, i.e., $\chi(PG(2, p)) = \{p\}$.*

So, for instance, you won't be able to find a Fano plane as a submatroid of $PG(2, 3)$. Since the projective geometry $PG(r, p)$ contains ($\begin{bmatrix} r \\ 2 \end{bmatrix}_p$ copies of) the projective plane $PG(2, p)$, we can make this sound even more impressive:

**Corollary 6.28.** *Let $p$ be prime and $r \geq 2$. Then the projective geometry $PG(r, p)$ is only representable as a matroid over fields of characteristic $p$, i.e., $\chi(PG(r, p)) = \{p\}$.*

Now suppose the rank $r$ matroid $M$ is representable over a field $\mathbb{F}$. Then we can view $M$ as a submatroid of the projective geometry $PG(r - 1, \mathbb{F})$. When will $M$ be *affine*, i.e., when can $M$ be embedded in the affine geometry $AG(r - 1, \mathbb{F})$?

For $M$ to be affine over $\mathbb{F}$, we must have an embedding of $M$ into the projective space $PG(r - 1, \mathbb{F})$ in such a way that $M$ *misses* some hyperplane of $PG(r - 1, \mathbb{F})$. This follows because $PG(r - 1, \mathbb{F}) - H = AG(r - 1, \mathbb{F})$ for any hyperplane $H$ of $PG(r - 1, \mathbb{F})$; see Proposition 5.23.

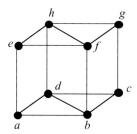

The "add 1" matroid $M_p$ is *not* affine over the field $\mathbb{F}_p$ because $M_p$ has a $(p+1)$-point line (in fact, it has two such lines), but every line in $AG(2, p)$ has exactly $p$ points.

It is worth pointing out that if $M$ is representable over $\mathbb{F}$, then $M$ is affine over some extension field of $\mathbb{F}$. The matroid $M_p$ will be affine over $\mathbb{F}_{p^2}$, for instance, because we can find an equation of the form $ax + by + cz = 0$ for $a, b, c \in \mathbb{F}_{p^2}$ that is not satisfied by any column vector in our matrix.

For $p > 3$ prime, the affine planes $AG(2, p)$ are still only representable over fields of characteristic $p$. That's worth making a fuss over:

**Theorem 6.29.** *If $p > 3$ is prime, then the affine plane $AG(2, p)$ is only representable over fields of characteristic $p$, i.e., $\chi(AG(2, p)) = \{p\}$.*

To prove Theorem 6.29, we would like to construct examples of rank 3 matroids affine over $\mathbb{F}_p$ that are only representable over characteristic $p$. This is a little tricky because we need smaller examples than our $M_p$ matroids. We invite the interested reader to work through one construction in Project P.5.

We conclude this chapter with a another famous non-representable matroid. The non-Pappus matroid has nine points. A smaller example was found by Vamos. Our proof that this matroid is not representable depends on properties of the ambient projective space.

**Example 6.30.** Let $M$ be the rank 4 matroid depicted in Figure 6.15. It's reasonably straightforward to show $M$ is not representable by attempting to find a representing matrix. See Exercise 24 for some help with this approach. Here we use a geometric argument.

The matroid has five four-point circuits: $abef, cdgh, adeh, bcfg$, and $bdfh$. Think of the first four of these as corresponding to the front, back, left and right faces of the cube. Then the circuit $bdfh$ corresponds to one "diagonal" plane of the cube. We will see this collection of four-point circuits forces the other diagonal $aceg$ to be coplanar in any ambient projective space. Note the top and bottom faces are not circuits in this matroid: $\{a, b, c, d\}$ and $\{e, f, g, h\}$ are bases.

Figure 6.16. In the Vamos cube, the six points $a, b, d, e, f, h$ form a prism. In any ambient projective space, the lines $\overline{ae}, \overline{bf}$ and $\overline{dh}$ meet at a common point $x$.

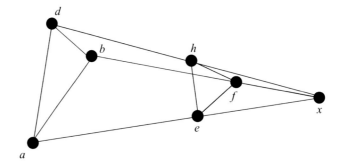

Figure 6.16. In the Vamos cube, the six points $a, b, d, e, f, h$ form a prism. In any ambient projective space, the lines $\overline{ae}, \overline{bf}$ and $\overline{dh}$ meet at a common point $x$.

First, why is this a matroid? One way to see this is to check the independent set axioms (I1), (I2) and (I3). As usual, (I1) and (I2) are trivial, and, since every subset of size 3 (or smaller) is independent, the only thing we need to check is the following: if $I$ and $J$ are independent with $|I| = 3$ and $|J| = 4$, then $I \cup x$ is independent for some $x \in J - I$. But, if $|I| = 3$, then it's easy to see there is at most one point $y$ with $I \cup y$ *not* independent. So $I \cup y$ is the only potential "bad" $J$, but such a $J$ would not be independent.

We'll now show $M$ is not representable over any field by attempting to embed $M$ in a projective space. Suppose $M$ is representable. Then, since the four points $a, b, e$ and $f$ are coplanar, the two lines $\overline{ae}$ and $\overline{bf}$ must meet in the ambient projective space (since every pair of coplanar lines meet in any projective space).

Now let $x = \overline{ae} \cap \overline{bf}$ be this point of intersection. Then $x$ must also be on the line $\overline{dh}$: this follows from the semimodular law (rank axiom (r3)) by setting $A = \{a, d, e, h, x\}$ and $B = \{b, d, f, h, x\}$. Then $r(A) = r(B) = 3$ and $r(A \cup B) = 4$, so $r(A \cap B) \leq 2$, so $d, h$ and $x$ are collinear. See Figure 6.16 for this configuration in a projective space.

We're almost done. We can make the same argument for the prism composed of the points $\{b, c, d, f, g, h\}$, i.e., the other half of the cube. This gives $c, g$ and $x$ collinear. But now $\overline{cgx}$ and $\overline{aex}$ are a pair of intersecting lines, so $r(\{a, c, e, g\}) = 3$, not 4, which is a contradiction.

The configuration in Figure 6.16 might look familiar. If the point $x$ is not on the line $\overline{dh}$, we get the Escher "matroid" configuration introduced in Figure 1.33 of Chapter 1 (see Exercise 11 of Chapter 1). This configuration, which is not a matroid, is redrawn in Figure 6.17.

Here is the takeaway message from this example:

The Vamos cube is a matroid, but is not representable because of properties of an ambient projective space. The Escher configuration is not a matroid because it violates fundamental matroid properties.

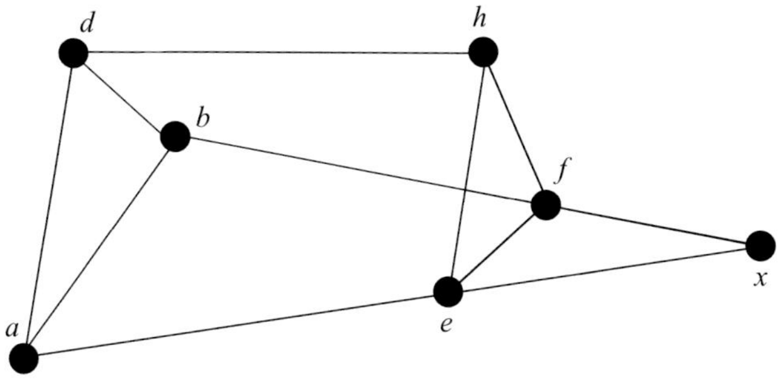

Figure 6.17. The Escher "matroid" appears in any embedding of the Vamos cube into projective space. Matroid axioms force $d$, $h$ and $x$ to be collinear.

## Exercises

### Section 6.1 – Matrices and matroids

(1) Consider the matroid $M[B]$ corresponding to the matrix:

$$B = \begin{array}{cc} \begin{array}{ccccc} a & b & c & d & e \end{array} \\ \left[ \begin{array}{ccccc} 1 & 0 & 1 & 1 & 2 \\ 0 & 1 & -1 & 2 & 0 \end{array} \right]. \end{array}$$

(a) What is the rank of $M[B]$?

(b) Draw the geometry of $M[B]$.

(2) Consider the matroid $M[C]$ on the columns of the matrix $C$.

$$C = \begin{array}{cc} \begin{array}{ccccc} a & b & c & d & e \end{array} \\ \left[ \begin{array}{ccccc} 1 & 0 & 1 & -1 & 0 \\ 1 & -1 & 0 & 0 & 0 \\ 0 & 0 & 1 & 2 & 2 \end{array} \right]. \end{array}$$

(a) Compute the determinant of the submatrices determined by the following sets of vectors

(i) $abc$     (ii) $abd$     (iii) $abe$     (iv) $cde$.

(b) Which of the following scts are bases of the matroid $M[C]$?

(i) $abc$     (ii) $abd$     (iii) $abe$     (iv) $cde$.

(c) Draw the geometry of $M[C]$.

(3) Let $A = \begin{array}{cc} \begin{array}{cccccccc} a & b & c & d & e & f & g & h \end{array} \\ \left[ \begin{array}{cccccccc} 2 & 0 & 0 & 1 & 2 & 3 & 3 & 1 \\ -1 & 1 & 3 & 1 & 2 & 1 & 0 & 2 \\ 0 & 1 & 3 & 0 & 0 & 1 & 0 & 1 \\ 1 & 0 & 0 & 0 & 0 & 1 & 1 & 0 \end{array} \right]. \end{array}$

(a) Use row operations to row reduce $A$ and then draw a picture of the column dependence matroid $M[A]$. (Hint: this is a rank 3 matrix.)

(b) Identify a circuit in your drawing and show that the corresponding columns of $A$ are linearly dependent.

(c) Now identify a basis from your drawing, and show the corresponding columns of $A$ are linearly independent.

(4) This exercise asks you to prove independent set axiom (I3) for linearly independent sets without using matrix row reduction, i.e., using direct linear independence arguments. (The proof still uses matrices, though.)

(a) Let $I = \{\bar{v}_1, \ldots, \bar{v}_m\}$ and $J = \{\bar{w}_1, \ldots, \bar{w}_k\}$ with $m < k$, and suppose the statement is false, i.e., each $\bar{w}_i$ can be written as a linear combination of $\{\bar{v}_1, \ldots, \bar{v}_m\}$. Show this gives

$$
\begin{bmatrix} a_{11} & \cdots & a_{1m} \\ \vdots & \ddots & \vdots \\ a_{k1} & \cdots & a_{km} \end{bmatrix} \begin{bmatrix} \bar{v}_1 \\ \vdots \\ \bar{v}_m \end{bmatrix} = \begin{bmatrix} \bar{w}_1 \\ \vdots \\ \bar{w}_k \end{bmatrix}.
$$

Rewrite this as $A \cdot \bar{V} = \bar{W}$, where $A$ is the $k \times m$ matrix given by the coefficients $(a_{ij})$, $\bar{V}$ is the $m \times 1$ array of vectors formed by the $\bar{v}_i$ and $\bar{W}$ is the $k \times 1$ array of vectors formed by the $\bar{w}_i$.

(b) Take the matrix transpose of the equation $A \cdot \bar{V} = \bar{W}$ and use the fact that $k > m$ to conclude there is a non-trivial solution $\bar{X}$ to the equation $A^T \cdot \bar{X} = \bar{0}$.

(c) If $\bar{X}$ is the non-trivial solution to the equation from part (b), take the transpose again and multiply by $\bar{V}$:

$$
\bar{X}^T \cdot A = \bar{0}_{1 \times m} \Rightarrow \bar{X}^T \cdot A \cdot \bar{V} = 0.
$$

Show how this contradicts the fact that $J$ is a linearly independent set.

(5) Verify the circuit exchange axiom (C3) directly for representable matroids $M[A]$. Let $C_1$ and $C_2$ be minimal linearly dependent sets, and let $x \in C_1 \cap C_2$. Show there is a minimal linearly dependent set $C_3 \subseteq C_1 \cup C_2 - x$.

(6) Prove Theorem 6.4 for matroids: if $I$ is independent and $S$ is a spanning set in a matroid $M$, with $I \subseteq S$, then there is a basis $B$ satisfying $I \subseteq B \subseteq S$.

(7) Let $A$ be the matrix $A =$

|   | a | b | ι | d | e | f | g | h |
|---|---|---|---|---|---|---|---|---|
|   | 1 | 0 | 0 | 0 | 0 | 0 | 0 | 0 |
|   | 1 | 1 | 0 | 0 | −1 | 1 | 0 | 0 |
|   | 1 | 0 | 1 | 0 | 0 | 1 | 1 | 0 |
|   | 1 | 0 | 0 | 1 | 0 | 0 | 1 | 0 |

(a) $M[A]$ has an isthmus and a loop. Find them.

(b) Draw a geometric picture of the matroid $M[A]$.

(c) Show that $M[A]$ is a graphic matroid by finding a graph $G$ with $M[A] = M(G)$. Then show that the vertex–edge incidence matrix (see Definition 4.16) for your graph is row equivalent to $A$.

(8) Prove the first part of Proposition 6.5: let $M[A]$ be the matroid defined on the matrix $A$, and suppose $e$ not an isthmus. Show the

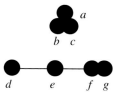

Figure 6.18. Matroid for
Exercise 11.

matroid $M[A] - e$ is represented by the matrix $A - e$, i.e., the
matrix obtained from $A$ by removing the column corresponding to
the point $e$.

(9) Let $A$ be a matrix and let $M[A]$ be the linear dependence matroid
on the columns of $A$.

(a) Show that $e$ is a loop of $M[A]$ if and only if $e$ corresponds to
the column vector $\overline{0}$ in $A$. Use this to show that the contraction
procedure for representable matroids from Proposition 6.5 fails
when $e$ is a loop.

(b) Show that the following isthmus recognition procedure is
valid:
  • Let $B$ be the matrix obtained from $A$ by performing ele-
    mentary row operations on $A$ so that there is exactly one
    non-zero entry in column $e$.
  • Let row $r$ of $B$ be the row containing the one non-zero entry
    of $e$.
  • Then $e$ is an isthmus of $M[A]$ if and only if the single non-
    zero entry of $e$ is also the only non-zero entry in row $r$.

(10) Prove Theorem 6.6: let $A = [I_{r \times r} \mid D]$ and let $A^* = [-D^T \mid I_{(n-r) \times (n-r)}]$. Show that the rows of $A^*$ form a basis for the null
space $N(A)$. Conclude that $M[A^*] = M[A]^*$.

### Section 6.2 – Finding representations

(11) Let $M$ be the matroid of Figure 6.18.

(a) Show that $A = \begin{array}{c} \begin{array}{ccccccc} a & b & c & d & e & f & g \end{array} \\ \begin{bmatrix} 1 & 2 & 3 & 1 & 0 & 1 & 1 \\ 1 & 2 & 3 & 0 & 1 & 1 & 1 \\ 2 & 4 & 6 & 0 & 1 & 1 & 1 \end{bmatrix} \end{array}$ is a representation of $M$.

(b) $M$ is a direct sum: $M = M_1 \oplus M_2$. Find matrices $A_1$ and $A_2$
that represent $M_1$ and $M_2$, respectively.

(c) Show the matrix $A$ from part (a) is row equivalent to
$\left( \begin{array}{c|c} A_1 & 0 \\ \hline 0 & A_2 \end{array} \right)$. (This is the representation of $M_1 \oplus M_2$ from
Exercise 19 of Chapter 3.)

(12) Find a matrix to represent the matroid $M$ from Figure 6.19. Show
$M$ is representable over any field except $\mathbb{F}_2$. (In fact, $M$ is affine
over all fields except $\mathbb{F}_2$.)

Figure 6.19. Matroid for
Exercise 12.

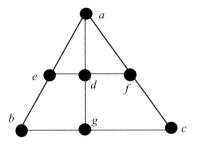

Figure 6.20. Matroid for
Exercise 13.

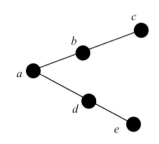

(13) Let $M$ be the matroid in Figure 6.20.
   (a) Use a coordinatizing path to find a projectively unique matrix to represent $M$ over all fields.
   (b) $M$ is a graphic matroid, so we can get a vertex–edge incidence matrix (see Definition 4.16) for the graph $G$. Find this matrix and show it represents $M$ over $\mathbb{F}_2$.
   (c) Now apply Proposition 6.9 to your incidence matrix to show it is equivalent to the matrix you found in part (a).
(14) Construct a proof of Proposition 6.9(1): let $A$ be an $r \times n$ matrix over a field $\mathbb{F}$ and let $B$ be an $r \times r$ invertible matrix. Then the column dependences of $A$ are the same as those of $BA$. Deduce the fact that a matrix and its reduced row echelon form have the same column dependences, i.e., Proposition 6.2. (Hint: show the null space (or *kernel*) of $A$ and $BA$ are the same: $Av = \bar{0}$ if and only if $(BA)v = \bar{0}$. Then interpret $Av = \bar{0}$ as a linear dependence of some of the columns of $A$.)
(15) Prove Proposition 6.9(2): let $A$ be an $r \times n$ matrix over a field $\mathbb{F}$ and let $D$ be an $n \times n$ invertible diagonal matrix. Show the column dependences of $A$ are the same as those of $AD$. (Hint: if $D$ is a diagonal matrix with entries $d_1, d_2, \ldots, d_n$ along the main diagonal and zeros everywhere else, then column $c_i'$ of $AD$ is obtained from the corresponding column $c_i$ of $A$ by multiplying every entry of $c_i$ by $d_i$.)
(16) (a) Find the basic circuit incidence matrix and bipartite graph we used in Example 6.11 to create the coordinatizing path.

Spanning circuit    (b) Suppose the rank $r$ matroid $M$ has a *spanning circuit* $C$, i.e., a circuit of size $r + 1$. Let $B$ be the basis obtained by deleting

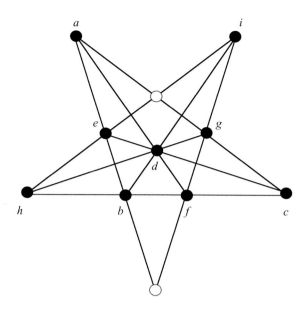

Figure 6.21. Betsy Ross matroid for Exercise 20.

some element $x$ of $C$, and let $A'$ be the basic circuit incidence matrix for $M$ with respect to this basis. Show there is a coordinatizing path in which $x$ corresponds to a column of 1's and the first non-zero entry of each additional column equals 1. (Note: for such a coordinatizing path, every vertex except $x$ on one side of the basic circuit bipartite graph $G$ will have degree 1, and the vertex labeled $x$ will have degree $r$.)

### Section 6.3 – Representations and field characteristics

(17) (a) Show the line $U_{2,n}$ is representable over all fields $\mathbb{F}$ with $|\mathbb{F}| \geq n - 1$.

    (b) Show the uniform matroid $U_{r,n}$ is representable over all field characteristics. (Hint: this can be done using transcendentals, as in Example 6.19.)

    Note: uniform matroids have been very easy to deal with – that is, up till now. It is an unsolved problem to determine precisely what fields the uniform matroid $U_{r,n}$ is representable over. This is closely related to classical problems of finding collections of *free* points in projective spaces.

(18) Suppose $M$ is a rank 2 matroid. Show that $M$ is representable over all field characteristics, i.e., $\chi(M) = \{0, 2, 3, 5, \ldots\}$.

(19) Suppose $M_1$ and $M_2$ are matroids on disjoint ground sets $E_1$ and $E_2$, respectively. Show the characteristic set $\chi(M_1 \oplus M_2) = \chi(M_1) \cap \chi(M_2)$. Conclude that the class of matroids representable over a given field is closed under the operation of direct sum.

(20) Let $M$ be the rank 3 matroid drawn in Figure 6.21. The matroid has nine points – ignore the white points for now.

Figure 6.22. Matroid $M[A]$ for Exercise 21.

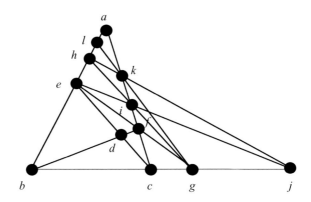

(a) Find a matrix to represent this matroid. If you do this carefully, you shouldn't need to introduce more than one indeterminate $x$.

(b) Show that if this matroid is representable over a field $\mathbb{F}$, then $\mathbb{F}$ contains $\sqrt{5}$.

(c) Show that if this matroid is embedded in an affine or projective space, then the two white points and the center point must be collinear.

(21) Here's a matrix:

$$
A = \begin{array}{c c c c c c c c c c c c}
 & a & b & c & d & e & f & g & h & i & j & k & l \\
\left[\begin{array}{c} \\ \\ \\ \end{array}\right. & \begin{array}{c} 1 \\ 0 \\ 0 \end{array} & \begin{array}{c} 0 \\ 1 \\ 0 \end{array} & \begin{array}{c} 0 \\ 0 \\ 1 \end{array} & \begin{array}{c} 1 \\ 1 \\ 1 \end{array} & \begin{array}{c} 1 \\ 1 \\ 0 \end{array} & \begin{array}{c} 1 \\ 0 \\ 1 \end{array} & \begin{array}{c} 0 \\ 1 \\ -1 \end{array} & \begin{array}{c} 1 \\ x \\ 0 \end{array} & \begin{array}{c} 1 \\ 0 \\ x \end{array} & \begin{array}{c} 0 \\ 1 \\ -x \end{array} & \begin{array}{c} 1 \\ 0 \\ x^2 \end{array} & \begin{array}{c} 1 \\ x^2 \\ 0 \end{array} & \left.\begin{array}{c} \\ \\ \\ \end{array}\right] .
\end{array}
$$

(a) Show that $M[A]$ is the matroid in Figure 6.22. The columns have been ordered so that, after the first four points, each subsequent point except $h$ is on the intersection of two lines (and so is projectively uniquely determined.)

(b) Now add the dependence $djl$ so that the three points $d$, $j$ and $l$ are collinear in the matroid. What equation does this force $x$ to satisfy?

(c) Use part (b) to show this matroid (with $djl$ dependent) cannot be represented over $\mathbb{Q}$.

(22) Generalize the matroid $M_5$ from Example 6.18 to define a matroid $M_p$ that is representable only over $\mathbb{F}_p$.

(23) Let $M'_5$ be the matroid in Figure 6.12. Show that $M'_5$ is representable over a field of characteristic $p$ if and only if $p > 5$. Generalize!

## Section 6.4 – Non-representable matroids

(24) Show the Vamos cube of Figure 6.15 is not representable by attempting to find a representing matrix. As a suggestion, show that if $A$ represents $M$ over some field, then we may assume $A$ has

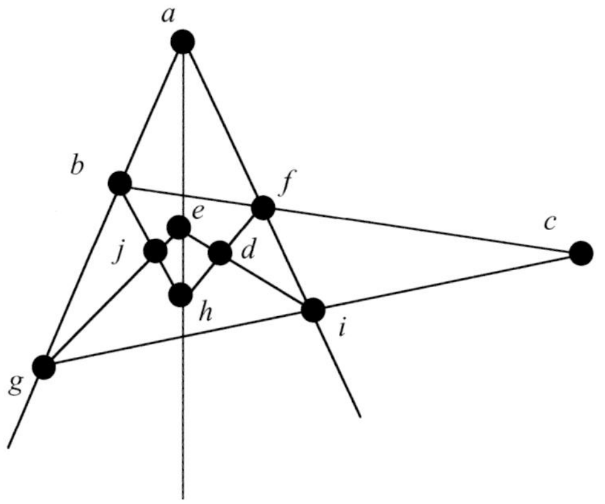

Figure 6.23. The non-Desargues configuration in Exercise 26.

the following form:

$$
A = \begin{array}{c c c c c c c c}
 & a & b & c & d & e & f & g & h \\
\left[\begin{array}{c c c c c c c c}
1 & 0 & 0 & 0 & 1 & x_1 & y_1 & z_1 \\
0 & 1 & 0 & 0 & 1 & x_2 & y_2 & z_2 \\
0 & 0 & 1 & 0 & 1 & 1 & y_3 & z_3 \\
0 & 0 & 0 & 1 & 1 & 1 & 1 & 1
\end{array}\right]
\end{array}.
$$

(We have used the circuit $abef$ to force the last two entries of $f$ to be equal. Use the remaining four circuits to show $x_1 = y_1, z_2 = z_3, z_1 = x_1 z_3$, and $y_2 = 1$. Now show the four points $a, c, e, g$ must be dependent.)

(25) Show that the matroid $M_p$ in Example 6.18 is not affine by showing every line of $PG(2, p)$ meets some point of $M_p$. (Recall: the lines in $PG(2, p)$ correspond to equations of the form $ax + by + cz = 0$, where $a, b, c \in \mathbb{F}_p$ and $(x, y, z)$ is a vector corresponding to a point.)

(26) Desargues Theorem in projective geometry concerns the configuration in Figure 6.10. The non-Desargues matroid is formed by removing the line $\overline{cdj}$ from the Desargues configuration, as in Figure 6.23. Show that the non-Desargues matroid is not representable over any field by trying to find a representing matrix and showing the columns corresponding to the points $c, d$ and $j$ must be linearly dependent, i.e., the points $c, d$ and $j$ must be collinear. (Hint: use the given alphabetical ordering of the ponts.)

(27) (Ingleton [18]) The proof of Theorem 6.21 shows the non-Pappus matroid $M$ is representable over a division ring with two non-commuting indeterminates. Let $M'$ be the matroid obtained from $M$ by adding the line $beh$ to the lines of $M$. Show $M'$ is not representable over any field or division ring. (Hint: this additional dependence forces $x = y$.)

Figure 6.24. Embed the
geometry $M_{\sqrt{-3}}$ in $AG(2, 3)$ –
Exercise 29.

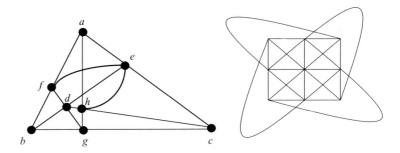

(28) Show that $M$ is a non-representable matroid if and only if the dual matroid $M^*$ is non-representable.

### Section 6.5 – Representations and geometry

(29) Show the matroid $M_{\sqrt{-3}}$ from Example 6.11 is affine over $\mathbb{F}_3$. Do this "directly" by finding a labeling for eight of the nine points of $AG(2, 3)$ so that the eight labeled points correspond to the eight points of $M_{\sqrt{-3}}$ (see Figure 6.24). (This exercise will be used in Project P.5, where we show $AG(2, 3)$ can be represented over any field characteristic.)

(30) (a) Show the graphic matroid $M(K_n)$ is never affine over $\mathbb{F}_2$ (for $n > 2$).
     (b) Show $M(K_4)$ is affine over all fields $\mathbb{F}$ with $|\mathbb{F}| \geq 4$.

(31) A *complete quadrangle* consists of seven points in the plane: four points $a$, $b$, $c$ and $d$, no three on a line, along with the three points of intersection $e = \overline{ab} \cap \overline{cd}$, $f = \overline{ac} \cap \overline{bd}$ and $g = \overline{ad} \cap \overline{bc}$. Suppose a matroid $M$ contains a complete quadrangle. Show that if $M$ is representable over $\mathbb{F}_2$ (i.e., $M$ is binary), then the points $e$, $f$ and $g$ are collinear.

(32) (*If you know some number theory.*) Show the matroid $M_{\sqrt{-3}}$ of Figure 6.3 is representable over a field $\mathbb{F}_p$ if and only if $p = 3$ or $p \equiv 1 \pmod 3$. (Hint: quadratic reciprocity.)

# 7

## Other matroids

Matroids are an important generalization of graphs and matrices; graphic and representable matroids have been the focus of much of the text. But other important combinatorial structures also have interpretations as matroids. In this chapter, we concentrate on two well-studied applications: transversal matroids (which arise from bipartite graphs) and hyperplane arrangements in $\mathbb{R}^n$, which are closely related to representable matroids.

### 7.1 Transversal matroids

Finding matchings in bipartite graphs is an extremely important and well-studied topic in combinatorics. For instance, the job assignment problem introduced in Example 1.20 in Chapter 1 asks you to determine which applicants to hire for a collection of jobs. This motivates the notion of a *transversal matroid*, a matroid associated with matchings in a bipartite graph. These matroids were defined in Example 1.20, and Theorem 7.2 asserts the collections of vertices that can be matched to satisfy the independent set axioms. In Chapter 1, we postponed the proof until "later." Now, it's later.[1]

In an effort to make this chapter self-contained (and to spare you the trouble of leafing back to Chapter 1), we remind you of all the relevant definitions. Let $G$ be a bipartite graph with vertex bipartition $V = X \cup Y$. Recall a subset of edges $N$ is called a *matching* if no two edges of $N$ share any vertex.[2] We write $N = (I, J)$ for $I \subseteq X$ and $J \subseteq Y$ for the vertices that the matching $N$ uses in $X$ and $Y$, respectively. We are interested in the sets $I$ that can be matched in $G$.

---

[1] When was it now?

[2] Notational comment: we use $N$ to denote a matching here instead of $M$, which we reserve for the matroid. For the bipartite graph $G$, we write $M_G$ for the transversal matroid. This is different from the graphic (cycle) matroid, which we denoted $M(G)$, which is different from the representable matroid based on the matrix $A$, which we denoted $M[A]$. We should all try to keep this straight.

Figure 7.1. Two different bipartite graphs may give the same transversal matroid.

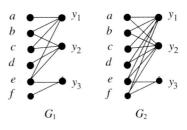

Figure 7.2. The usual geometric representation of the (identical) transversal matroids $M_{G_1}$ and $M_{G_2}$.

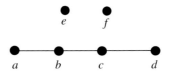

**Example 7.1.** Consider the bipartite graphs $G_1$ and $G_2$ in Figure 7.1. We show the associated transversal matroids $M_{G_1}$ and $M_{G_2}$ on the ground set $X = \{a, b, c, d, e, f\}$ are the same by finding the subsets that can be matched in each bipartite graph. These will be the independent sets of the associated transversal matroid.

First, it's obvious that every singleton in $\{a, b, c, d, e, f\}$ can be matched in both bipartite graphs. You can also check every subset of size 2 can be matched, and so these subsets are also independent in $M_{G_1}$ (and in $M_{G_2}$).

Since $|Y| = 3$ in both graphs, the maximum size of any matching is 3. Then the subsets of size 3 that can be matched will be the bases of the matroid. Notice that the four elements $a, b, c$ and $d$ are only joined to $y_1$ and $y_2$ in $G_1$ (and also in $G_2$). That means no subset of size 3 taken from the set $\{a, b, c, d\}$ can be matched. But this is the only restriction in either bipartite graph, so $G_1$ and $G_2$ have the same bases. See Figure 7.2 for a picture of this matroid. Since every subset of size 3 taken from $\{a, b, c, d\}$ is dependent, we have $\{a, b, c, d\}$ is a four-point line.

We now prove the subsets of $X$ that can be matched form the independent sets in a matroid.

**Theorem 7.2.** *Let $G$ be a bipartite graph with vertex partition $X \cup Y$. Let $\mathcal{I}$ be the collection of subsets $I \subseteq X$ such that the vertices $I$ are precisely those vertices of $X$ in some matching $N = (I, J)$ of $G$. Then $\mathcal{I}$ satisfies the independent set axioms (I1), (I2) and (I3).*

*Proof Theorem 7.2.* We give a proof that uses matrices. This will take advantage of the fact that we've already proven the linearly independent subsets of columns of a matrix satisfy the matroid independent set axioms (Theorem 6.1).

Here's the key idea: given a bipartite graph $G$ with vertex bipartition $V = (X, Y)$, form an incidence matrix $A$ as follows: label the rows by

$Y$ and the columns by $X$, and define

$$a_{i,j} = \begin{cases} z_{ij} & \text{if an edge joins } x_j \text{ to } y_i \\ 0 & \text{otherwise,} \end{cases}$$

where the $z_{ij}$ are indeterminates (think of these as independent variables – the technical term is "independent transcendentals").[3]

For instance, in Example 7.1, the bipartite graph $G_1$ has incidence matrix

$$A = \begin{array}{c} \\ y_1 \\ y_2 \\ y_3 \end{array} \begin{array}{cccccc} a & b & c & d & e & f \\ \left[\begin{array}{cccccc} z_{11} & z_{12} & z_{13} & 0 & z_{15} & 0 \\ 0 & z_{22} & z_{23} & z_{24} & z_{25} & 0 \\ 0 & 0 & 0 & 0 & z_{35} & z_{36} \end{array}\right] \end{array}$$

where we have coded $a = x_1, b = x_2$, etc.

Then we know $M[A]$ is a matroid. The remainder of the proof is just checking the linearly independent subsets of columns of $M[A]$ are precisely the subsets of $X$ that can be matched in the bipartite graph $G$. There are two things to check:

- Every linearly independent subset of columns of $M[A]$ corresponds to a subset that can be matched.
- Every subset of $X$ that can be matched in $G$ corresponds to a linearly independent subset of columns in $M[A]$.

First things first: let $I$ be a linearly independent subset of columns in $M[A]$. Now row reduce the submatrix formed by those $|I|$ column vectors. This will give us an identity matrix of size $|I| \times |I|$ with some rows of zeros below (possibly). Thus, there must have been some collection of exactly $|I|$ of our $z_{ij}$ entries in our submatrix, with no two entries in the same row or same column. Those entries give us a matching of the elements of $I$.

Second things second: suppose $I \subseteq X$ can be matched in the bipartite graph $G$. We need to show the columns of $M[A]$ indexed by $I$ are linearly independent. Reorder the rows and columns so there is a non-zero entry in position $(k, k)$ in the submatrix formed by $I$. Then the determinant of the submatrix formed by the columns of $I$ is non-zero since it includes the term $\prod_{i=1}^{|I|} z_{ii}$, which can't be cancelled by other terms since we are using independent indeterminates. So these columns are linearly independent. □

An alternate proof that uses the *augmenting path algorithm* is outlined in Exercise 13. This algorithm is important in matching theory, and it also allows a direct proof that maximal matchings all have the same size – Exercise 14.

---

[3] We did this in reverse in Section 6.2.3, where coordinatizing paths were associated with a spanning tree in a bipartite graph derived from an incidence matrix.

Figure 7.3. The matroid $M$ is not a transversal matroid.

$a\ b$        $c\ d$        $e\ f$

*Not all matroids are transversal.*[4] So, before going any further, let's look at an example of a matroid that is not transversal.

**Example 7.3.** Let $M$ be the rank 2 matroid of Figure 7.3. Then, if $M$ were transversal, we can find a representing bipartite graph $G$ with $|X| = 6$ and $|Y| = 2$ (see Proposition 7.9).

But, in any such bipartite graph, the points $a$ and $b$ must be joined to one vertex of $Y$ and the points $c$ and $d$ must be joined to the other vertex of $Y$. Now it is impossible for $e$ and $f$ to be adjacent to one vertex of $Y$. Thus $M$ is not transversal.[5]

## 7.2  Transversal matroids, matching theory and geometry

### 7.2.1  Matching theory and systems of distinct representatives

Matching theory is a well-developed area of combinatorics, and many authors prefer using *set systems* to bipartite graphs. We prefer graphs, but we also want to present another way to view matchings, along with some sort of dictionary to translate between the two approaches.

Instead of talking about matchings in a bipartite graph, you can define *systems of distinct representatives* (SDRs) for a family of subsets of a set. Here is the official definition:

**Definition 7.4.** Let $X$ be a finite set and let $\mathcal{F} = \{S_1, S_2, \ldots, S_k\}$ be a family of subsets of $X$. Then the subset $B = \{x_1, x_2, \ldots, x_k\} \subseteq X$ is a
*system of distinct representatives* (SDR) for $\mathcal{F}$ provided $x_i \in S_i$ for $1 \le i \le k$. $B$ is also called a *transversal* of the set system $\{S_1, S_2, \ldots, S_k\}$.

System of distinct representatives

Transversal

**Example 7.5.** Let $X = \{a, b, c, d, e, f\}$ and let $S_1 = \{a, b, c, e\}$, $S_2 = \{b, c, d, e\}$ and $S_3 = \{e, f\}$. Then $\{a, d, e\}$ is a system of distinct representatives because $a \in S_1$, $d \in S_2$ and $e \in S_3$.

When this material is presented in a generic combinatorics text,[6] you might see the following interpretation for an SDR: $X$ is a group of people, the $S_i$ are clubs, and we wish to choose one person from each club to represent that club, where nobody represents more than one club.

Given a set system $\mathcal{F} = \{S_1, S_2, \ldots, S_k\}$ based on a ground set $X$, it's very easy to get a bipartite graph $G$ that corresponds to that set system: let $X$ be the ground set and let $Y = \{S_1, S_2, \ldots, S_k\}$, and connect an

---

[4]  At this point in the text, this statement should have lost all of its shock value.
[5]  This example appeared as Exercise 21 in Chapter 1.
[6]  Assignment: find a generic book and figure out what makes it generic.

Table 7.1. *Set systems and transversals.*

| Matroid | Bipartite graph | Set system |
|---|---|---|
| Independent set | Matching | Partial transversal |
| Basis | Maximal matching | SDR or transversal |

element $x_i$ to a set $Y_j$ precisely when $x_i \in Y_j$. In Example 7.5, you can check the bipartite graph you get is identical to the bipartite graph $G_1$ of Example 7.1.

Table 7.1 is a dictionary to help you translate from bipartite graphs to set systems.

The fundamental question in matching theory is the following:

- Given a bipartite graph $G$ with vertices $V = (X, Y)$ and $A \subseteq X$, when can $A$ be matched into $Y$?

This question was answered by Philip Hall in 1935: Hall's "Marriage"[7] Theorem gives necessary and sufficient conditions that tell us when a subset $A \subseteq X$ can be matched into $Y$. Before stating the theorem we need a bit of notation.

Let $A \subseteq X$ in the bipartite graph $G = (X, Y)$, and define $R(A)$ as follows:

$$R(A) = \{y \in Y \mid xy \text{ is an edge of } G \text{ for some } x \in A\}.$$

We can also define the inverse relation: for $Z \subseteq Y$, let

$$R^{-1}(Z) = \{x \in X \mid xy \text{ is an edge of } G \text{ for some } y \in Z\}.$$

If $|R(A)| < |A|$, then it's easy to see the set $A$ cannot be matched – this follows from the pigeonhole principle. For instance, in the graph $G_1$ in Figure 7.1, we have $R(abc) = \{y_1, y_2\}$, so there is no matching of the set $\{a, b, c\}$ into $Y$ (and, consequently, the set $\{a, b, c\}$ is dependent in the transversal matroid $M$). In fact, more is true: if $|R(S)| < |S|$ for some subset $S \subseteq A$, then it's also impossible to match $A$ into $Y$. But Hall's Theorem says this obvious necessary condition is also sufficient:

**Theorem 7.6.** *[Hall's Theorem] Let $G$ be a bipartite graph with vertices $V = (X, Y)$ and let $A \subseteq X$. Then the set $A$ can be matched into $Y$ if and only if $|R(S)| \geq |S|$ for all $S \subseteq A$.*

You can find a proof of Hall's Theorem in your generic combinatorics text.[8] An application appears in Exercises 16 and 17.

---

[7] This theorem is often referred to in the literature as the Marriage Theorem, in which the bipartition of the vertices consists of women and men; edges correspond to mutual compatibility and a matching is a set of marriages. Enough said.

[8] One popular reference is [36], where you can also find lots of nice applications of the theorem.

Figure 7.4. Two presentations
of the same transversal matroid.

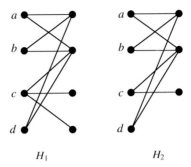

A generalization of Hall's Theorem relates the matroid rank directly
to the bipartite graph $G$. Define the *deficiency* $\delta(A)$ of a subset $A \subseteq X$
to be

$$\delta(A) = \max_{S \subseteq A} (|S| - |R(S)|).$$

Since $\delta(\emptyset) = 0$, $\delta(A) \geq 0$ for all sets $A$. For instance, in the graph $G_1$ in
Figure 7.1, we see $\delta(abc) = 1$, $\delta(e) = 0$ and $\delta(S) = 3$.

**Theorem 7.7.** *[Ore's Theorem] Let $G$ be a bipartite graph with vertices
$V = (X, Y)$ and let $A \subseteq X$. The size of the largest subset of $A$ that can
be matched into $Y$ is $|A| - \delta(A)$.*

Thus, the rank of $A \subseteq X$ in the transversal matroid $M_G$ is equal
to $|A| - \delta(A)$. In the transversal matroid $M_{G_1}$ associated with $G_1$ in
Figure 7.1, we note $r(abc) = 2$, $r(e) = 1$ and $r(S) = 3$.

$r(A) = |A| - \delta(A)$

### 7.2.2  Finding nice bipartite graphs

Suppose you are given a transversal matroid. What can you assume
about the bipartite graph $G$ that generates $M$? We saw in Example 7.1
that the bipartite graph need not be unique. We say the bipartite graphs
$G_1$ and $G_2$ in Example 7.1 are different *presentations* of the transversal
matroid $M$.

Both $G_1$ and $G_2$ in that example have the attractive property that the
size of the largest subset that can be matched is equal to the size of $Y$.
Thus, for instance, when $r(M) = |Y|$, our incidence matrix (from the
proof of Theorem 7.2) has full row rank.

Unfortunately, this need not always be true. Let's look at an example.

**Example 7.8.**  Consider the bipartite graph $H_1$ on the left in Figure 7.4.
Then the largest subset of $X$ that can be matched in $H_1$ has size 3, but
$|Y| = 4$. Note that the presentation for $H_2$ has $|Y| = 3$, and these two
bipartite graphs have the same subsets that can be matched: $M_{H_1} = M_{H_2}$.

What's the problem with the bipartite graph $H_1$ in Figure 7.4? Note
that the point $c$ is an isthmus, and the problem with $H_1$ is that $c$ is

represented with some redundancy. This is easy to fix, and, it turns out, this is the only problem we will ever face in constructing a nice bipartite graph.

**Proposition 7.9.** *If M is a transversal matroid, then there is a presentation in which* $|Y| = r(M)$.

In order to prove Proposition 7.9, we will need a lemma. Recall a flat $F$ in a matroid $M$ is *cyclic* if $F$ is a union of circuits. This means that both $r(F - x) = r(F)$ for all $x \in F$ and $r(F \cup x) = r(F) + 1$ for all $x \notin F$.

**Lemma 7.10.** *Suppose M is a transversal matroid.*

(1) *If C is a circuit in M, then* $|R(C)| = |C| - 1$.
(2) *If F is a rank k cyclic flat of M, then, in any bipartite graph presentation for M, we must have* $|R(F)| = k$.

*Proof Lemma 7.10.*

(1) This is Exercise 7. Do it.
(2) Suppose we have a presentation and let $B = \{x_1, x_2, \ldots, x_k\}$ be a basis for $F$. Then there is a matching of $B$ into $Y$. We may assume $x_1$ is matched to $y_1$, $x_2$ to $y_2$, and so on. Let $B' = \{y_1, y_2, \ldots, y_k\}$. Then we claim:

*Claim 1.* If $x \in F - B$, then $R(x) \subseteq B'$. This is immediate, since if $x$ is joined to some $y \notin B'$, we could match $\{x, x_1, x_2, \ldots, x_k\}$ into $Y$, and so $r(F) > k$, a contradiction.

*Claim 2.* $R(B) = B'$. Again, suppose not. Then some element of $B$ is joined to some $y \notin B'$. Say $x_1$ is the culprit. Since $F$ is a cyclic flat,[9] $x_1$ is in some circuit. In fact, we can claim more: $x_1$ is in a *basic* circuit $C$ with respect to the basis $B$. (See Exercise 15.)
Then $C - B = \{x\}$, where $x \notin B$, so $C = \{x, x_1, x_2, \ldots, x_m\}$, say, and, by part (1), we know $|R(C - x)| = m$. But we already know $R(\{x_1, x_2, \ldots, x_m\}) \supseteq \{y, y_1, y_2, \ldots, y_m\}$ since $x_1$ is joined to $y \notin B'$. This contradiction finishes the claim.
Thus, $R(F) = B'$, so $|R(F)| = k$, which is what we were after.
□

We can now prove Proposition 7.9.

*Proof Proposition 7.9.* First, let $M$ be a transversal matroid, and (temporarily) ignore any isthmuses $M$ might have. Then $M$ is a cyclic flat, since every point is in some circuit. Thus, in *any* bipartite presentation, we must have $r(M) = |Y|$. Now return the isthmuses to $M$ and note we can simply add one new element $y$ to $Y$ for each isthmus $x$ of $M$, joining

---

[9] It's always nice to see the hypothesis used.

Table 7.2. *The subsets $R(x)$ for $x \in X$ for the graph $G$ of Figure 7.5.*

| $x$ | $a$ | $b$ | $c$ | $d$ | $e$ | $f$ | $g$ |
|---|---|---|---|---|---|---|---|
| $R(x)$ | $y_1$ | $y_1, y_2$ | $y_1, y_2$ | $y_1, y_2, y_4$ | $y_1, y_2, y_4$ | $y_2, y_3$ | $y_1, y_4$ |

Figure 7.5. The bipartite graph $G$ and the associated free simplicial picture of the transversal matroid $M_G$. The points labeled $y_2$, $y_3$ and $y_4$ are *not* points of the matroid.

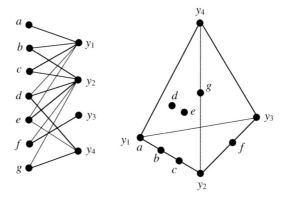

$x$ to $y$. This bipartite graph will then be a valid presentation for $M$ and it satisfies $r(M) = |Y|$, so we're done.    $\square$

**Minimal and maximal presentations**    We can say a little more about the presentations of a transversal matroid. Define a presentation to be *minimal* if the removal of any edge changes the subsets of $X$ that can be matched (and hence the associated matroid). Similarly, a presentation is *maximal* if no edges can be added without changing the matroid. In Example 7.1, you can check that the presentation $G_1$ is minimal and $G_2$ is maximal. Both of these viewpoints are useful, depending on the particular application under consideration.

### 7.2.3 Transversal matroids and geometry

There is a nice, easy, direct way to represent the geometry of a transversal matroid. We describe the procedure in the next example.

**Example 7.11.**    Let $G$ be the bipartite graph on the left in Figure 7.5. In this example, our drawing will be based on a *simplex* in three-dimensional Euclidean space, i.e., a tetrahedron. Then we draw the matroid using the following steps.

- First, draw a tetrahedron $T$ with four vertices, and label the vertices of $T$ with $y_1$, $y_2$, $y_3$ and $y_4$, as in the picture on the right in Figure 7.5. Think of your tetrahedron $T$ as a *scaffolding* upon which we'll add the points of the matroid.
- For each $x \in X$, make a list of the $y_i$'s your $x$ is joined to in the bipartite graph, as in Table 7.2.
- Now each non-empty subset $S \subseteq Y$ determines a *face* of $T$, where a "face" can be a single point (a corner of $T$), an edge, a triangular face or all of $T$ (if $S = Y$).

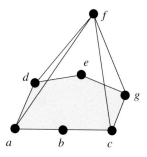

Figure 7.6. The transversal matroid associated to the bipartite graph in Figure 7.5, redrawn. Note that $f$ is an isthmus.

Now build the picture of your matroid by placing the point $x$ on the interior of the face determined by the points of $R(x)$. For example, since $R(c) = \{y_1, y_2\}$, we place $c$ "freely" on the edge $\overline{y_1 y_2}$ of $T$. This procedure places the points $d$ and $e$ freely in the interior of the triangle $\triangle y_1 y_2 y_4$, and it also places $a$ at the corner of $T$ labeled by $y_1$.

This procedure will always produce a valid picture of our transversal matroid, although it may not be the first picture you might think to draw. In this example, a more natural drawing of this rank 4 matroid appears in Figure 7.6.

Here's how the procedure for drawing a transversal matroid from Example 7.11 works in general. We need a few definitions:

The *convex hull* of a set of points in Euclidean space is the smallest convex set containing all the points. An $(n + 1)$-*simplex* is the convex hull of $n + 1$ points in general position in $\mathbb{R}^n$. Thus, a 2-simplex is a triangle, a 3-simplex is a tetrahedron, etc.

Convex hull

**Definition 7.12.** Let $S$ be a simplex in Euclidean space. We say a set of points $E$ in $\mathbb{R}^n$ is *free-simplicial* if all the points of $E$ are freely placed on the faces of various dimensions of $S$ (including the "face" corresponding to the interior of $S$).

Free-simplicial

Given a free-simplicial set, we get a matroid structure by simply using affine dependence in $\mathbb{R}^n$ – see Section 5.2 of Chapter 5. If $M$ is a free-simplicial matroid with all its points on a simplex $S$, we can describe the circuits easily. If $C$ is a circuit and $|C| = 2$, then both points of $C$ must be located at a vertex of $S$. If $|C| = 3$, then all three points of $C$ are located on an edge of $S$. If $|C| = 4$, then the four points of $C$ all lie on a two-dimensional face of $S$, and so on.

The key observation that makes this work is just this: there are no "accidental" dependences[10] in $M$; for instance, three points positioned in the interior of some face of dimension at least 2 must be independent. Let's state the general result as a theorem.

---

[10] This is also what made our use of transcendentals in the incidence matrix proof of Theorem 7.2 work. You can think of the free-simplicial structure as a geometric realization of the transcendentals in our matrix.

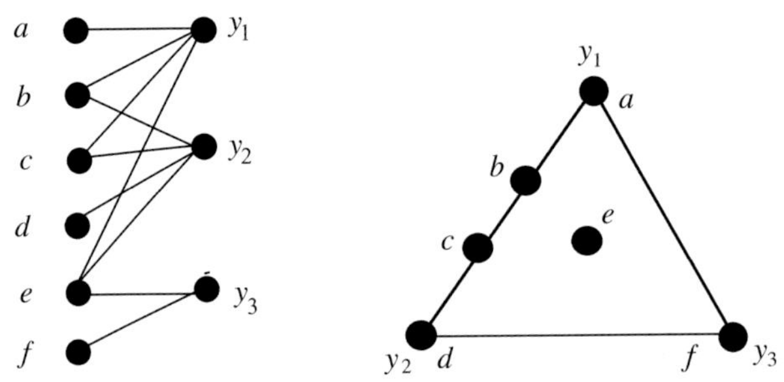

Figure 7.7. The bipartite graph $G_1$ is a minimal presentation of the matroid $M$. The free simplicial picture associated with this presentation is on the right.

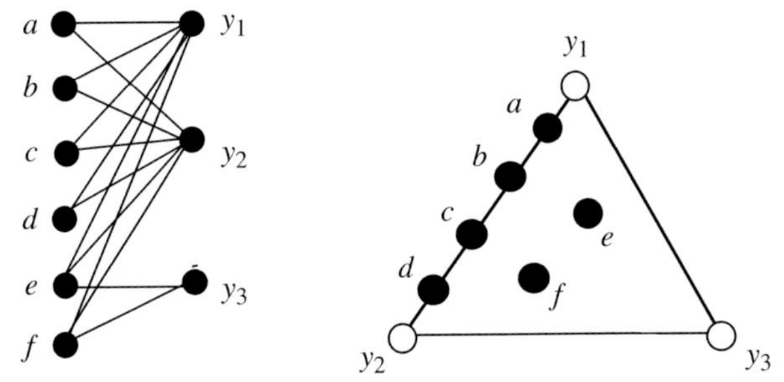

Figure 7.8. The bipartite graph $G_2$ and its associated free-simplicial picture. $G_2$ is a maximal presentation of $M$.

**Theorem 7.13.** *[Brylawski] A matroid M is a transversal matroid if and only if M is isomorphic to a free-simplicial set in Euclidean space.*

*Proof Theorem 7.13.* If $M$ is transversal, use Proposition 7.9 to choose a bipartite presentation in which $|Y| = r(M)$. Now simply use the procedure described in Example 7.11 to create a free-simplicial set based on a simplex $S$ in Euclidean space. Then a set of $d$ points in your free-simplicial set will be independent if and only if that collection of points is on a face of the simplex of dimension at least $d - 1$. But, by the way we've placed the points on our simplex, this happens precisely when that collection of points can be matched in our bipartite graph, i.e., is independent in the transversal matroid.

For the converse, note that we can take a free-simplicial set and reverse the procedure described in Example 7.11. This immediately gives us a bipartite graph, and the transversal matroid associated to that graph is clearly isomorphic to our free-simplicial set in Euclidean space. □

The difference between minimal and maximal presentations is visible in our free-simplicial drawings of the matroid. For instance, using the minimal presentation in the bipartite graph $G_1$ on the left in Figure 7.8, we have the matroid drawn on the right.

The maximal presentation $G_2$ on the left in Figure 7.8 of the same matroid gives us the representation of $M$ shown on the right in that figure.

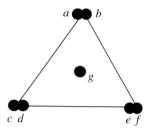

Figure 7.9. A transversal matroid $M$ that produces a non-transversal dual.

We emphasize both of these pictures yield the same matroid: a rank 3 matroid consisting of a four-point line and two points not on that line (Figure 7.2).

Finally, Theorem 7.13 can also be used to tell us a matroid is not transversal. For example, the matroid in Figure 7.3 is not free-simplicial, and no redrawing of this matroid will make it so, so this matroid is (still) not transversal.

### 7.2.4 Transversal matroids, deletion, contraction and duality

Here's a question that may have occurred to you:

• Is the dual of a transversal matroid always transversal?

We've looked at duality for subclasses of matroids before. In particular, we've seen the dual of a graphic matroid need not be graphic (in fact, $M(G)^*$ is graphic if and only if $G$ is planar – this is Theorem 3.27). On the other hand, the dual of a representable matroid is always representable (Theorem 6.6).

Let's start with an easy result involving deletion. You are asked to prove this in Exercise 3.

**Proposition 7.14.** *Suppose M is transversal and e is not an isthmus. Then $M - e$ is also transversal.*

Why don't we look at an example?[11]

**Example 7.15.** Let $M$ be the matroid given in Figure 7.9. Then $M$ is clearly a free-simplicial matroid, so it's transversal. (For some quick practice, draw a bipartite graph $G$ that shows this matroid $M$ is transversal.)

We claim this example produces a matroid whose dual is not transversal. Here is a clever, indirect way to see this.

• Step 1: contract the point $g$. You can check this gives the matroid $M$ in Figure 7.3.
• Step 2: from Example 7.3, we know $M/g$ is not transversal.

---

[11] This is a rhetorical question. Don't try to answer it.

- Step 3: suppose the dual matroid $M^*$ is transversal. Then so is $M^* - g$ (by Proposition 7.14).
- Step 4: now $M^* - g$ is transversal, but its dual $(M^* - g)^* = M/g$ is not.

Thus, we must have one of the following two alternatives:

- $M$ is transversal, but $M^*$ is not, or
- $M^* - g$ is transversal, but $(M^* - g)^*$ is not.

Whichever case holds, we have produced an example of a matroid whose dual is not transversal. (It turns out the second alternative is true; $M^* - g$ and $M^*$ are both transversal – see Exercises 8 and 9.)

There are a few observations that follow from Example 7.15. First, note that $M$ is transversal, but $M/g$ is not. Thus,

- The class of transversal matroids is not closed under contraction, i.e., $M$ transversal does not imply $M/e$ transversal.

Secondly, the example creates a transversal matroid whose dual is not transversal, but we can't tell (from the logical structure of the argument) what that matroid is. This is reminiscent of the following result from number theory:

- There are irrational numbers $a$ and $b$ such that $a^b$ is rational.

Here's an indirect proof: Consider $a = b = \sqrt{2}$. Then, if $\sqrt{2}^{\sqrt{2}}$ is rational, we're done. Otherwise, set $a = \sqrt{2}^{\sqrt{2}}$ and note $a^b = (\sqrt{2}^{\sqrt{2}})^{\sqrt{2}} = 2$. Thus, $a^b$ is rational for some irrational $a$ and $b$, but this argument can't distinguish between the two alternatives.[12]

In summary, we note the class of transversal matroids is closed under deletion, but not contraction or duality. Some of this follows from general matroid theory: if a class of matroids is closed under deletion and duality, then it must be closed under contraction since $M/e = (M^* - e)^*$. (This is one of the two missing cases in Exercise 42 from Chapter 3.)

### 7.2.5  Representations of transversal matroids

When is a transversal matroid representable over a field $\mathbb{F}$? It turns out that it's not too hard to answer this question, at least from a theoretical viewpoint.

First, we can use our free-simplicial embedding of the points of $M$ to guess that transversal matroids are always representable over $\mathbb{R}$. Here's why: your simplex in $\mathbb{R}^n$ has no accidental dependences, and each point

---

[12] It turns out $\sqrt{2}^{\sqrt{2}}$ is irrational, but that follows from a deep result in number theory.

is located at an ordered $n$-tuple in $\mathbb{R}^n$. We know these coordinates give us our affine representation, and that's all we need.

In fact, it should be clear that we can represent any transversal matroid over $\mathbb{Q}$, the rational numbers. Just place the vertices of your simplex at rational points in $\mathbb{R}^n$ and make sure you choose rational coordinates for all the remaining points. An approach that uses some ideas from abstract algebra gives us a bit more. In fact, a consequence of our proof of Theorem 7.2 is that transversal matroids are representable over all field characteristics.

**Theorem 7.16.** *Let M be a transversal matroid. Then there is some positive integer N such that if $|\mathbb{F}| > N$, then M is representable over the field $\mathbb{F}$.*

*Proof Theorem 7.16.* Let $\mathbb{F}$ be any field. The idea behind the proof is pretty simple: first, find a bipartite graph that corresponds to the matroid $M$. Now create an incidence matrix $A$ to represent $M$ as in the proof of Theorem 7.2. In general, the matrix $A$ will involve several transcendentals over the field $\mathbb{F}$. Thus, $A$ represents the matroid $M$ over the extension field $\mathbb{F}(x_1, x_2, \ldots, x_k)$.

To get a representation over the field $\mathbb{F}$, we have to get rid of all the transcendentals. But, in our matrix $A$, the transcendentals serve the purpose of ensuring certain column vectors are linearly independent. Since there are only a finite number of these dependences to avoid, there are only a finite number of subdeterminants that we need to ensure are non-zero. Then, provided the size of the field $|\mathbb{F}|$ is large enough, we can find elements $e_i \in \mathbb{F}$ so that setting $x_i = e_i$ for $1 \leq i \leq k$ does not change any of our linear dependences in $A$. □

## 7.3 Hyperplane arrangements and matroids

Matroids arise from geometry in a variety of ways. One of the most striking is through collections of lines in the plane, planes in $\mathbb{R}^3$, and, more generally, collections of $(d-1)$-dimensional hyperplanes in $\mathbb{R}^d$.

Hyperplanes in $\mathbb{R}^d$ are the solutions to linear equations:

$$H = \{a_1x_1 + a_2x_2 + \cdots + a_dx_d = b \mid a_i, b \in \mathbb{R}\}.$$

Given a collection $\mathcal{A}$ of hyperplanes in $\mathbb{R}^d$, our goal is to define a matroid $M$ with ground set $\mathcal{A}$.

### 7.3.1 Central arrangements and matroids

Let's start with a special type of hyperplane arrangement in Euclidean space, one in which all the hyperplanes have a common intersection.

Central arrangement
Essential arrangement
**Definition 7.17.** An arrangement $\mathcal{A}$ of hyperplanes is *central* if its intersection is non-empty: $\bigcap_{H \in \mathcal{A}} H \neq \emptyset$. A central arrangement is *essential* if $\bigcap_{H \in \mathcal{A}} H = P$, where $P$ is a point in $\mathbb{R}^d$.

As a simple example, the three planes $x = y$, $x = z$ and $y = z$ form a central arrangement. In fact, each of these planes contains the line $x = y = z$ in $\mathbb{R}^3$, so this arrangement is not essential. Note that the three normal vectors $(1, -1, 0)$, $(1, 0, -1)$ and $(0, 1, -1)$ do not span $\mathbb{R}^3$. This is true in general for inessential arrangements:

**Proposition 7.18.** *A central arrangement $\mathcal{A}$ of hyperplanes in $\mathbb{R}^d$ is essential if and only if the collection of normal vectors spans $\mathbb{R}^d$.*

We leave the proof to you – see Exercise 21. To associate a matroid to a central arrangement $\mathcal{A}$ of hyperplanes in $\mathbb{R}^d$, we need to define the independent sets. Define a subset $I$ of $k < d$ hyperplanes to be independent if

$$\dim \left( \bigcap_{H \in I} H \right) = d - k.$$

**Theorem 7.19.** *Let $\mathcal{A}$ be a central arrangement of hyperplanes in $\mathbb{R}^d$. Then $\mathcal{A}$ is the ground set of a matroid, denoted $M(\mathcal{A})$, whose independent sets are those subsets $I$ of $k < d$ hyperplanes with $\dim \left( \bigcap_{H \in I} H \right) = d - k$.*

*Proof Theorem 7.19.* Let $I = \{H_1, H_2, \ldots, H_k\}$ be a collection of $k$ hyperplanes. We claim $I$ is independent if and only if the corresponding $k$ normal vectors are linearly independent. It will then follow that these are the independent sets of a matroid because linearly independent subsets of vectors are the independent sets of a matroid – Theorem 6.1.

We assume the hyperplanes $H_1, H_2, \ldots, H_k$ all go through the origin. This is not a restriction; since the hyperplanes have non-empty intersection, we could just translate all the hyperplanes back to the origin without changing the dimension of the intersection.

Let $x = (x_1, x_2, \ldots, x_n)$. Then the hyperplane $H_i$ has equation $v_i \cdot x = 0$, where $v_i \in \mathbb{R}^d$ is the normal vector for $H_i$ (and $\cdot$ is the ordinary dot product). Let $B$ be the $k \times d$ matrix whose rows are the $k$ normal vectors $v_1, v_2, \ldots, v_k$. Then $\bigcap_{i=1}^{k} H_i = N(B)$, the null space of the matrix $B$. Then, by the Rank–Nullity Theorem,[13] $n(B) + r(B) = d$, where $n(B)$ is the nullity of $B$ (the dimension of the null space) and $r(B)$ is the matrix rank of $B$.

Thus $n(B) = d - k$ if and only if $r(B) = k$, i.e., if and only if the rows of $B$ are linearly independent.    □

---

[13]  See Section 5.1.2 of Chapter 5 for a review.

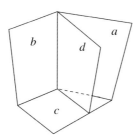

Figure 7.10. Central
hyperplane arrangement
in $\mathbb{R}^3$

Note: the word "hyperplane" has two meanings here. It can be a
$(d-1)$-dimensional subspace of $\mathbb{R}^d$, or it can be a matroid hyperplane,
i.e., a maximal non-trivial flat of the matroid. This is unfortunate and
unavoidable, but it shouldn't trouble us here: throughout this section,
we are using hyperplanes in the former, not the latter, sense. (It's worth
observing that the matroid hyperplanes of a central, essential hyperplane arrangement are *lines* in $\mathbb{R}^d$ corresponding to intersections of
hyperplanes in the arrangement.)

The matroid rank of a subset of hyperplanes in $\mathbb{R}^d$ is just the *codi-
mension* of the intersection of that subset:

**Proposition 7.20.** *Let A be a collection of hyperplanes in a central
arrangement $\mathcal{A}$. Then*

$$r(A) = codim \left( \bigcap_{H \in A} H \right) = d - \dim \left( \bigcap_{H \in A} H \right).$$

The proof follows from our interpretation via the normal vectors –
the proof of Theorem 7.19. Here's an easy example.

**Example 7.21.** Consider the four planes in $\mathbb{R}^3$ with equations $x =
0, y = 0, z = 0$ and $x - y = 0$. Then the corresponding normal vec-
tors are $(1, 0, 0)$, $(0, 1, 0)$, $(0, 0, 1)$ and $(1, -1, 0)$. We abuse notation by
using $a, b, c$ and $d$ to mean three different things here: they label the
hyperplanes themselves, the corresponding normal vectors, and also the
elements of the matroid. (To be a bit pedantic, we should probably say
the hyperplane $H_a$ has normal vector $v_a$, and corresponds to the matroid
element $a$. We hope this does not lead to any confusion.[14])

We give a picture of the planes in $\mathbb{R}^3$ in Figure 7.10. Then the normal
vectors give the matroid $M(\mathcal{A})$ shown in Figure 7.11. Note that $c$ is an
isthmus – in fact, this normal vector is orthogonal to the plane determined
by the other three normals.

By the way, note that the matroids associated to central hyperplane
arrangements are *simple*, i.e., they have no loops or multiple points. This

---

[14] Well, any *additional* confusion. Some confusion is tolerated, at all times, in all math
books.

Figure 7.11. Normal vectors and a picture of the associated matroid $M(\mathcal{A})$ for the arrangement $\mathcal{A}$ in Figure 7.10. The vectors $a$, $b$ and $d$ are coplanar, and $c$ is orthogonal to the plane determined by the vectors $a$, $b$ and $d$.

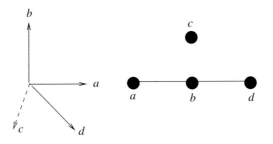

Figure 7.12. The intersection lattice for the hyperplanes of Figure 7.10 is the geometric lattice for $M(\mathcal{A})$. The order is *reverse inclusion* for the intersection of the hyperplanes, where $abd$ denotes the intersection $a \cap b \cap d$, and so on.

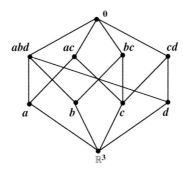

is not a deep observation – loops would correspond to hyperplanes with normal vector $(0, 0, \ldots, 0)$[15] and multiple points would correspond to hyperplanes with parallel normals, i.e., identical hyperplanes.

One more easy connection with matroids: the *flats* of the matroid associated to a central arrangement correspond to the various intersections of hyperplanes. This makes it very easy to interpret the geometric lattice of flats for these matroids. For the arrangement in Figure 7.10, we get the lattice of Figure 7.12. (The poset of intersections of the hyperplanes, ordered by reverse inclusion, is a well-studied object in combinatorial topology.)

What if a central arrangement of hyperplanes in $\mathbb{R}^d$ is not essential? Then the intersection of all the hyperplanes is a line, plane, or some higher-dimensional subspace. You still get a matroid, but the rank of the matroid is less than $d$. See Exercise 20.

One consequence of the normal vector–hyperplane arrangement connection: the matroids associated to central hyperplane arrangements are just the matroids representable over the reals. In fact, the entire theory is contained in the theory of representable matroids. So, not to make too much of this, why are we bothering you with hyperplane arrangements?

Here are a few reasons we made up in response to this question:

(1) Hyperplane arrangements are important in the study of *oriented matroids*, an important field that is closely related to topology in $\mathbb{R}^n$.

---

[15] Which is just plain silly.

Table 7.3. *The nine equations and normal vectors for planes of symmetry for a cube.*

| Equation | Normal | Equation | Normal | Equation | Normal |
|---|---|---|---|---|---|
| $x = 0$ | $(1, 0, 0)$ | $x + y = 0$ | $(1, 1, 0)$ | $x - y = 0$ | $(1, -1, 0)$ |
| $y = 0$ | $(0, 1, 0)$ | $x + z = 0$ | $(1, 0, 1)$ | $x - z = 0$ | $(1, 0, -1)$ |
| $z = 0$ | $(0, 0, 1)$ | $y + z = 0$ | $(0, 1, 1)$ | $y - z = 0$ | $(0, 1, -1)$ |

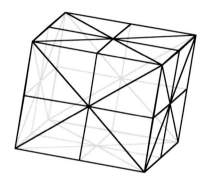

Figure 7.13. The nine planes of symmetry for a cube.

(2) The complement of an arrangement of hyperplanes $\mathbb{R}^d - \bigcup_{H \in \mathcal{A}}$ is an open subset of $\mathbb{R}^d$, and topologists are interested in studying these arrangements. The Orlik–Solomon algebra is, for example, an important invariant associated to an arrangement of hyperplanes.

(3) Arrangements arise in enumeration problems: the following problem is the prototype:

> Into how many pieces can one divide a cube of cheese using $n$ straight cuts with a knife?

We will study some generalizations of this problem in Section 7.4.

(4) Several interesting groups have associated Cayley graphs whose geometric duals are the open chambers of $\mathbb{R}^d - \bigcup_{H \in \mathcal{A}}$ for some central arrangement. Studying the arrangements can give insight into the average word length in a Coxeter group, for instance.

(5) Hyperplane arrangements can give you some very pretty pictures.

To put an exclamation point on the last reason, here's something pretty to look at.

**Example 7.22.** Consider the cube in Figure 7.13. Then you can check that there are nine planes in $\mathbb{R}^3$ that act as planes of symmetry for a cube. These symmetry planes can be thought of as *planes of reflection* or *mirrors* for the cube; for us, they are a central hyperplane arrangement.

If we place our cube in $\mathbb{R}^3$ so that the eight vertices are at the points $(\pm 1, \pm 1, \pm 1)$, then we can write down equations for our planes. For each such plane, we give its equation and a normal vector in Table 7.3.

Figure 7.14. The matroid $M(\mathcal{A})$ is built from the complete graph $K_4$ by adding points $a$, $b$ and $c$ as intersections of lines in $M(K_4)$.

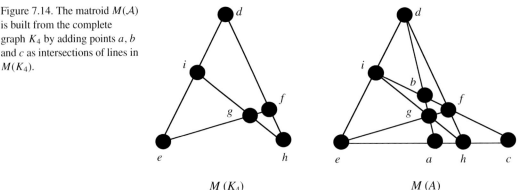

$$M(K_4) \qquad\qquad\qquad M(\mathcal{A})$$

The associated matroid $M(\mathcal{A})$ is the matroid represented by the matrix of normals:

$$
\begin{array}{ccccccccc}
a & b & c & d & e & f & g & h & i
\end{array}
$$
$$
\begin{bmatrix}
1 & 0 & 0 & 1 & 1 & 0 & 1 & 1 & 0 \\
0 & 1 & 0 & 1 & 0 & 1 & -1 & 0 & 1 \\
0 & 0 & 1 & 0 & 1 & 1 & 0 & -1 & -1
\end{bmatrix}.
$$

We draw the matroid $M(\mathcal{A})$ as follows. First, note that the column vectors corresponding to the six vectors $d, e, f, g, h, i$ correspond to the graphic matroid $M(K_4)$. Now locate the points $a, b$ and $c$ as points of intersection of lines in that matroid, as in Figure 7.14:

$$a = \overline{eh} \cap \overline{dg} \qquad b = \overline{fi} \cap \overline{dg} \qquad c = \overline{eh} \cap \overline{fi}.$$

It's interesting to compare the symmetry of the cube with the symmetry of this matroid. See Exercise 25 for one such connection. For some fun with the icosahedron, see Exercise 26.

Let $S$ be the set of nine normal vectors in Example 7.22. Then the 18 vectors $\pm S$ form the *root system* $B_3$. Root systems[16] are collections of vectors that have beautiful symmetry properties, and they are important in certain algebraic classification problems.

### 7.3.2 Non-central arrangements

We can use central arrangements to obtain non-central arrangements whose representations are a bit more efficient. Here's the main idea.

- Given a central arrangement of hyperplanes $\mathcal{A}$ in $\mathbb{R}^d$, assume all of your hyperplanes contain the origin.

---

[16] Research question: find all botanical references in this text.

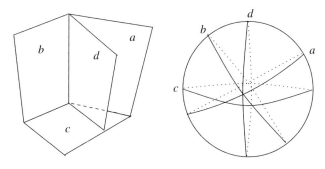

Figure 7.15. A central arrangement (left) and the corresponding spherical arrangement (right).

- Now add the unit sphere[17] in $\mathbb{R}^d$ to your collection of hyperplanes. Each hyperplane will meet the unit $d$-dimensional sphere in a $(d-1)$-dimensional sphere.
- Throw away the hyperplanes and keep all the $(d-1)$-dimensional spheres of intersection.

This simple procedure gives a *spherical arrangement*. In $\mathbb{R}^3$, you get a collection of great circles on an ordinary sphere. Figure 7.15 shows the spherical arrangement corresponding to the central hyperplane arrangement given in Figure 7.10.

Spherical arrangement

Since the central arrangement gave us a matroid, the spherical arrangement does, too. But, we can simplify this picture even more. Note that there is a two-fold redundancy in the spherical arrangement: we can slice the sphere into two hemispheres, and the picture of intersections is exactly the same in both hemispheres. Thus, all the matroid information is encoded in just one hemisphere. We accomplish this geometrically with one more step:

- Given the spherical arrangement from above, find a hyperplane (a "hyperplane at infinity") that is not in $\mathcal{A}$ which divides the sphere into two hemispheres. Then just keep one of the hemispheres and identify antipodal points on the boundary.

This final result is a *projective arrangement*. The procedure outlined here is entirely analogous to the procedure we used in Chapter 5 to define the projective space $PG(n, q)$ from the affine space $AG(n+1, q)$. In Figure 7.16, the hyperplane at infinity is the bounding circle.

Projective arrangement

Finally,[18] if we are given a projective arrangement in some hemisphere of $S^{d-1}$, we can "flatten" this hemisphere out to get a non-central arrangement in $\mathbb{R}^{d-1}$. This last arrangement is called an *affine arrangement*.

Affine arrangement

---

[17] The unit sphere $S^{d-1} = \{v \in \mathbb{R}^d \mid ||v|| = 1\}$. So $S^2$ is the surface of a ball in $\mathbb{R}^3$, for example.

[18] This time we mean it.

Figure 7.16. Left: the projective
arrangement corresponding to
the arrangement in Figure 7.10,
with the "hyperplane at
infinity" the bounding circle.
Right: the affine arrangement is
a flattened version which is a
non-central arrangement in $\mathbb{R}^2$.

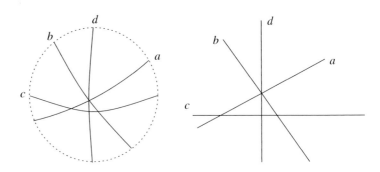

Figure 7.16. Left: the projective
arrangement corresponding to
the arrangement in Figure 7.10,
with the "hyperplane at
infinity" the bounding circle.
Right: the affine arrangement is
a flattened version which is a
non-central arrangement in $\mathbb{R}^2$.

Here's the punch line: we started with a central arrangement in $\mathbb{R}^d$ and ended with a non-central arrangement in $\mathbb{R}^{d-1}$. These arrangements are equivalent in the sense that they define the same matroid.

$$\text{Central in } \mathbb{R}^d \mapsto \text{Spherical in } S^{d-1}$$
$$\mapsto \text{Projective in } PG(d, \mathbb{R}) \mapsto \text{Affine in } \mathbb{R}^{d-1}.$$

When can this process be reversed? The answer: sometimes. If the non-central arrangement has no *parallel intersections*, we can build a central arrangement in $\mathbb{R}^d$ that corresponds to a given non-central arrangement in $\mathbb{R}^{d-1}$.

Parallel intersections    **Definition 7.23.** A collection of hyperplanes $\mathcal{A}$ in $\mathbb{R}^d$ has *parallel intersections* if the intersection of some subset of hyperplanes is a subspace $W$ with $\dim(W) \geq 1$ and $W$ is parallel to $H$ for some $H \in \mathcal{A}$, i.e., $W \cap H = \emptyset$.

Then it's clear that if you have a parallel intersection in your non-central arrangement, you cannot find a central arrangement to lift to. On the other hand, any non-central arrangement that avoids these parallel intersections will lift. We summarize the discussion with a proposition, and we leave the proof to you – Exercise 23.

**Proposition 7.24.**

(1) *Suppose $\mathcal{A}$ is a central arrangement of hyperplanes in $\mathbb{R}^d$. Then the associated affine arrangement $\mathcal{A}'$ in $\mathbb{R}^{d-1}$ has no parallel intersections.*

(2) *Conversely, suppose $\mathcal{A}'$ is a non-central arrangement of hyperplanes in $\mathbb{R}^{d-1}$ with no parallel intersections. Then there is a central arrangement of hyperplanes $\mathcal{A}$ in $\mathbb{R}^d$ so that $\mathcal{A}'$ is the affine arrangement associated to $\mathcal{A}$.*

In particular, this allows us to define a matroid structure on a family of lines in the plane, provided every pair of lines intersect. We'll look at line configurations in the next section.

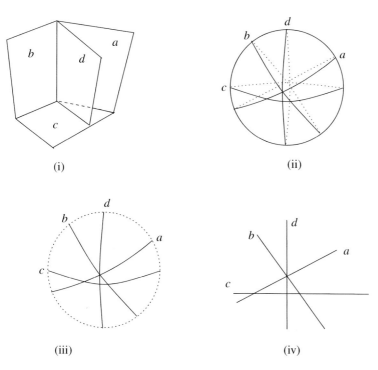

Figure 7.17. Equivalent hyperplane arrangements: (i) central; (ii) spherical; (iii) projective; (iv) non-central affine.

**Remark.** The difference between projective and affine arrangements is consistent with our equations for hyperplanes in $PG(n, q)$ and $AG(n, q)$ from Chapter 5. In particular, $H$ is a projective hyperplane if and only if $H$ satisfies an equation of the form $a_1 x_1 + a_2 x_2 + \cdots + a_d x_d = 0$, while $H$ is an affine hyperplane if and only if $H$ satisfies an equation of the form $a_1 x_1 + a_2 x_2 + \cdots + a_d x_d = b$, for some real scalars $a_1, a_2, \ldots, a_d$ and $b$. (See the comments preceding Table A.5.)

Figure 7.17 gives a tidy summary of this discussion.

## 7.4 Cutting cheese; counting regions via deletion and contraction

We start with a question that's a favorite for mathematics contests.

- What is the largest number of regions produced when $n$ lines are drawn in the plane?

This problem is also useful when introducing mathematical induction.

The three-dimensional version appeared as Problem E554 in the *American Mathematical Monthly* in 1943, where J. L. Woodbridge of Philadelphia asked:

- Show that $n$ cuts can divide a cheese into as many as $(n + 1)(n^2 - n + 6)/6$ pieces.

Figure 7.18. Five lines in
general position: no two are
parallel, and no three meet at a
point.

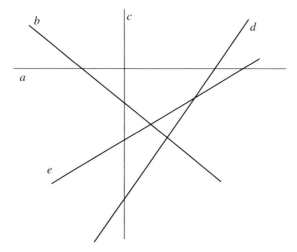

Figure 7.18. Five lines in general position: no two are parallel, and no three meet at a point.

Both of these questions are answered by a beautiful, general formula[19] discovered by L. Schläfli[20] (published posthumously in 1901):

- The largest number of regions produced when $n$ hyperplanes are drawn in $d$-dimensional space equals $\displaystyle\sum_{k=0}^{d} \binom{n}{k}$.

Our goal in this section is to give a matroid interpretation to this formula, use a variation to count *bounded* regions, and then to adapt it to count regions in *central* arrangements. We concentrate on affine, non-central arrangements, but give detailed outlines for the counts for several other cases in Exercises 27, 28, 29 and 30. Throughout, assume $\mathcal{A}$ is an affine, non-central arrangement, with no parallel intersections. Let's start with lines in the plane.

**Example 7.25.** Consider the five lines in *general position* (so no two lines are parallel and no three meet at a point) in Figure 7.18. This is an affine arrangement of hyperplanes in $\mathbb{R}^2$, so it gives us a matroid $M$ on five points. What is the matroid?

Well, lifting to a central arrangement in $\mathbb{R}^3$, every line corresponds to a plane through the origin. Then three observations will tell us who $M$ is.

(1) Since the lines are not parallel to each other, the five planes are all distinct.

---

[19] This formula is sometimes attributed to R. C. Buck, who published a short paper on this topic in the *American Mathematical Monthly* in 1943. Buck is best known for his *Advanced Calculus* text, which is still an excellent resource.

[20] Ludwig Schläfli (1814–1895) did pioneering work in geometry, discovering the six regular solids in four dimensions by 1855. Schläfli symbols are still used to classify regular polytopes.

(2) No three lines meet at a point. That means, given any three lines, the normal vectors for the corresponding planes are linearly independent.
(3) Since the normal vectors for all the planes live in $\mathbb{R}^3$, any subset of four (or more) normals will be linearly dependent. Thus $r(M) = 3$.

Now $M$ can honestly say "I am a rank 3 matroid on five points, and every subset of size 3 is independent. Who am I?"

You should be able to answer this question with a question:[21] "You must be the uniform matroid $U_{3,5}$, mustn't you?"

It turns out that our old friends, deletion and contraction, will play an important role here. We wish to describe how to delete and contract a hyperplane from an arrangement; this should give us two new arrangements, each with one fewer hyperplane. Then we can use induction to solve our number-of-regions problem for the original arrangement.

As we did for graphs and matrices, if $x \in M(\mathcal{A})$ is a point in the matroid (and so corresponds to a hyperplane $H_x$ in the arrangement), we would like to represent $M(\mathcal{A}) - x$ and $M(\mathcal{A})/x$ as hyperplane arrangements. In particular, we seek arrangements $\mathcal{A}'$ and $\mathcal{A}''$ so that

$$M(\mathcal{A}') = M(\mathcal{A}) - x \qquad \text{and} \qquad M(\mathcal{A}'') = M(\mathcal{A})/x.$$

Deletion and contraction in hyperplane arrangements

As usual, deletion is easy – just erase: $\mathcal{A}' = \mathcal{A} - H_x$ will satisfy $M(\mathcal{A}') = M(\mathcal{A}) - x$. (Question: can erasing the hyperplane $H_x$ reduce the matroid rank? Answer: not unless the normal vector for $H_x$ is an isthmus in the matroid $M(\mathcal{A})$. We assume $x$ is not an isthmus in this case. For an example, see Figure 7.10, where the hyperplane $c$ is an isthmus.)

For the contraction $M(\mathcal{A})/x$, note that every hyperplane $H \in \mathcal{A} - H_x$ intersects $H_x$ in a $(d-2)$-dimensional flat (translate of a subspace) of $\mathbb{R}^d$. This gives a new hyperplane arrangement $\mathcal{A}'' = \{H \cap H_x \mid H \in \mathcal{A}\}$. This time, we think of $H_x$ as the $(d-1)$-dimensional home of the arrangement, and $\mathcal{A}''$ as an affine hyperplane arrangement in $H_x$. This will give us $M(\mathcal{A}'') = M(\mathcal{A})/x$. (Note that the rank drops by one in this construction, which is what we expected, since $r(M/x) = r(M) - 1$.)

We summarize this in the next proposition; the proof is left to you in Exercise 24.

**Proposition 7.26.** *Let $\mathcal{A}$ be an affine arrangement of hyperplanes in $\mathbb{R}^d$, and let $H_x \in \mathcal{A}$. Let $\mathcal{A}' = \mathcal{A} - H_x$ and $\mathcal{A}'' = \{H \cap H_x \mid H \in \mathcal{A}\}$. Then*

$$M(\mathcal{A}') = M(\mathcal{A}) - x \qquad \text{and} \qquad M(\mathcal{A}'') = M(\mathcal{A})/x.$$

---

[21] Answering questions with questions is a fun, occasionally annoying game. It has (almost) no place in a math text.

Figure 7.19. Two arrangements
derived from the arrangement
$\mathcal{A}$ in Figure 7.18. Left: deleting
$a$. Right: contracting $a$.

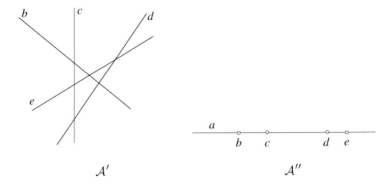

$\mathcal{A}'$

$\mathcal{A}''$

Returning to Example 7.25, we get the arrangements $\mathcal{A}'$ and $\mathcal{A}''$ in
Figure 7.19. You can check the associated matroids $M(\mathcal{A}') = U_{3,4}$ and
$M(\mathcal{A}'') = U_{2,4}$, which are $U_{3,5} - a$ and $U_{3,5}/a$, respectively.

Back to counting regions. Note that every region of $\mathcal{A}$ either includes
part of the line $a$ along its boundary, or it doesn't.[22] We group the regions
determined by $\mathcal{A}$ into three categories.

I.  The regions that have $a$ as part of their boundary, all having $a$ on
the same side. (This means any line from the interior of one of these
regions to another such region will not pass through the hyperplane $a$.)

II.  These regions are also bounded by $a$, but on the opposite side of $a$
from the regions in I. (So any line from the interior of a region in
this class to the interior of a region in class I *must* pass through $a$.)

III.  These regions do not have the hyperplane $a$ along their boundary.

In Figure 7.20, we show these three classes for the arrangement from
Figure 7.18. Note the number of regions in class I is the same as the
number in class II. In fact, we can't tell the difference between these
two classes, but that won't matter; these two classes will always have
the same number of regions. Once we place a region in class I, say,
everything else is uniquely placed.

Now let $c(\mathcal{A})$ be the number of regions determined by the hyperplanes
in the arrangement $\mathcal{A}$. We also let $c_1$, $c_2$ and $c_3$ denote the number of
regions of type I, II and III, respectively.

Let's count! Here are three relations that must hold in any affine
arrangement:

$$c(\mathcal{A}) = c_1 + c_2 + c_3 \qquad c(\mathcal{A}') = c_2 + c_3 \qquad c(\mathcal{A}'') = c_1.$$

---

[22] This observation is not deep; every statement is either true or its negation is true.
Aristotle formulated this as the Law of the Excluded Middle, and all of logic is based
on it.

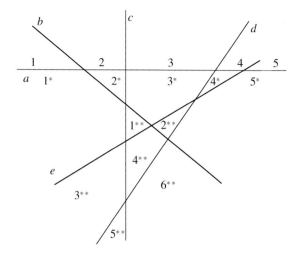

Figure 7.20. The three kinds of regions in an arrangement. Regions $\{1, 2, 3, 4, 5\}$ are bounded by $a$, and are all above $a$. Regions $\{1^*, 2^*, 3^*, 4^*, 5^*\}$ are also bounded by $a$, but on the opposite side of $a$ from the regions in I. Regions $\{1^{**}, 2^{**}, \ldots, 6^{**}\}$ are not bounded by $a$.

Why is this always true? Well, it should be clear that every region of $\mathcal{A}''$ can be thought of as a region in class I, since we could choose one side of $a$ and expand each region in $\mathcal{A}''$ to a region in class I. Thus, $c(\mathcal{A}'') = c_1$.

For $c(\mathcal{A}')$, we note that each region of $\mathcal{A}'$ is either entirely disjoint from $a$ (and so is in class III), or has $a$ divide its interior into two pieces. This means the hyperplane $a$ adds $c_1$ new regions to $\mathcal{A}'$, so $c(\mathcal{A}') = c(\mathcal{A}) - c_1 = c_2 + c_3$.

Evidently, we just proved the following nice recursive formula.

**Theorem 7.27.** *Let $c(\mathcal{A})$ be the number of regions determined by the affine hyperplane arrangement $\mathcal{A}$, and let $\mathcal{A}'$ and $\mathcal{A}''$ be defined as above. Then $c(\mathcal{A}) = c(\mathcal{A}') + c(\mathcal{A}'')$.*

In Figure 7.20, we have $c(\mathcal{A}) = 16$, $c(\mathcal{A}') = 11$ and $c(\mathcal{A}'') = 5$. An interpretation of Theorem 7.27 in terms of the *Tutte polynomial* appears in Exercise 22 of Chapter 9.

We now return to the task of finding a general formula for the number of regions. Let's first find a formula in $\mathbb{R}^2$: we want the maximum number of regions $n$ lines can divide the plane into. Some more notation: let $c(n, d)$ be the maximum number of regions determined by $n$ hyperplanes in $\mathbb{R}^d$. (The hyperplanes will be in *general position*, i.e., there will be no parallel intersections, and the intersection of any $d + 1$ hyperplanes will be $\emptyset$.)

We want the formula for $c(n, 2)$. Let's first make a chart with some small values of $n$ – see Table 7.4.

How does Theorem 7.27 help us figure out a formula for the number of regions? According to the theorem, we should have

$$c(n, 2) = c(n - 1, 2) + c(n - 1, 1).$$

Table 7.4. *The number of regions determined by n lines in general position in the plane.*

| $n$ | 0 | 1 | 2 | 3 | 4 | 5 |
|---|---|---|---|---|---|---|
| # regions | 1 | 2 | 4 | 7 | 11 | 16 |

Now $c(n-1, 1)$ is just the number of line segments determined by $n-1$ points on a line, so $c(n-1, 1) = n$. Thus, we get $c(n, 2) = c(n-1, 2) + n$. (This is pretty easy to guess without the theorem – just stare at the sequence $1, 2, 4, 7, 11, 16$.)

Finally, how can we solve this recursive formula to get a nice expression for $c(n, 2)$? There are lots of ways to solve recurrence relations, and a thorough treatment would take us far afield. For now, let's just use *back substitution* to solve this recursion:

$$c(n, 2) = c(n-1, 2) + n$$
$$= c(n-2, 2) + (n-1) + n$$
$$\vdots$$
$$= c(0, 2) + 1 + 2 + \cdots + (n-1) + n.$$

Now $c(0, 2) = 1$, so our formula is $c(n, 2) = \dfrac{n^2 + n + 2}{2}$. You can check this agrees with Schläfli's formula:

$$\frac{n^2 + n + 2}{2} = \binom{n}{0} + \binom{n}{1} + \binom{n}{2}.$$

Theorem 7.27 can also be used to *prove* a formula is correct, provided we already have the formula. We give a careful inductive proof, just for practice.

**Proposition 7.28.** *Suppose n lines are drawn in the plane, with no two parallel and no three lines coincident. Then these n lines divide the plane into $c(n, 2) = \dfrac{n^2 + n + 2}{2}$ regions.*

*Proof Proposition 7.28.* By induction on $n$. When $n = 1$, you can check $\frac{1^2 + 1 + 2}{2} = 2$, which is correct: one line divides the plane into two regions.

Now assume the result is true for any configuration of $n-1$ lines in the plane, no three of which are coincident, and suppose $\mathcal{A}$ is a collection of $n$ lines in the plane, no two parallel and no three coincident.

Then, by Theorem 7.27, we have $c(\mathcal{A}) = c(\mathcal{A}') + c(\mathcal{A}'')$. From Proposition 7.26 and the definition of $c(n, d)$, we know $c(\mathcal{A}) = c(n, 2)$,

$c(A') = c(n - 1, 2)$ and $c(A'') = c(n - 1, 1)$. Thus,

$$c(n, 2) = c(n - 1, 2) + c(n - 1, 1)$$
$$= c(n - 1, 2) + n$$
$$= \frac{(n - 1)^2 + (n - 1) + 2}{2} + n \qquad \text{(by induction)}$$
$$= \frac{n^2 + n + 2}{2}.$$

$\square$

The rest of the region counting is left for your pleasure in the exercises.[23] In Exercise 27, you are asked to prove Schläfli's formula for general $n$ and $d$. Exercise 28 gives the outline for two proofs that the number of *bounded* regions is $\binom{n-1}{d}$. A different approach to region counting for general non-central arrangements appears in Exercise 22 of Chapter 9.

Exercise 29 shows how to adapt these counting methods to count the number of regions in a central arrangement of hyperplanes in general position. (We interpret *general position* for a central arrangement all containing the origin in $\mathbb{R}^d$ to mean the intersection of every set of $d$ hyperplanes is the origin.)

Finally, in Exercise 30, we (ask you to) show that the number of *unbounded* regions in a non-central arrangement equals the *total* number of regions in an assoiciated central arrangement (where both arrangements are assumed to be in general position). Enjoy!

## Exercises

### Section 7.1 – Transversal matroids

(1) In the bipartite graph in Figure 7.21, give two examples of each of the following. All of your answers should be subsets of the ground set $X = \{a, b, \ldots, g\}$. Give short explanations as you see fit.

 (a) $I$ is independent.

 (b) $B$ is a basis.

 (c) $C$ is a circuit.

 (d) $F$ is a flat.

 (e) $H$ is a hyperplane.

(2) For the bipartite graph in Figure 7.21, find a different bipartite graph that represents the same matroid $M$.

(3) [Proposition 7.14] Suppose $M$ is transversal and $e$ is not an isthmus. Show that the deletion $M - e$ is also transversal by finding a bipartite graph that represents $M - e$.

---

[23] This will be much more fun, and the arguments and results will stick with you longer, if you work through the details yourself. As with everything mathematical. And non-mathematical.

Figure 7.21. Bipartite graph for
Exercises 1 and 2.

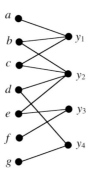

(4) Let $G$ be a bipartite graph with vertices $V = (X, Y)$ and let $M$ be
the transversal matroid based on $G$.
  (a) Show $x$ is an isthmus of $M$ if and only if $R^{-1}(y) = \{x\}$ for
    some $y \in Y$.
  (b) Show $x$ is a loop of $M$ if and only if $R(x) = \emptyset$.

## Section 7.2 – Hall's Theorem and free-simplicial representations

(5) Show that any matroid on five or fewer points is transversal, justi-
fying the claim that the matroid in Figure 7.3 is a smallest example
of a non-transversal matroid.

(6) Are the Fano and non-Fano matroids transversal? If so, find bipartite presentations; if not, explain why not.

(7) Suppose $M$ is a transversal matroid on the bipartite graph $G$ with
vertex partition $V = (X, Y)$.
  (a) Use Hall's Marriage Theorem to show $C$ is a circuit in $M$ if
    and only if $|R(C)| = |C| - 1$, and, for all $x \in C$, $C - x$ can
    be matched.
  (b) Suppose $|R(C)| = |C| - 1$, and, for all $x \in C$, $|R(C - x)| =$
    $|C| - 1$. Give an example to show $C$ need not be a circuit
    of $M$.

(8) Let $M$ be the non-transversal matroid in Figure 7.3. Show that the
matroid $M^*$ is transversal by finding a bipartite graph representing
$M^*$. Note this gives a direct argument that resolves the uncertainty
in Example 7.15.

(9) Let $M$ be the matroid in Figure 7.22. (This is the same matroid we
considered in Example 7.15.)
  (a) Show $M$ is transversal by constructing a bipartite presentation
    for $M$.
  (b) Show $M$ is graphic, and can be represented by a planar
    graph $G$.
  (c) Find the dual graph $G^*$ with $M(G^*) = M^*$, and show $M^*$ is
    transversal by finding an appropriate bipartite graph.

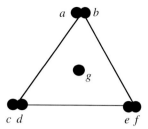

Figure 7.22. $M$ and $M^*$ are both transversal, but can be used to produce a transversal matroid with a non-transversal dual. See Exercise 9.

Note: this implies the deletion $M^* - g$ is also transversal. Then the argument given in Example 7.15 shows $M^* - g$ and $(M^* - g)^* = M/g$ are a pair of matroids where one is transversal and the other is not. (This agrees with Exercise 8.)

(10) Find an example of a graphic matroid that is not transversal. Then find an example of a transversal matroid that is not graphic.[24]

(11) Suppose $G$ is a bipartite graph with vertices $V = (X, Y)$, and suppose the transversal matroid $M$ has $r(M) = |Y|$.
   (a) Prove that if $F = R^{-1}(S)$ for some subset $S \subseteq Y$, then $F$ is a flat of the transversal matroid $M$. Show the converse is false by finding a counterexample.
   (b) Prove that if $H = R^{-1}(Y - y)$ for some $y \in Y$, then $H$ is a hyperplane of $M$. Find a counterexample to show the converse is false.
   (Hint: these are straightforward when you think about free-simplicial representations of $M$.)

(12) Suppose $G$ is a bipartite graph with vertices $V = (X, Y)$ and transversal matroid $M$. Let $A \subseteq X$.
   (a) Show $A \subseteq R^{-1}(R(A))$.
   (b) Show $\overline{A} \subseteq R^{-1}(R(A))$.
   (c) Suppose $r(A) = |R(A)|$, i.e., the matroid rank of $A$ equals the number of elements that are joined to $A$ in $G$. Show $\overline{A} = R^{-1}(R(A))$.
   (d) Suppose $r(A) < |R(A)|$. Show that the conclusion in part (c) need not be true.
   (e) Use part (c) and Lemma 7.10(2) to prove that if $F$ is a cyclic flat, then $F = R^{-1}(R(F))$.

(13) Let $G$ be a bipartite graph with vertices $V = (X, Y)$. In this problem, you will show the subsets of $X$ that can be matched satisfy the independent set axiom (I3) "directly" from the bipartite graph. This will give us another proof that transversal matroids are matroids. Let $I$ and $J$ be subsets of $X$ that can be matched, with $|I| < |J|$.

---

[24] You can do these in either order. How would anyone know?

Let $E_I$ and $E_J$ be the edges of some matching of $I$ and $J$, resp., and consider the subgraph $G'$ with edges $E_I \cup E_J$.

(a) Show that every vertex in $G'$ has degree 1 or 2.

(b) Show that each connected component of $G'$ is a path or a cycle.

(c) Use the fact that $|J| > |I|$ to show that there must be at least one component of $G'$ that consists of a path $x_1 \rightarrow y_1 \rightarrow x_2 \rightarrow y_2 \rightarrow \cdots \rightarrow x_n \rightarrow y_n$, where
  - $x_1 \in J - I$, and
  - $x_i \in I \cap J$ for $2 \le i \le n$.

(d) Now use your path to show $I \cup x_1$ can be matched in $G'$. Conclude that $J$ can be matched. This is the *augmenting path algorithm*.

(14) Let $G$ be a bipartite graph with vertices $V = (X, Y)$ and suppose $S$ and $T$ are maximal subsets of $X$ that can be matched into $Y$.

(a) Show that $|S| = |T|$. (This tells you that *maximal* and *maximum* are the same for matchings. Hint: think about matroid basis axiom (B2').)

(b) Give a direct proof (i.e., one that does not rely on the matroid basis properties) that the maximal transversals of a bipartite graph all have the same size. (Hint: see the procedure described in Exercise 13.)

(15) Let $F$ be a cyclic flat in a matroid $M$, i.e., $F$ is a flat and $F$ is a union of circuits of $M$. Let $B$ be a basis for $F$ and let $x \in B$. Show that $x$ is in some basic circuit with respect to the basis $B$. (Hint: show $|F - \overline{B - x}| \ge 2$.) [Note: we used this fact in our proof of Lemma 7.10(2).]

### The next two problems are concerned with matchings, not matroids

(16) (Due to Laszlo Babai) Beyond the seven seas there is a tiny island, six square miles in all. The island is inhabited by six native tribes and by six turtle species. Each tribe and each turtle species occupies exactly one square mile of territory. The territories of the tribes don't overlap with one another; nor do the territories of the different turtle species. Each tribe wishes to select a totem animal from among the turtle species found in the tribe's territory; and each tribe must have a different totem animal. Prove that such a selection is always possible. (Hint: Hall's Theorem.)

(17) (Due to Laszlo Babai) A *permutation matrix* is an $n \times n$ matrix in which every row and every column has exactly one entry equal to 1, and all the other entries are zero. An $n \times n$ matrix is *doubly stochastic* if all entries are non-negative and all row sums and all column sums are 1. A *convex combination of vectors* is a linear combination where each coefficient is non-negative and the sum of the coefficients is 1.

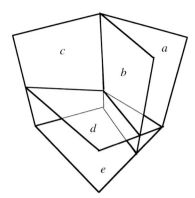

Figure 7.23. An arrangement of five planes for Exercises 18 and 19.

(a) How many $n \times n$ permutation matrices are there?

(b) Show that all permutation matrices are doubly stochastic and that any convex combination of doubly stochastic matrices is doubly stochastic.

(c) Prove that every doubly stochastic matrix is a convex combination of permutation matrices. (Hint: think about the turtles from Exercise 16.)

### Section 7.3 – Hyperplane arrangements

(18) Consider the five planes $\{a, b, c, d, e\}$ of Figure 7.23. In the arrangement, the planes $a, b, c$ all meet along a line, as do planes $a, d, e$. Show the arrangement is central, and draw a picture of the associated matroid. Is this matroid graphic? Have you seen this matroid before?[25]

(19) For the (central) arrangement of Figure 7.23, draw the associated affine (non-central) arrangement of five lines in the plane. Find the number of regions in both arrangements. (Hint for all parts: get a tennis ball and some rubber bands.)

(20) Suppose $\mathcal{A}$ is a central arrangement of hyperplanes in $\mathbb{R}^d$, but is not essential. Show that this still gives a matroid, but the rank of the matroid is $r(M) = r(A) = d - \dim\left(\bigcap_{H \in \mathcal{A}} H\right)$, where $A$ is the matrix whose rows are the normal vectors of the hyperplanes of $\mathcal{A}$.

(21) Prove Proposition 7.18: a central arrangement of hyperplanes in $\mathbb{R}^d$ is essential if and only if its collection of normal vectors spans $\mathbb{R}^d$.

(22) If $\mathcal{A}$ is a central, inessential hyperplane arrangement in $\mathbb{R}^d$, it is possible to define an essential arrangement from $\mathcal{A}$ as follows: set $W = \bigcap_{H \in \mathcal{A}} H$ and suppose $\dim(W) = s$. Then replace $H_i \mapsto H_i \cap W^\perp$, where $W^\perp$ is the orthogonal complement of $W$ (so $W^\perp = \{v \in \mathbb{R}^d \mid v \cdot w = 0 \text{ for all } w \in W\}$). This is called the *essentialization* of $\mathcal{A}$. Show that the new collection of hyperplanes

---

[25] The answer to the last question is a most emphatic "Yes!".

Figure 7.24. The planes of symmetry of an icosahedron give a central hyperplane arrangement in $\mathbb{R}^3$. See Exercise 26.

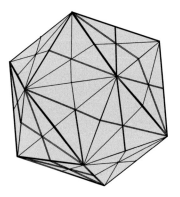

is an essential arrangement in the vector space $W^\perp$, and the rank of this matroid is $d - s$.

(23) Prove Proposition 7.24: first show central arrangements of hyperplanes in $\mathcal{R}^d$ give rise to non-central arrangements in $\mathbb{R}^{d-1}$ with no parallel intersections. Then show the converse also holds by describing a procedure to "lift" the affine hyperplanes in $\mathbb{R}^{d-1}$ having no parallel intersections to $(d-1)$-dimensional subspaces in $\mathbb{R}^d$.

(24) Prove Proposition 7.26: let $\mathcal{A}$ be an affine arrangement of hyperplanes in $\mathbb{R}^d$, and let $H_x \in \mathcal{A}$. Let $\mathcal{A}' = \mathcal{A} - H_x$ and $\mathcal{A}'' = \{H \cap H_x \mid H \in \mathcal{A}\}$. Then $M(\mathcal{A}') = M(\mathcal{A}) - x$ and $M(\mathcal{A}'') = M(\mathcal{A})/x$.

(25) (*If you know some group theory.*) Consider the matroid $M[A]$ for the matrix $A$ in Example 7.22.

    (a) Every symmetry of the cube corresponds to an automorphism of the matroid $M[A]$. Find the automorphism of $M[A]$ corresponding to reflection in the plane $x = 0$.

    (b) There are 48 isometries of the cube, and the structure of this group is $H = \mathbb{Z}_2^3 \rtimes S_3$. Show $Aut(M[A]) \cong \mathbb{Z}_2^2 \rtimes S_3$ is an index 2 subgroup of $H$. (Note that if $z$ represents the map that sends each point in $\mathbb{R}^3$ to its antipodal point, then the pair of isometries of the cube $h$ and $zh$ both correspond to the same matroid automorphism.)

(26) The icosahedron is one of the five Platonic solids. When placed with center at the origin, the planes of symmetry are a central hyperplane arrangement consisting of 15 hyperplanes through the origin. Figure 7.24 shows the planes of symmetry.

    The matrix of normal vectors is

| $a$ | $b$ | $c$ | $d$ | $e$ | $f$ | $g$ | $h$ | $i$ | $j$ | $k$ | $l$ | $m$ | $n$ | $o$ |
|---|---|---|---|---|---|---|---|---|---|---|---|---|---|---|
| 1 | 0 | 0 | 1 | $-1$ | 1 | 1 | $\tau$ | $-\tau$ | $\tau$ | $\tau$ | $\tau^2$ | $\tau^2$ | $-\tau^2$ | $\tau^2$ |
| 0 | 1 | 0 | $\tau$ | $\tau$ | $-\tau$ | $\tau$ | $-\tau^2$ | $\tau^2$ | $\tau^2$ | $\tau^2$ | 1 | $-1$ | 1 | 1 |
| 0 | 0 | 1 | $\tau^2$ | $\tau^2$ | $\tau^2$ | $-\tau^2$ | 1 | 1 | $-1$ | 1 | $-\tau$ | $\tau$ | $\tau$ | $\tau$ |

where $\tau = \frac{1+\sqrt{5}}{2}$ is the golden ratio.

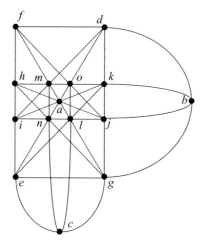

Figure 7.25. The rank 3 matroid associated to the hyperplane arrangement for an icosahedron. See Exercise 26.

(a) In the matroid, we find 15 three-point lines and 10 five-point lines. Interpret these results "geometrically" by trying to convince yourself that the three planes that bound a triangle of the icosahedron correspond to a three-point line in the matroid. For the five-point lines, look at the five planes that pass through the center of each triangle.

(b) Show that every point is on two three-point lines and two five-point lines. (This is immediate if you believe the picture in Figure 7.25. Try using a geomatric argument as in part (a).)

(c) Show that the 15 points of the matroid can be partitioned into five subsets such that, for the three points $xyz$ in a block of the partition, the matroid has three two-point lines $xy$, $xz$ and $yz$:

$$\{abc\} \quad \{dhl\} \quad \{ejm\} \quad \{fkn\} \quad \{gio\}.$$

(Hint: again, this follows easily from the picture in Figure 7.25. For a proof that uses the planes of symmetry, the three points in a block correspond to three mutually orthogonal normal vectors.) This is closely related to the *compound of five cubes* associated with an icosahedron.

### Section 7.4 – Counting regions

(27) Let $\mathcal{A}$ be a collection of $n$ hyperplanes in $\mathbb{R}^d$ in general position (so the intersection of any $d + 1$ is $\emptyset$, and there are no parallel intersections).

(a) Show $M(\mathcal{A}) = U_{d,n}$, the uniform rank $d$ matroid on $n$ points.

(b) Use Theorem 7.27 and induction to prove Schläfli's formula for the number of regions:

$$c(n, d) = \binom{n}{d} + \binom{n}{d-1} + \cdots + \binom{n}{0}.$$

(There are two variables here, $n$ and $d$. It's probably easiest to do induction on $n + d$.)

(c) If $n \le d$, then Schläfli gives $c(n, d) = 2^n$. Explain this answer combinatorially.

(d) Show that if $n < d$, then $\mathcal{A}$ is central and inessential.

(28) As in Exercise 27, assume $\mathcal{A}$ is a collection of $n$ hyperplanes in $\mathbb{R}^d$ in general position. Then we wish to show the number of *bounded* regions is $\binom{n-1}{d}$.

(a) Let $b(n, d)$ be the number of bounded regions. Modify the proof of Theorem 7.27 to show $b(n, d) = b(n - 1, d) + b(n - 1, d - 1)$. Then use Proposition 7.26 and induction to prove the formula. (Hint: there is some hyperplane in $\mathcal{A}$ so that all the bounded regions are on the same side of $\mathcal{A}$. Now delete and contract that hyperplane.)

(b) For an alternate proof, first choose a hyperplane $H \in \mathcal{A}$.

(i) Call a point $x \in \mathbb{R}^d$ a *point of $\mathcal{A}$* if $x$ is the intersection of exactly $d$ hyperplanes of $\mathcal{A}$. Show there are precisely $\binom{n-1}{d}$ points of $\mathcal{A}$ not on the hyperplane $H$.

(ii) For each point $x$ of $\mathcal{A}$ not on $H$, show there is a unique bounded region of $\mathcal{A}$ that has $x$ as the point in that region *farthest* from $H$.

(iii) Conclude that the number of bounded regions is $\binom{n-1}{d}$.

(29) Let $\mathcal{A}$ be a central hyperplane arrangement in $\mathbb{R}^d$, and assume all the hyperplanes contain the origin. We also assume the hyperplanes are placed as freely as possible, so that the intersection of any subset of $d$ of them is also just the origin (and, in particular, $M(\mathcal{A})$ is a uniform matroid). We are interested in the number of regions $c(\mathcal{A})$ (all of which will be unbounded).

(a) Intersect $\mathcal{A}$ with the unit sphere $S$ in $\mathbb{R}^d$, as usual, to create a spherical arrangment. Then let $\mathcal{A}_1$ be the (non-central) affine arrangement in $\mathbb{R}^{d-1}$ derived from $\mathcal{A}$, except, this time, use one of the hyperplanes in the arrangement as the "hyperplane at infinity," then remove that hyperplane. Show that $\mathcal{A}_1$ is a collection of $n - 1$ hyperplanes *in general position* in $\mathbb{R}^{d-1}$. (Hint: if you use rubber bands on a tennis ball as a model for the spherical arrangement, then pick one rubber band and just restrict to the half of the ball on one side of that rubber band. Then flatten.)

(b) Show that $c(\mathcal{A}) = 2c(n - 1, d - 1)$, where $c(n - 1, d - 1)$ is the number of regions in the Schläfli formula from Exercise 27. Conclude that

$$c(\mathcal{A}) = 2\left( \binom{n-1}{d-1} + \binom{n-1}{d-2} + \cdots + \binom{n-1}{0} \right).$$

This formula was found by Jakob Steiner in 1826.

(30) We can find the number of *unbounded* regions in an arrangement, too. Let $u(n, d)$ be the number of unbounded regions in a non-central arrangement of $n$ hyperplanes in $\mathbb{R}^d$ in general position.

(a) Use the formulas from Exercises 27 (Schläfli's formula) and 28 to get a formula for $u(n, d)$.

(b) Let $\mathcal{A}$ be a *central* arrangement of $n$ hyperplanes in $\mathbb{R}^d$ in general position (as defined in Exercise 29). Use your formula from part (a) and the formula from part (b) of Exercise 29 to prove that $u(n, d) = c(\mathcal{A})$.

(c) Give a geometric argument that shows $u(n, d) = c(\mathcal{A})$. (Hint: when $d = 2$, the number of unbounded regions is $2n$. Draw $n$ lines, no two parallel, on a piece of paper, then hold the paper really really far from your face.)

# 8

# Matroid minors

## 8.1 Examples, excluded minors and the Scum Theorem

When we created new matroids from old matroids in Chapter 3, the operations of deletion and contraction played a central role. Combining and iterating these operations produces a *minor* of the original matroid. That's the theme of this section, and it's also the central theme of a very active research program.

Let $M$ be a matroid on the ground set $E$, and let $A$ and $B$ be two disjoint subsets of $E$. Then $M/A - B$, the matroid formed by deleting all the elements of $B$ and contracting all the elements of $A$, is called a *minor* of $M$. Note that Proposition 3.8 allows us to ignore the order in which we apply the deletion and contraction operations. We do need to make sure that we never deleted an isthmus or contracted a loop along the way, so we make sure $A$ is an independent set of $M$ and $B$ contains no isthmuses after all the elements of $A$ are contracted.

**Example 8.1.** Let's show that the non-Fano matroid (see Figure 8.1) $F_7^-$ has the four-point line $U_{2,4}$ as a minor. Since the non-Fano plane has seven points, we need to find three points to delete or contract. Furthermore, since $r(F_7^-) = 3$ (and since contraction reduces rank), we know that we will need to contract one point (since a four-point line has rank 2) and delete two points.

Let's contract $e$: the independent sets in $F_7^-/e$ are all subsets of $\{a, b, c, d, f, g\}$ of size at most 2 *except* $\{a, b\}$ and $\{c, d\}$. (This follows from the fact that $abe$ and $cde$ are the only dependent sets of size 3 containing the point $e$.) Thus, $F_7^-/e$ is a rank 2 matroid with two double points: $ab$ and $cd$. A picture of this matroid is given in Figure 8.2.

Now we're almost done. If we delete one point from each of the two double points, we get $U_{2,4}$. You have some freedom here; for instance, deleting $b$ and $d$ gives $F_7^-/e - \{b, d\} \cong U_{2,4}$.

How much freedom did we have in Example 8.1? By symmetry, we could have contracted either $e$, $f$ or $g$ initially, then we had two

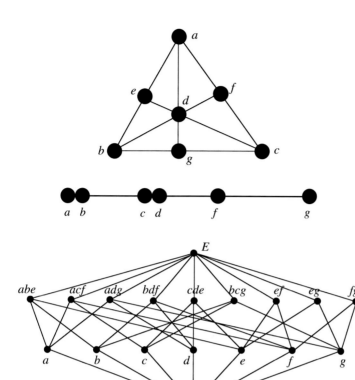

Figure 8.1. The non-Fano matroid $F_7^-$.

Figure 8.2. The contraction $F_7^-/e$.

Figure 8.3. The geometric lattice for the non-Fano plane $F_7^-$.

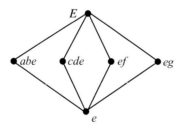

Figure 8.4. The geometric lattice for the four-point line $U_{2,4}$ minor is visible in the geometric lattice for $F_7^-$.

independent choices to make for the two points to delete. This gives a total of $3 \cdot 2 \cdot 2 = 12$ four-point line minors of $F_7^-$.

You can "see" the $U_{2,4}$ minor in the geometric lattice of flats. The lattice for the non-Fano plane is shown in Figure 8.3. Note there are seven atoms (corresponding to the seven points) and nine hyperplanes (six three-point lines and three two-point lines).

Then if $N = F_7^-/e$ is the minor formed by contracting $e$ (from Figure 8.2), the flats of $N$ correspond precisely to the flats of $F_7^-$ that contain $e$. There are four flats that cover $e$; they are $abe, cde, ef$ and $eg$. See Figure 8.4.

This trick always works, in the following sense. First, we can always form our minor by first contracting, then deleting (by commutativity of these operations). Next, since we never contract loops, the set we

Figure 8.5. The restriction of the non-Fano plane to the points $\{a, b, d, f\}$.

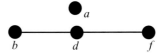

contracted must have been an independent set in $M$ (see Exercise 10). Then the flats in the minor $N = M/A - B$ are in one-to-one correspondence with the flats of $M$ that contain $\overline{A}$, the closure of $A$ (which is always a flat).

This argument justifies (without offering a careful proof) the next result, known as the "Scum Theorem."

Scum Theorem

**Theorem 8.2.** *[Scum Theorem] Let N be a minor of M on the ground set E. Assume $A \subseteq E$ is the subset of points contracted in creating the minor N. Then the geometric lattice for N is (isomorphic to) the upper interval $[\overline{A}, E]$ in the geometric lattice for M.*

The name "Scum Theorem" is a bit unfortunate, and reflects some sort of prejudice on the part of matroid theorists (at least, it tells you what they think of minors[1]). The minor is thought of as the scum, which rises to the top[2] of the geometric lattice (which is a pond in this analogy, we suppose) as the upper interval $[F, E]$ for some flat $F$.

The last step in Example 8.1 involved deleting two points to get rid of the multiple points that popped up in the contraction $F_7^-/e$. Then the four-point line $U_{2,4}$ is just the simplification of $F_7^-/e$. (Recall that if $M$ is a matroid with loops and multiple points, then the *simplification* of $M$ is obtained by deleting all the loops of $M$ and replacing each of the multiple points by a single point.)

Submatroid

If $S \subseteq E$, we can get a *submatroid* by simply restricting our attention to $S$, ignoring everything outside of $S$. This is the operation of *matroid*

Restriction

*restriction,* which we defined in Definition 3.39 of Chapter 3. For an easy example, consider the non-Fano matroid in Figure 8.1. If we let $S = \{a, b, d, f\}$, we get the restriction $M|_S$ pictured in Figure 8.5.

In this example, $M|_S$ has the same rank as $M$. It's apparent that $M|_{\{a,b,d,f\}}$ is what you would get if you had simply deleted everything else in the matroid: $M - \{c, e, g\}$. This is always true when the restriction has full rank.

**Proposition 8.3.** *Suppose M is a matroid on E, $S \subseteq E$ and $r(M) = r(M|_S)$. Then*

$$M|_S = M - (E - S).$$

The proof is obvious, and we omit it. As a simple[3] application, note that if $si(M)$ denotes the simplification of $M$, then we can obtain $si(M)$

---

[1] Underage readers may elect to skip this discussion.
[2] Anyone for the "Cream Theorem"?
[3] Pun intended.

as a restriction of $M$. We point out $r(si(M)) = r(M)$, i.e., the rank of the simplification is the same as the rank of the original matroid.

**Corollary 8.4.** *Suppose $M$ is a matroid on $E$ and let $si(M)$ be the simplification of $M$. Then $si(M) = M|_S = M - (E - S)$ for some $S \subseteq E$.*

What happens when the rank of the restriction is smaller than the rank of $M$? Here's a quick example:

**Example 8.5.** As before, let $M = F_7^-$ be the non-Fano plane, as in Figure 8.1. This time, let $S = \{b, d, f\}$. Then the restriction $M|_S$ is just the three-point line $bdf$. You can get this from $M$ by first deleting $\{c, e, g\}$, then contracting the isthmus $a$. (This is easy to see in Figure 8.5: the three points $\{c, e, g\}$ have already been deleted, so all that remains is the contraction of $a$.)

This is true generally: any restriction can be obtained from $M$ by deleting or contracting everything else in the matroid. We state this as a theorem, pointing out that the number of things you need to contract is exactly the difference between the rank of $M$ and the rank of the restriction.

**Theorem 8.6.** *Let $M$ be a matroid on the ground set $E$ and let $S \subseteq E$. Then*

$$M|_S = (M - A)/B$$

*where $A \cup B = E - S$, $A \cap B = \emptyset$, and $|B| = r(M) - r(M|_S)$.*

Again, we omit the proof, which is still obvious.[4] So restrictions are minors. The converse is false, however: in Example 8.1, we were able to get a four-point line as a minor of $F_7^-$, the non-Fano plane. But no restriction of $F_7^-$ is a four-point line because all the lines ("rank 2 flats" would be the matroid way to say this) of $F_7^-$ have either two or three points.

*All restrictions are minors*

There is a close connection between minors of $M$ and minors of the dual matroid $M^*$. For practice with your intuition about minors, cover up everything after the words *if and only if* in the statement of the next result. Then guess the conclusion.

**Proposition 8.7.** *Let $M$ and $M^*$ be a pair of dual matroids. Then $N$ is a minor of $M$ if and only if $N^*$ is a minor of $M^*$.*

*Proof Proposition 8.7.* $N$ is a minor of $M$ if and only if $N = M/A - B$ for appropriate disjoint subsets $A$ and $B$. Then, from Proposition 3.21, $N^* = M^*/B - A$. (It's worth a quick comment here that, by Proposition 3.22, the prohibition on deleting loops and contracting isthmuses is mirrored in the operations in the dual.) □

---

[4] Just saying something is obvious over and over again is not a good pedagogical technique.

There are lots of very deep and important results in matroid theory that have the form:

> A matroid $M$ is —————————— if and only if $M$ has no minor isomorphic to ——————————, —————————— or ——————————.

The first blank can be filled in with adjectives like *binary, regular* or *graphic,* and the other blanks are filled with a list of so-called *excluded* or *forbidden* minors. (Recall that binary matroids are those matroids representable over $\mathbb{F}_2$, while regular matroids are those representable over all fields.) Theorems of this form are very attractive because they give a short characterization of an important class of matroids.

We give an easy example of such a result in the next proposition. Recall that a matroid in which every subset is independent is called a *Boolean algebra*. A Boolean algebra cannot contain a loop as a minor, and a loop is precisely the unique minor preventing a matroid from being a Boolean algebra:

**Proposition 8.8.** *A matroid $M$ is a Boolean algebra if and only if $M$ does not contain a loop as a minor.*

*Proof Proposition 8.8.* We need to show two things: first, if $M$ is a Boolean algebra, then $M$ does not have a loop as a minor; second, if $M$ is not a Boolean algebra, then $M$ does have a loop as a minor.

So assume $M$ is a Boolean algebra on $n$ points. Then every point in $M$ is an isthmus, so minors are created solely through contraction, and every minor is just a collection of some number of isthmuses. This means $M$ does not have a loop as a minor.

For the converse, suppose $M$ is not a Boolean algebra. We need to show that $M$ does contain a loop as a minor. Since $M$ is not a Boolean algebra, $M$ has a circuit $C = \{e_1, \ldots, e_k\}$. So first form the minor $M|_C$ by restricting to $C$, then contract the independent set $\{e_2, \ldots, e_k\}$. That leaves a one-point matroid on the element $e_1$, and $e_1$ will be a loop. $\square$

It's only slightly harder to prove the following characterization of uniform matroids. See Exercise 3.

**Proposition 8.9.** *Let $M = U_{1,1} \oplus U_{0,1}$ be the matroid with two points – a loop and an isthmus. Then a matroid is uniform if and only if it does not have $M$ as a minor.*

Hereditary class    If a class $\mathcal{M}$ of matroids is closed under the operations of deletion and contraction, then we say $\mathcal{M}$ is a *hereditary* class. For instance, the class of all graphic matroids is hereditary, as is the class of all matroids representable over a field $\mathbb{F}$. See Exercise 2 for some examples of hereditary and non-hereditary classes.

If a class $\mathcal{M}$ of matroids is hereditary, then it makes sense to talk about the minimal *excluded* or *forbidden* minors for $\mathcal{M}$. Here's the definition.

**Definition 8.10.** Let $\mathcal{M}$ be a hereditary class of matroids. Then a matroid $M$ is an *excluded* or *forbidden* minor for the class $\mathcal{M}$ if $M \notin \mathcal{M}$, but every *proper* minor of $M$ is in $\mathcal{M}$.

We see from Definition 8.10 that an excluded or forbidden minor is always minimal. As a preview of Section 8.3 and an indication of some of the depth of matroid theory, we state the following excluded minor characterization of the class of graphic matroids, proved by Tutte in 1958 [35].

**Theorem 8.11.** *[Tutte] A matroid M is graphic if and only if M has no minor isomorphic to any of the following matroids:*

$$U_{2,4}, F_7, F_7^*, M(K_5)^*, M(K_{3,3})^*.$$

We know all of these matroids: $U_{2,4}$ is the four-point line, $F_7$ is the Fano plane, and the two matroids $M(K_5)$ and $M(K_{3,3})$ are the graphic matroids you get from the complete graph $K_5$ and the complete bipartite graph $K_{3,3}$.

One direction of Theorem 8.11 is easy – showing none of these matroids is graphic. It's also not hard to show that any proper minor of any of these five matroids is graphic – see Exercise 9. The truly difficult part of the theorem lies in showing that this is the complete list: *any* non-graphic matroid has one of these five matroids as a minor.

We revisit this theorem and several other excluded minor characterizations in Section 8.3.

## 8.2 Binary matroids

Binary matroids are matroids representable over $\mathbb{F}_2$, the two-element field $\{0, 1\}$. That means every such matroid can be identified as the column dependences of a matrix of 0's and 1's, with all the dependences modulo 2. Because these matrices are rather simple[5] objects, binary matroids have lots of special structure not shared by the collection of all matroids.

### 8.2.1 Graphs and binary matroids

Let's start by recalling a nice connection between graphs and matrices, introduced in Example 4.17.

---

[5] Don't be fooled. There are several deep results in matroid theory concerning binary matroids.

Figure 8.6. (Example 8.12)
Left: the graph $G$. Right: the
graphic matroid corresponding
to $G$.

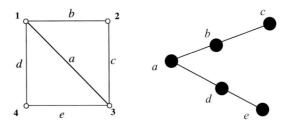

**Example 8.12.** Let $G$ be the graph of Figure 8.6.

As in Example 4.17, form the vertex–edge incidence matrix as follows
(see Definition 4.16):

- Label the rows by the vertices and the columns by the edges.
- Set $a_{i,j} = 1$ if vertex $i$ is incident to edge $j$, and set $a_{ij} = 0$
  otherwise.

$$
A_G = \begin{array}{c} \\ 1 \\ 2 \\ 3 \\ 4 \end{array}
\begin{array}{c} \begin{array}{ccccc} a & b & c & d & e \end{array} \\
\begin{bmatrix} 1 & 1 & 0 & 1 & 0 \\ 0 & 1 & 1 & 0 & 0 \\ 1 & 0 & 1 & 0 & 1 \\ 0 & 0 & 0 & 1 & 1 \end{bmatrix} \end{array}.
$$

There are two things we want to observe about this matrix. First, the
last row is the sum of the first three rows, modulo 2. That mean the
row-rank of the matrix is smaller than the number of rows. In this case,
the rank of the matrix (and the matroid) is 3:

$$
\begin{array}{c} \begin{array}{ccccc} a & b & c & d & e \end{array} \\
\begin{bmatrix} 1 & 1 & 0 & 1 & 0 \\ 0 & 1 & 1 & 0 & 0 \\ 1 & 0 & 1 & 0 & 1 \\ 0 & 0 & 0 & 1 & 1 \end{bmatrix} \end{array}
\quad \text{row reduces to} \quad
\begin{array}{c} \begin{array}{ccccc} a & b & c & d & e \end{array} \\
\begin{bmatrix} 1 & 0 & 1 & 0 & 1 \\ 0 & 1 & 1 & 0 & 0 \\ 0 & 0 & 0 & 1 & 1 \\ 0 & 0 & 0 & 0 & 0 \end{bmatrix} \end{array}.
$$

Secondly, and more importantly, the matrix $A_G$ (and the row-reduced
version) represents the graphic matroid over $\mathbb{F}_2$. (This is the content of
Exercise 4(c) of Chapter 4.) It's easiest to see this by looking at the
circuits of the matroid. In this case, there are three: $abc$, $ade$ and $bcde$.
Then the corresponding columns of $A_G$ are:

$$
abc \leftrightarrow \begin{array}{c} \begin{array}{ccc} a & b & c \end{array} \\
\begin{bmatrix} 1 & 1 & 0 \\ 0 & 1 & 1 \\ 1 & 0 & 1 \\ 0 & 0 & 0 \end{bmatrix} \end{array}
\quad ade \leftrightarrow \begin{array}{c} \begin{array}{ccc} a & d & e \end{array} \\
\begin{bmatrix} 1 & 1 & 0 \\ 0 & 0 & 0 \\ 1 & 0 & 1 \\ 0 & 1 & 1 \end{bmatrix} \end{array}
\quad bcde \leftrightarrow \begin{array}{c} \begin{array}{cccc} b & c & d & e \end{array} \\
\begin{bmatrix} 1 & 0 & 1 & 0 \\ 1 & 1 & 0 & 0 \\ 0 & 1 & 0 & 1 \\ 0 & 0 & 1 & 1 \end{bmatrix} \end{array}.
$$

Note that these submatrices have exactly zero or two 1's in each row.
This means the sum of the column vectors in each circuit is the zero
vector, modulo 2. For instance, $a + b + c = \bar{0}$ (over the field $\mathbb{F}_2$). That

means the column vectors corresponding to circuits are linearly dependent sets. That's good.

We also need to check every circuit in the matroid $M(A_G)$ (the matroid defined from the matrix, over $\mathbb{F}_2$) corresponds to a cycle in the graph. In this example, you can check that the columns $abc$, $ade$ and $bcde$ are the only circuits in $M(A_G)$.

The most important take-away message from Example 8.12 is this:

**Theorem 8.13.** *Graphic matroids are binary.*

*Proof Theorem 8.13.* (This is Exercise 4(c) of Chapter 4.) Let $G$ be a graph and $A_G$ its vertex–edge incidence matrix. We will show the cycles of $G$ correspond precisely to the circuits in the represented binary matroid $M(A_G)$.

First, let's dispense with loops. A loop in the graph $G$ will correspond to a column vector of all 0's in the matrix $A_G$, i.e., a linearly dependent set. (One way to justify this is that the vertex incident to the loop will generate a 2 in the vertex–edge incidence matrix, but $2 \equiv 0$ (modulo 2).)

Now suppose $C$ is a cycle in $G$, with $C = \{c_1, c_2, \ldots, c_m\}$, where $m > 1$. Then we can reorder and/or rename the rows and columns of $A_G$ so that the first $m$ rows and columns of the matrix look like:

$$
\begin{array}{cccccc}
c_1 & c_2 & c_3 & \cdots & c_{m-1} & c_m \\
\end{array}
$$
$$
\begin{bmatrix}
1 & 0 & 0 & \cdots & 0 & 1 \\
1 & 1 & 0 & \cdots & 0 & 0 \\
0 & 1 & 1 & \cdots & 0 & 0 \\
0 & 0 & 1 & \cdots & 0 & 0 \\
\vdots & \vdots & \vdots & \vdots & \vdots & \vdots \\
0 & 0 & 0 & \cdots & 1 & 0 \\
0 & 0 & 0 & \cdots & 1 & 1 \\
\end{bmatrix}.
$$

Thus, there are 1's along the main diagonal of this $m \times m$ square submatrix, and 1's below the main diagonal, and a single 1 in the upper right-hand corner.

Then it is clear that the sum of the columns is $c_1 + c_2 + \cdots + c_m = 0$ in $\mathbb{F}_2$ since there are exactly two 1's in each row (and all the entries in rows $m + 1, m + 2 \ldots$ are 0). Thus, these columns are linearly dependent.

It remains to show that removing any column leaves a linearly independent set. First, since any of these columns can be transformed into any other column through linear operations, we may as well remove the last column. Then row reducing the remaining $m - 1$ columns leaves an $(m - 1) \times (m - 1)$ identity matrix, so they are linearly independent. Thus, the columns corresponding to $C$ form a circuit in $M(A_G)$.

For the other direction, we assume $C$ is a circuit in the binary matroid $M(A_G)$ and show $C$ corresponds to a cycle in $G$. So let

$C = \{c_1, c_2, \ldots, c_m\}$ be a circuit in $M(A_G)$. Then these column vectors are linearly dependent, but any proper subset of them is linearly independent.

Let $A_C$ be the submatrix formed by the $m$ columns of the circuit $C$. Since these columns are linearly dependent, we know there are an even number of 1's in each row of $A_C$. Recall that the rows of $A_G$ correspond to the vertices of $G$. Then we can find a cycle contained in $C$ as follows:

- First, find an entry of $A_C$ that equals 1 (since $m > 1$, there is some non-zero entry). Assume this entry is in the $a_{1,1}$ position in $A_C$. In the graph $G$, this corresponds to choosing a vertex $v_1$ and an edge $c_1$ incident to that vertex in $G$.
- Since there are two 1's in each column, there must be some other non-zero entry in the first column; let's say the $a_{2,1}$ entry is 1. In $G$, this corresponds to choosing the other vertex incident to $c_1$; let's say this vertex is $v_2$.
- Now, looking at the second row of $A_C$, there must be some other non-zero entry (since there are an even number of entries in every row). We assume this entry appears in position $a_{2,2}$, corresponding to an edge from $c_2$ incident to $v_2$.

$$
\begin{array}{c} \\ v_1 \\ \\ v_2 \\ \\ v_3 \\ \vdots \end{array}
\begin{array}{c}
\overset{c_1}{\phantom{x}} \quad \overset{c_2}{\phantom{x}} \quad \overset{c_3}{\phantom{x}} \quad \overset{\cdots}{\phantom{x}} \\
\left[
\begin{array}{cccc}
1 & 0 & 0 & \cdots \\
\downarrow & & & \\
1 \;\rightarrow\; & 1 & 0 & \cdots \\
& \downarrow & & \\
0 & 1 \;\rightarrow\; & 1 & \cdots \\
\vdots & \vdots & \vdots & \cdots
\end{array}
\right]
\end{array}.
$$

We continue in this way, at each stage choosing a non-zero entry in a row or a column.

When will this process terminate? When one of the vertical steps lands us in a row we've already visited, i.e., when we complete a cycle in the graph. Then the cycle we created must include all the edges of $C$; otherwise, this cycle would be a linearly dependent proper subset of $C$.

Thus, if $C$ is a circuit in the binary matroid $M(A_G)$, then the edges corresponding to $C$ in the graph $G$ form a cycle. □

## 8.2.2 Equivalent conditions for binary matroids

In light of Theorem 8.13, it makes sense to see if theorems that hold for graphic matroids also hold for binary matroids. Here are two prototypical graph theorems:

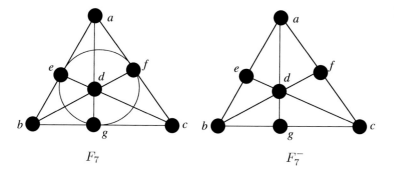

Figure 8.7. Left: Fano plane $F_7$. Right: non-Fano plane $F_7^-$.

- Proposition 4.11: Suppose $C$ is a circuit and $C^*$ is a cocircuit in the graphic matroid $M(G)$. Then $|C \cap C^*|$ is even.
- Theorem 4.5: Let $C_1$ and $C_2$ be circuits in $M(G)$. Then the symmetric difference $C_1 \triangle C_2$ is a disjoint union of circuits. (Recall: the *symmetric difference is $A \triangle B = (A - B) \cup (B - A)$*.)

Do these results hold for binary, non-graphic matroids? What about non-binary matroids? As usual, it's best to try to build some intuition by looking at an example. Let's examine some circuits and cocircuits in the Fano plane $F_7 = PG(2, 2)$ (which is a binary matroid) and the non-Fano plane $F_7^-$ (which isn't). (See Propositions 6.16 and 6.17 for proofs that $F_7$ is binary and $F_7^-$ is not.)

**Example 8.14.** The Fano plane and the non-Fano plane are shown in Figure 8.7.

For the Fano plane, you can check both results remain true: $|C \cap C^*|$ is even for any circuit $C$ and cocircuit $C^*$, and the symmetric difference $C_1 \triangle C_2$ will always be a disjoint union of circuits (in fact, $C_1 \triangle C_2$ will always be a circuit in the Fano plane). For example, let $C = \{a, b, e\}$ and let $C^* = \{b, d, e, g\}$, so $C$ is a circuit and $C^*$ is a cocircuit in $F_7$. Then $C \cap C^* = \{b, e\}$, so $|C \cap C^*| = 2$. For two circuits $C_1 = \{e, f, g\}$ and $C_2 = \{a, d, e, f\}$ in $F_7$, we have $C_1 \triangle C_2 = \{a, d, g\}$, which is a circuit.

Are either of these results valid in the non-Fano plane? (You might enjoy pausing to search for counterexamples here.) Both results fail in $F_7^-$. Here are two counterexamples:

- Let $C = \{d, e, f, g\}$ and $C^* = \{a, d, e, f\}$. Then $C$ is a circuit and $C^*$ is a cocircuit, and we have $|C \cap C^*| = 3$.
- Let $C_1 = \{d, e, f, g\}$ and $C_2 = \{a, d, e, f\}$. Then $C_1$ and $C_2$ are both circuits, and $C_1 \triangle C_2 = \{a, g\}$ which is an independent set. (Note that $\{a, d, e, f\}$ is both a circuit and a cocircuit.)

It turns out each of these results – $|C \cap C^*|$ is always even, and $C_1 \triangle C_2$ is always a disjoint union of circuits – is *always true* for binary matroids, and *always false*[6] for non-binary matroids. In other words, these two properties each characterize binary matroids.

The next theorem gives two equivalent ways to say "My matroid is binary." (The symmetric difference $C_1 \triangle C_2 \triangle \cdots \triangle C_k$ is just the set of points in an *odd* number of the $C_i$.)

**Theorem 8.15.** *Let M be a matroid. Then the following are equivalent.*

(1) *M is binary.*
(2) *Suppose C is a circuit and $C^*$ is a cocircuit. Then $|C \cap C^*|$ is even.*
(3) *Let $C_1, C_2, \ldots, C_k$ be circuits. Then the symmetric difference*

$$C_1 \triangle C_2 \triangle \cdots \triangle C_k$$

*is a disjoint union of circuits (possibly empty).*

Condition (3) is a stronger version of Theorem 4.5, which used the symmetric difference of two circuits. We leave the proof of the various implications in Theorem 8.15 to the exercises, but provide detailed hints in examples. In Example 8.16 and Exercise 16, we give outlines for showing (1) $\Rightarrow$ (2). Showing (2) $\Rightarrow$ (3) is the focus of Example 8.17 and Exercise 18. Finally, Example 8.18 and Exercise 19 offer a proof of (3) $\Rightarrow$ (1). Although we don't need it for the logical structure of our proof, we offer a proof of (1) $\Rightarrow$ (3) in Exercise 20.

For now, let's look at three examples that illustrate some of the ideas used in proving the implications of Theorem 8.15.

**Example 8.16.** Let $M(A)$ be the binary matroid on the 11 column vectors of the following matrix:

Demonstrating
(1) $\Rightarrow$ (2)

$$A = \begin{array}{c} \\ \\ \\ \\ \\ \end{array} \begin{array}{ccccccccccc} e_1 & e_2 & e_3 & e_4 & e_5 & e_6 & e_7 & e_8 & e_9 & e_{10} & e_{11} \\ \left[\begin{array}{ccccccccccc} 1 & 1 & 1 & 1 & 0 & 1 & 1 & 0 & 0 & 0 & 0 \\ 1 & 1 & 0 & 0 & 0 & 1 & 0 & 1 & 0 & 0 & 0 \\ 0 & 0 & 1 & 0 & 1 & 1 & 0 & 0 & 1 & 0 & 0 \\ 0 & 1 & 0 & 0 & 1 & 1 & 0 & 0 & 0 & 1 & 0 \\ 1 & 0 & 1 & 1 & 1 & 1 & 0 & 0 & 0 & 0 & 1 \end{array}\right] \end{array}.$$

This example is designed to give a matrix-theory approach to proving (1) $\Rightarrow$ (2) in Theorem 8.15 – see Exercise 16. You can check the first five columns form a circuit $C$ (over $\mathbb{F}_2$). Also note that the hyperplane $H$ with equation $x_1 = 0$ consists of columns $e_5, e_8, e_9, e_{10}$ and $e_{11}$. Thus, $E - H = C^* = \{e_1, e_2, e_3, e_4, e_6, e_7\}$ forms the cocircuit with equation $x_1 = 1$, i.e., the first coordinate of each of these columns is 1. Note that $C \cap C^* = \{e_1, e_2, e_3, e_4\}$.

We would like to give a matrix-theory interpretation of the circuit $C$ and the cocircuit $C^*$. For the circuit, note that because the sum of

---

[6] *Always false* is interpreted mathematically: for instance, in any non-binary matroid, there is a circuit $C$ and a cocircuit $C^*$ such that $|C \cap C^*|$ is odd.

the first five columns of $A$ is the zero vector, we have $Av = \bar{0}$, where $v = (1, 1, 1, 1, 1, 0, 0, 0, 0, 0, 0)^t$. (The $^t$ is the transpose; we need a column vector for this matrix multiplication to make sense.) You can think of $v$ as the incidence vector for the circuit $C$.

For the cocircuit $C^*$, note that $uA = (1, 1, 1, 1, 0, 1, 1, 0, 0, 0, 0)$, where $u = (1, 0, 0, 0, 0)$ is a row vector. This matrix multiplication just places a 1 in position $i$ if the $i$th column of $A$ is in the cocircuit $C^*$, i.e., $uA$ is the incidence vector for $C^*$.

How can we determine $|C \cap C^*|$? We have two $1 \times 11$ incidence vectors, so their dot product will be the size of the intersection:

$$(uA) \cdot v = |C \cap C^*|.$$

But, since matrix multiplication is associative,[7] $(uA)v = u(Av) = u \cdot \bar{0} = 0$, where $u \cdot \bar{0}$ is the dot product of two $1 \times 5$ vectors, and the final 0 is just the scalar 0. Since this last computation is done in $\mathbb{F}_2$, we have $|C \cap C^*| \equiv 0 \mod 2$, which is condition (2).

**Example 8.17.** This example illustrates the implication (2) $\Rightarrow$ (3) in Theorem 8.15. Let $M$ be the binary matroid $M = M(A)$ based on the matrix $A$ from Example 8.16. We assume $|C \cap C^*|$ is even for any circuit $C$ and any cocircuit $C^*$. (Note: we aren't allowed to use the fact that $M(A)$ is a binary matroid in constructing a general argument, and we won't be using this in the example. Stay alert!)

*Demonstrating (2) $\Rightarrow$ (3)*

Here are four circuits: $C_1 = \{e_2, e_3, e_4, e_6, e_{11}\}$, $C_2 = \{e_3, e_4, e_9\}$, $C_3 = \{e_1, e_4, e_8\}$ and $C_4 = \{e_2, e_3, e_6, e_7\}$. We need to show the symmetric difference

$$D = C_1 \triangle C_2 \triangle C_3 \triangle C_4 = \{e_1, e_3, e_4, e_7, e_8, e_9, e_{11}\}$$

is a disjoint union of circuits. (You can check $D = C' \cup C''$ is partitioned into two circuits: $C' = \{e_4, e_7, e_{11}\}$ and $C'' = \{e_1, e_3, e_8, e_9\}$.)

It will be easier to do the computations (and more useful for a general proof) to recode the circuits as incidence vectors:

$$C_1 \leftrightarrow (0, 1, 1, 1, 0, 1, 0, 0, 0, 0, 1)$$
$$C_2 \leftrightarrow (0, 0, 1, 1, 0, 0, 0, 0, 1, 0, 0)$$
$$C_3 \leftrightarrow (1, 0, 0, 1, 0, 0, 0, 1, 0, 0, 0)$$
$$C_4 \leftrightarrow (0, 1, 1, 0, 0, 1, 1, 0, 0, 0, 0).$$

Use these four vectors to form the rows of a matrix:

$$N = \begin{pmatrix} 0 & 1 & 1 & 1 & 0 & 1 & 0 & 0 & 0 & 0 & 1 \\ 0 & 0 & 1 & 1 & 0 & 0 & 0 & 0 & 1 & 0 & 0 \\ 1 & 0 & 0 & 1 & 0 & 0 & 0 & 1 & 0 & 0 & 0 \\ 0 & 1 & 1 & 0 & 0 & 1 & 1 & 0 & 0 & 0 & 0 \end{pmatrix}.$$

---

[7] There are many examples of non-trivial theorems whose proofs depend on the associativity of matrix multiplication.

Let $D = C_1 \triangle C_2 \triangle C_3 \triangle C_4$, and note that the incidence vector for $D$ is just the mod 2 sum of the four rows of $N$. Let's first show $D$ is dependent (this is obvious here since $D$ contains seven points in a rank 5 matroid, but we give a more general argument). So suppose $D$ is independent. Then extend $D$ to a basis $B$, and let $C^* = E - B - e_1$. Then $C^*$ is a cocircuit of the matroid. By construction of $C^*$, we know $C^* \cap D = \{e_1\}$. By (2), we know $C^*$ meets each of the $C_i$ in an even number of points.

We claim $C^*$ also meets $D$ in an even number of points – this is the key observation in this implication. A nice way to see this uses some simple matrix multiplication over $\mathbb{F}_2$. Let $v = (1, 0, 1, 1, 0, 0, 1, 1, 1, 0, 1)^t$ be the incidence vector for the cocircuit $C^*$, written as a column vector. Then $Nv \equiv \overline{0}$ (modulo 2), so $v$ is in the null space of the matrix $N$. Thus, $v$ is orthogonal to the entire row space of $N$. Now the row space includes the sum of the rows, so $v$ is also orthogonal to the sum of the rows of $N$. But that sum is the incidence vector for the symmetric difference; thus $C^*$ meets $D$ in an even number of points.

This contradicts the fact that $|C^* \cap D| = 1$, so $D$ must be dependent. If $D$ is a circuit, we're done. If it isn't, let $C' \subsetneq D$ be a circuit and repeat the above argument on $D - C'$.

Demonstrating
$(3) \Rightarrow (1)$

**Example 8.18.** In this example, we examine the implication $(3) \Rightarrow (1)$ from Theorem 8.15. The general case is treated in Exercise 19. Let $M$ be the Fano plane. With the labeling from Figure 8.7, there are 14 circuits:

| $C_1$ | $abe$ | $C_8$ | $cdfg$ |
|-------|-------|-------|--------|
| $C_2$ | $acf$ | $C_9$ | $bdeg$ |
| $C_3$ | $adg$ | $C_{10}$ | $bcef$ |
| $C_4$ | $bdf$ | $C_{11}$ | $aceg$ |
| $C_5$ | $bcg$ | $C_{12}$ | $adef$ |
| $C_6$ | $cde$ | $C_{13}$ | $abfg$ |
| $C_7$ | $efg$ | $C_{14}$ | $abcd$ |

We are assuming (3): the symmetric difference of any collection of circuits is a disjoint union of circuits. We are interested in showing $M$ must be binary. We use an approach that takes advantage of some basic results from linear algebra.

Now for each $C_i$, we make a $1 \times 7$ incidence vector $v_i$ with

$$v_i(j) = \begin{cases} 1 & \text{if } j \in C_i, \\ 0 & \text{otherwise.} \end{cases}$$

For instance, for $C_1$ above, we have $v_1 = (1, 1, 0, 0, 1, 0, 0)$.

Then the vectors $v_1, v_2, \ldots, v_{14}$ generate a subspace $W$ of $\mathbb{F}_2^7$. A basis for $W$ consists of four vectors, which we've assembled as the rows of the matrix

$$
\begin{array}{c}
\\
C_1 \\
C_2 \\
C_3 \\
C_4
\end{array}
\begin{array}{ccccccc}
a & b & c & d & e & f & g \\
\end{array}
\left[
\begin{array}{ccccccc}
1 & 1 & 0 & 0 & 1 & 0 & 0 \\
1 & 0 & 1 & 0 & 0 & 1 & 0 \\
1 & 0 & 0 & 1 & 0 & 0 & 1 \\
0 & 1 & 0 & 1 & 0 & 1 & 0 \\
\end{array}
\right].
$$

(This matrix represents the dual matroid $F_7^*$.) Since the sum of any subset of vectors in $W$ corresponds to the symmetric difference of the corresponding subsets, we see (by (3)) that every element of $W$ corresponds to a disjoint union of circuits of the Fano plane.

Now form the subspace $W^\perp$ consisting of all vectors orthogonal to everything in $W$. The subspace $W^\perp$ is three dimensional. A basis is given by the rows of the matrix

$$
\begin{array}{ccccccc}
a & b & c & d & e & f & g \\
\end{array}
\left[
\begin{array}{ccccccc}
1 & 0 & 1 & 0 & 1 & 0 & 1 \\
0 & 1 & 1 & 0 & 1 & 1 & 0 \\
1 & 1 & 1 & 1 & 0 & 0 & 0 \\
\end{array}
\right].
$$

Then you can check this matrix represents $M$ over $\mathbb{F}_2$. For instance, the column vectors $e$, $f$ and $g$ all have last coordinate 0, so they are dependent (over any field). To see why this procedure always works, see Exercise 19.

We don't need to assume the symmetric difference of *any* collection of circuits is a disjoint union of circuits. As we remarked above, conditions (1), (2) and (3) in Theorem 8.15 are also equivalent to the apparently weaker condition on symmetric differences:

> If $C_1, C_2 \in \mathcal{C}$ with $C_1 \neq C_2$, then $C_1 \triangle C_2$ is a disjoint union of elements of $\mathcal{C}$.

In fact, we can actually use the condition on symmetric differences to provide a cryptomorphism for the circuits of a binary matroid.

**Corollary 8.19.** *Let $\mathcal{C}$ be a family of subsets of a finite set $E$. Then $\mathcal{C}$ is the collection of circuits of a binary matroid if and only if*

(1) $\emptyset \notin \mathcal{C}$;
(2) *if $C_1, C_2 \in \mathcal{C}$ and $C_1 \subseteq C_2$, then $C_1 = C_2$;*
(3) *if $C_1, C_2 \in \mathcal{C}$ with $C_1 \neq C_2$, then $C_1 \triangle C_2$ is a disjoint union of elements of $\mathcal{C}$.*

We omit the proof of Corollary 8.19. Most matroid texts prove this as part of a longer list of equivalent conditions. An encyclopedic treatment of binary matroids appears in Oxley's text [26].

Since the dual of a binary matroid is also binary (Theorem 6.6), we also get:

Figure 8.8. Left: the Petersen
graph $G$. Right: two edges of
the Petersen graph have been
deleted.

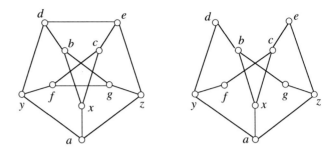

**Corollary 8.20.**   *$M$ is binary if and only if $C_1^* \triangle C_2^*$ is a disjoint union of cocircuits for any two cocircuits $C_1^*$ and $C_2^*$.*

It's worth looking back at Theorem 4.13, where we showed how to construct cocircuits in graphic matroids.

## 8.3  Summary of excluded minor results

One of the classic theorems of graph theory is Kuratowski's characterization of planarity. We mentioned this briefly in Theorem 3.27. In a footnote to a sketch of the proof, we defined a *topological minor* of a graph.

**Definition 8.21.**   A subgraph $H$ is a *topological minor* of a graph $G$ if you can obtain $H$ by deleting edges of $G$ and also contracting edges incident to vertices of degree 2.

Then the complete graph $K_5$ and the complete bipartite graph $K_{3,3}$ are the unique minimal obstructions for a graph to be planar. When phrased in terms of graph minors, this theorem is attributed to Wagner.

**Theorem 8.22.**   *[Wagner] A graph $G$ is planar if and only if it does not contain either $K_{3,3}$ or $K_5$ as a topological minor.*

**Example 8.23.**   A standard (and very pretty) application of this theorem can be used to show the *Petersen* graph of Figure 4.15 is not planar (compare this method with the proof based on Euler's formula given in Exercise 10 in Chapter 4).

We can show the Petersen graph has $K_{3,3}$ as a topological minor. Referring to Figure 8.8, we first delete the edges $de$ and $fg$. After deleting these two edges, each of the four vertices $d, e, f$ and $g$ will have degree 2, so we can contract the four edges $bd, ce, fy$ and $gz$. (This is the same as "erasing" the four vertices $e, f, g$ and $h$.) Then we get the graph in Figure 8.9, which is clearly isomorphic to $K_{3,3}$, with each of the three vertices $a, b$ and $c$ adjacent to each of $x, y$ and $z$.

Table 8.1. *Excluded minors for various classes of matroids.*

| Class | Excluded minors | Reference |
|---|---|---|
| Graphic | $U_{2,4}, F_7, F_7^*, M(K_5)^*, M(K_{3,3})^*$ | Tutte (1959) |
| Cographic | $U_{2,4}, F_7, F_7^*, M(K_5), M(K_{3,3})$ | Tutte (1959) |
| Planar graphic | $U_{2,4}, F_7, F_7^*, M(K_5), M(K_{3,3}),$ $M(K_5)^*, M(K_{3,3})^*$ | Tutte (1959) |
| Regular | $U_{2,4}, F_7, F_7^*$ | Tutte (1958) [35] |
| Binary | $U_{2,4}$ | Tutte (1958) [35] |
| Ternary | $U_{2,5}, U_{3,5}, F_7, F_7^*$ | Reid (1971 unpublished), Bixby (1979) [5], Seymour (1979) [31] |

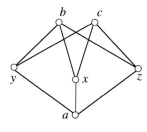

Figure 8.9. $K_{3,3}$ is a topological minor of the Petersen graph.

Note that if $H$ is a topological minor of $G$, then $M(H)$ is a matroid minor of $M(G)$. So we can translate Kuratowski's Theorem into a matroid setting:

**Theorem 8.24.** *Suppose $G$ is a graph. Then $G$ is planar if and only if the cycle matroid $M(G)$ has no minor isomorphic to $M(K_5)$ or $M(K_{3,3})$.*

We saw in Theorem 3.27 that $G$ is planar if and only if the dual matroid $M(G)^*$ is also graphic. That means $M(K_5)^*$ and $M(K_{3,3})^*$ are excluded minors for the class of graphic matroids. See Table 8.1.

Another way to say "$G$ is a planar graph" is "$G$ can be drawn on a sphere with no edge crossings." Then Kuratowski's Theorem tells you there are exactly two excluded topological minors for graphs on the sphere.

It's pretty easy to see that *every* graph can be drawn on some surface without edge crossings. For instance, although $K_5$ and $K_{3,3}$ are not planar, they can both be drawn on a *torus* without edge crossings. In general, for an arbitrary graph $G$, first draw it on a sphere, possibly with edge crossings. Then, for each such crossing, add a "handle" to your sphere, with one edge using the handle to avoid a crossing. Then it's clear we can add enough handles (one per crossing will do) to embed $G$ in some surface with no edge crossings. (By the way, the minimum number of handles needed is called the *genus* of the graph.)

How many excluded minors are there for a given surface? Could you have an infinite family of graphs $G_1, G_2, \ldots$, all of which were excluded minors for a given surface? More generally, could you have any infinite collection of graphs $G_1, G_2, \ldots$ with no $G_i$ a minor of $G_j$?[8]

The answer is no, and the proof, which spans more than 20 papers and hundreds of pages, is one of the monumental achievements in all of mathematics in the last 20 years. Robertson and Seymour have developed an intricate theory based on a structure theorem for minor-closed classes of graphs that shows there are no infinite collections of non-comparable (in the minor sense) graphs.

**Theorem 8.25.** *[Graph Minor Theorem – Robertson & Seymour] Any infinite family of graphs contains a pair G and H with G a minor of H.*

Returning to matroids, we saw in Theorem 8.13 that graphs are binary matroids. In fact, graphs are representable over all fields, and we can do so by a *unimodular* matrix.

Unimodular matrix

**Definition 8.26.** A matrix $A$ with integer entries is *unimodular* if every $k \times k$ submatrix has determinant 0, 1 or $-1$. A matroid is *unimodular* or *regular* if it can be represented over $\mathbb{Q}$ by a unimodular matrix.

Regular matroid

Since $1 \times 1$ submatrices are included in our definition, every entry in a unimodular matrix must be 0, 1 or $-1$. Unimodular matrices are important in many applications of linear algebra; the fact that the submatrices have determinants 0, $\pm 1$ tells you that the inverses of submatrices, when they exist, have integer entries. (You can define a unimodular matrix over an arbitrary ring, with the condition that the determinants of the square submatries are either 0 or units in the ring.)

**Theorem 8.27.** *The matroid M is regular if and only if M can be represented over every field.*

We omit the proof of Theorem 8.27, but one direction is quite easy. Suppose $M$ has rank $r$ and $n$ points. If $M$ is represented over $\mathbb{Q}$ by an $r \times n$ unimodular matrix $A$, then we can consider the same matrix $A$ over any field characteristic. Then the determinant of an $r \times r$ submatrix will be non-zero over $\mathbb{Q}$ precisely when it is non-zero modulo $p$, for any prime $p$ (since the only non-zero determinants possible are $\pm 1$). So the bases are the same, regardless of the field.

For the converse, we would need to produce a unimodular matrix. Although we omit the details, this can be accomplished by finding a matrix that simultaneously represents $M$ over $\mathbb{F}_2$ and $\mathbb{F}_3$. Once this is done, we need to check all the determinants of $k \times k$ submatrices are 0, 1 or $-1$. This takes some work, and is generally done inductively. See [26] for the details.

Returning briefly to graphs, we state the fact that graphic matroids are unimodular as a theorem.

---

[8]  And, is it possible to compose a paragraph solely from questions? Well, is it?

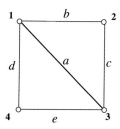 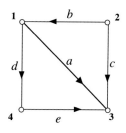

Figure 8.10. Left: the graph *G*. Right: an orientation of *G*.

**Theorem 8.28.** *Graphic matroids are regular.*

The proof is closely related to the proof of Theorem 8.13. As in the proof of that theorem, begin with the vertex–edge incidence matrix. Now, just change one of the two 1's in each column to −1. (In terms of the graph, this corresponds to putting an *orientation* on the edges of the graph, as in Figure 8.10.)

$$
\begin{array}{c} \\ 1 \\ 2 \\ 3 \\ 4 \end{array}
\begin{array}{cccccc} a & b & c & d & e \\ \left[\begin{array}{ccccc} 1 & 1 & 0 & 1 & 0 \\ 0 & 1 & 1 & 0 & 0 \\ 1 & 0 & 1 & 0 & 1 \\ 0 & 0 & 0 & 1 & 1 \end{array}\right] \end{array}
\longrightarrow
\begin{array}{c} \\ 1 \\ 2 \\ 3 \\ 4 \end{array}
\begin{array}{cccccc} a & b & c & d & e \\ \left[\begin{array}{ccccc} 1 & -1 & 0 & 1 & 0 \\ 0 & 1 & 1 & 0 & 0 \\ -1 & 0 & -1 & 0 & -1 \\ 0 & 0 & 0 & -1 & 1 \end{array}\right]. \end{array}
$$

Then this new matrix will represent the graphic matroid $M(G)$ over every field. This requires some proof, of course, but is straightforward. See Exercise 28.

Tutte provided the first excluded minor characterization in 1959.

**Theorem 8.29.** *A matroid is binary if and only if it does not contain* $U_{2,4}$ *as a minor.*

We can restate Theorem 8.29 as follows:

> $U_{2,4}$ is the only excluded minor for the class of binary matroids.

We can use our characterization of binary matroids from Theorem 8.15 to prove Theorem 8.29

*Proof Theorem 8.29 – Sketch.* To show $U_{2,4}$ is an excluded minor, we must show $U_{2,4}$ is not binary, but any minor of $U_{2,4}$ is binary. But this is trivial.

The difficult part of the proof is showing that $U_{2,4}$ is the *only* excluded minor for the class of binary matroids. So, suppose $M$ is a rank $r$ matroid that is not binary. We must show $M$ has $U_{2,4}$ as a minor. We use Theorem 8.15(2): a matroid is binary if and only if $|C \cap C^*|$ is even for every circuit $C$ and cocircuit $C^*$.

Restating this equivalence, if $M$ is not binary, then there is a circuit $C$ and cocircuit $C^*$ with $|C \cap C^*|$ odd. Since $|C \cap C^*| \neq 1$ for all matroids (by Proposition 4.8 or Exercise 9 of Chapter 2), we must have $|C \cap C^*| \geq 3$.

So suppose $C \cap C^* \supseteq \{x, y, z\}$. Since $|C| \geq 3$, we know $r(M) \geq 2$. Then $H = E - C^*$ is a hyperplane, $r(H) \geq 1$, and $H$ contains a rank $r - 2$ flat $F$.

We now assume the flat $F$ can be chosen so that $C \cap F$ is a basis for $F$. This means we assume $C$ is a spanning circuit of $M$, with $C = \{x, y, z, a_1, a_2, \dots, a_{r-2}\}$, where $\{a_1, a_2, \dots, a_{r-2}\}$ is a basis for $F$.[9]

We now claim $F$ is covered by the following four hyperplanes: $H, \overline{F \cup \{x\}}, \overline{F \cup \{y\}}$ and $\overline{F \cup \{z\}}$. If we can prove this claim, we'll be done. For then we could contract a basis for the flat $F$, leaving a rank 2 matroid. Deleting everything else that was in $F$ (they will all be loops at this point) leaves four (or more) rank 1 flats. Then it's clear $U_{2,4}$ is a minor, and we will be finished.

So it remains to show $H, \overline{F \cup \{x\}}, \overline{F \cup \{y\}}$ and $\overline{F \cup \{z\}}$ are four distinct hyperplanes. Since none of $x$, $y$ and $z$ are in $H$, we know the rank of each of these flats is $r - 1$, i.e., they are hyperplanes.

Why are these four hyperplanes all distinct? Well, suppose $\overline{F \cup \{x\}} = \overline{F \cup \{y\}}$, for instance. Then let $B = \{x, y, a_1, a_2, \dots, a_{r-2}\}$. Then $B$ is a basis for the matroid $M$ since it's a proper subset of a circuit and it contains exactly $r$ elements. Thus $y \notin \overline{F \cup \{x\}}$, so $\overline{F \cup \{x\}} \neq \overline{F \cup \{y\}}$. □

Note that if $M$ has a $U_{2,4}$ minor, then $M$ must have a (corank 2) flat $F$ covered by (at least) four distinct hyperplanes. This is immediate from the Scum Theorem (Theorem 8.2). We needed the converse of this fact here: we had the four hyperplanes covering the flat $F$, and needed to create $U_{2,4}$ as a minor. We restate this scummy interpretation as a corollary.

**Corollary 8.30.** *A rank $r$ matroid $M$ is binary if and only if $M$ has no flat of rank $r - 2$ that is covered by four hyperplanes.*

The rest of this section is devoted to summarizing some of the most important excluded minor characterizations.

Note that all of the excluded minors are rather small matroids; the largest is $M(K_5)$ (and its dual), which have 10 points. The entire list of excluded minors is also known for the class of *quaternary* matroids (published in 2000 [14]); these are the matroids representable over the field $\mathbb{F}_4$, the field with four elements. That list includes exactly seven matroids, the largest of which has eight points.

Recall that if $\mathbb{F}$ is a finite field, then $\mathbb{F}$ has $p^k$ elements for some prime $p$ and some positive integer $k$, and there is a unique such field for every

---

[9] We have not justified this step – this is why this is a proof sketch. See Exercise 15 for an example of why this step is crucial.

$p$ and $k$. G.-C. Rota conjectured that the number of excluded minors for the class of matroids representable over a finite field is always finite.

**Rota's conjecture** Rota's conjecture

The number of excluded minors for the class of matroids representable over a finite field is finite.

Then we know Rota's conjecture is true for fields with two, three or four elements. The conjecture is open for all larger fields. This is quite striking: the (presumably) easiest open case is the class of matroids representable over $\mathbb{F}_5$, and it is not known if the number of excluded minors is finite in this case.

It's easy to prove $U_{2,q+2}$ (and, therefore, its dual $U_{q,q+2}$) is an excluded minor for representability over $\mathbb{F}_q$; see Exercise 21. It's much more difficult to prove other uniform matroids are excluded minors for a general finite field $\mathbb{F}_q$. In fact, $U_{r,p+2}$ is an excluded minor for $\mathbb{F}_p$ representability (for $p$ prime) for all $r$ between 2 and $p$. The proof of this result is quite recent.[10]

A closely related area of research is directly related to the Graph Minors Theorem of Robertson and Seymour (Theorem 8.25).[11] Suppose $M_1, M_2, \ldots$ is an infinite collection of matroids representable over $\mathbb{F}_q$ for some prime power $q = p^k$. Then the question analogous to the Robertson–Seymour Theorem is open.

**Well-quasi-ordered conjecture**

Suppose $M_1, M_2, \ldots$ is an infinite collection of matroids representable over $\mathbb{F}_q$. Then $M_i$ is a minor of $M_j$ for some $i \neq j$.

This has recently been shown to be true for binary matroids by Geelen, Gerards and Whittle; their result is part of a larger project attempting to extend the Robertson–Seymour approach to matroids represented over $\mathbb{F}_q$.

Finally, we note the well-quasi-ordered conjecture is false for matroids represented over infinite fields. The following counterexample is due to Brylawski.

**Example 8.31.** Let $M_n$ be a rank 3 matroid on $2n$ points $\{a_1, a_2, \ldots, a_n, b_1, b_2, \ldots, b_n\}$ whose three-point circuits are $\{a_i, b_i, a_{i+1}\}$, where indices are computed mod $n$, as in Figure 8.11.

Then no $M_i$ is a minor of $M_j$ for any two matroids $M_i$ and $M_j$ in the family, but each $M_i$ is representable over the rational numbers. See Exercise 27.

[10] S. Ball, On large subsets of a finite vector space in which every subset of a basis size is a basis, to appear in *J. Eur. Math. Soc.* (2012).

[11] Technically, Robertson and Seymour proved the class of all graphs is *well-quasi-ordered*. That means the partially ordered set of all graphs, ordered by the minor relation, has no infinite *antichains*.

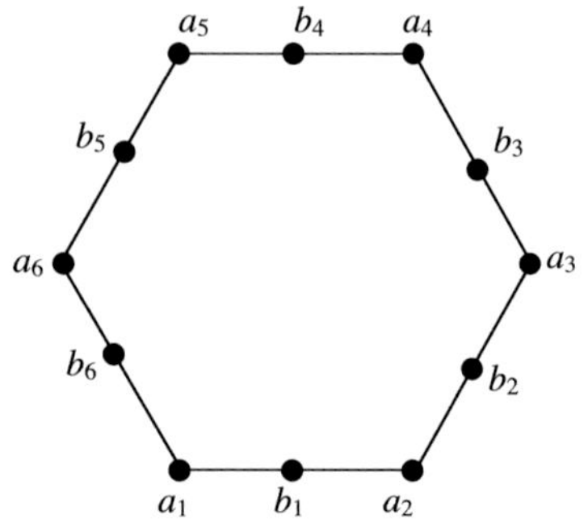

Figure 8.11. The matroid $M_6$ is a member of an infinite family of matroids representable over $\mathbb{Q}$. No matroid in the family is a minor of another matroid in the family.

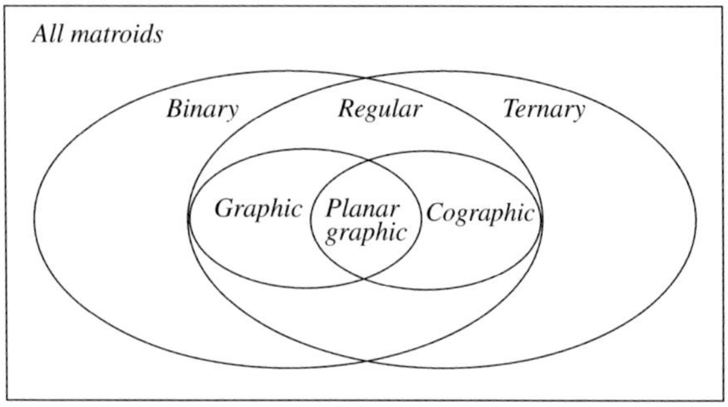

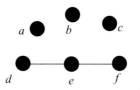

Figure 8.12. The various containment relations among six hereditary classes of matroids.

Figure 8.13. Matroid for Exercise 1.

A summary of the relationships among the matroid classes considered here is shown in Figure 8.12.

## Exercises

### Section 8.1 – Examples, excluded minors and the Scum Theorem

(1) Let $M$ be the matroid in Figure 8.13.
  (a) Show that $M$ has a $U_{2,5}$ minor. How many $U_{2,5}$'s can you find?
  (b) Show that $U_{3,5}$ is also a minor of $M$, and count these, too.
  (c) $M$ is self-dual (see Exercise 21 in Chapter 3). Explain your answers to parts (a) and (b) in light of this fact.
(2) For this problem, we use the following classes of matroids from Exercise 42 in Chapter 3.
  • $C_1$ = all matroids on 10 elements.
  • $C_2$ = all matroids on 10 or fewer elements.
  • $C_3$ = all rank 4 matroids on 10 elements.
  • $C_4$ = all matroids of rank at most 5.
  • $C_5$ = all graphic matroids.

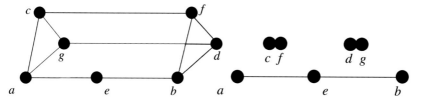

Figure 8.14. A matroid and its dual – see Exercises 6, 7, 8 and 13.

- $C_6$ = all connected matroids.
- $C_7$ = all matroids having no isthmuses.
- $C_8$ = all simple matroids, i.e., matroids having no loops or multiple points.

For each class, determine if the class is hereditary, and, if it is, describe the (minimal) excluded minors for the class.

(3) Prove Proposition 8.9: A matroid is uniform if and only if it does not contain the two-point matroid $U_{1,1} \oplus U_{0,1}$ as a minor.

(4) Suppose every (proper) minor of a matroid $M$ is a uniform matroid $U_{m,n}$ for some values of $m$ and $n$. Must $M$ be a uniform matroid? If so, prove it; if not, find a specific counterexample.

(5) This exercise is a version of Proposition 8.8 for simple matroids. Recall a matroid is *simple* if it has no loops or multiple points. A simple matroid is also called a *combinatorial geometry*. Show that a simple matroid $M$ is a Boolean algebra if and only if $M$ has no three-point line as a minor.

(6) Let $M$ be the rank 4 matroid on the left in Figure 8.14.
   (a) Show that the rank 3 matroid in Figure 8.13 is a minor of $M$.
   (b) Show explicitly that $M$ has a $U_{2,4}$ minor. Then find a rank 2 flat $F$ covered by four distinct hyperplanes. (This verifies the conclusion of the Scum Theorem – Theorem 8.2.)

(7) Recall the *simplification si(M)* of a matroid $M$ is obtained from $M$ by removing all loops and replacing each multiple point by a single point.
   (a) Let $M$ be the matroid on the right in Figure 8.14. Draw $si(M)$.
   (b) Show $si(M)$ is always a restriction (and hence, a minor) of $M$.
   (c) Show that $si(M)$ is obtained by repeatedly performing the following operations:
      - Delete all loops of $M$.
      - For all two-point circuits $\{x, y\}$, delete one of $x$ or $y$.

(8) The *cosimplification co(M)* of a matroid $M$ is obtained from $M$ by repeatedly performing these operations:
   - Contract all isthmuses of $M$.
   - For all two-point cocircuits $\{x, y\}$, contract one of $x$ or $y$.
   (a) Let $M$ be the matroid on the left in Figure 8.14. Draw $co(M)$. (Hint: there are no isthmuses and two 2-point cocircuits.)
   (b) Show $co(M)$ is always a minor of $M$.

(c)  Show $co(M) = (si(M^*))^*$, where $si(M)$ is the simplification of $M$ from Exercise 7. Check this equality for the two
dual matroids in Figure 8.14.

(9)  Prove the easy direction of Tutte's Theorem (Theorem 8.11) characterizing graphic matroids in terms of excluded minors: show
that each of the matroids $U_{2,4}$, $F_7$, $F_7^*$, $M(K_5)^*$ and $M(K_{3,3})^*$ is
an excluded minor for the class of graphic matroids, i.e., for each
matroid $M$ on the list, $M$ is not graphic, but every proper minor of
$M$ is.

(10)  Suppose $N = M/A - B$ is a minor of $M$, and $r(N) = r(M) - a$,
for some $a \geq 0$. Show that $A$ is independent in $M$, and $|A| = a$.

(11)  (a)  Suppose $\mathcal{M}$ is a hereditary class of matroids, and let $\mathcal{M}^* =
\{M^* \mid M \in \mathcal{M}\}$. Show $\mathcal{M}^*$ is also a hereditary class.

(b)  Suppose a matroid $M$ is an excluded minor for a hereditary
class of matroids $\mathcal{M}$. Use Proposition 8.7 to show $M^*$ is an
excluded minor for $\mathcal{M}^*$.

(c)  Recall a matroid $M$ is *cographic* if $M^*$ is graphic. Show that
a matroid $M$ is cographic if and only if $M$ does not contain
$U_{2,4}$, $F_7$, $F_7^*$, $M(K_5)$ or $M(K_{3,3})$ as a minor.

### Section 8.2 – Binary matroids

(12)  Show that the affine plane $AG(2, 3)$ and the projective plane
$PG(2, 3)$ are not binary by finding $U_{2,4}$ minors. (In fact, $AG(n, q)$
and $PG(n, q)$ are not representable over characteristic 2 when $q$
is odd and $n \geq 2$. This follows from Theorems 6.27 and 6.29 in
Chapter 6.)

(13)  Show that the two matroids of Figure 8.14 are not binary by finding
a $U_{2,4}$ minor. Are they ternary?

(14)  Let $G$ be a graph with $k$ components and $v$ vertices, and let $A_G$ be
the vertex–edge incidence matrix. Then show the matrix rank (over
$\mathbb{F}_2$) is given by $r(A_G) = v - k$. (This follows immediately from
Theorem 8.13 and the description of the rank of a graphic matroid
given in Section 4.1. Try to construct a matrix-based proof that
interprets the components of $G$ in terms of the incidence matrix
$A_G$.)

(15)  Suppose $M$ is a binary matroid with a circuit $C$ and cocircuit
$C^*$ where $|C \cap C^*| = 4$. Then the argument given in our sketch
of Theorem 8.29 might lead you to believe we could find *five*
hyperplanes that covered a given rank $r - 2$ flat. You would be
wrong: let $F_7$ be the Fano plane.

(a)  Show that every cocircuit $C^*$ has four elements, and that every
cocircuit is also a circuit. Conclude $|C \cap C^*| = 4$ when we
choose $C = C^*$.

(b)  Choose a four-point circuit $C$ in $F_7$, so $C$ is also a cocircuit.
Then show the proof sketch of Theorem 8.29 breaks down
where we assumed we could choose the flat $F$ so that $C \cap F$

is a basis for $F$. (In fact, in this case, $C \cap F = \emptyset$ for any $F$, which is just a point.)

(16) This exercise gives an outline of a proof that if $M$ is a binary matroid, then $|C \cap C^*|$ is even for any circuit $C$ and any cocircuit $C^*$. (This is implication (1) $\Rightarrow$ (2) in Theorem 8.15.) The approach is modeled on Example 8.16.

Let $M$ be a rank $r$ binary matroid on $n$ points and let $A$ be an $r \times n$ matrix with $M = M(A)$, where column dependences are taken modulo 2. Suppose $C$ is a circuit and $C^*$ a cocircuit. Suppose

$$C \cap C^* = \{c_1, c_2, \ldots, c_k\},$$

where $C = \{c_1, c_2, \ldots, c_k, d_1, d_2, \ldots, d_l\}$ and $C^* = \{c_1, c_2, \ldots,$ $c_k, e_1, e_2, \ldots, e_m\}$. Reorder the columns of $A$, if necessary, so that the first $k + l + m$ columns of $A$ correspond to $\{c_1, \ldots, c_k, d_1, \ldots,$ $d_l, e_1, \ldots, e_m\}$.

(a) Let $v$ be the $n \times 1$ column vector with 1's in the first $k + l$ positions and 0's elsewhere, i.e., $v$ is the incidence vector for the circuit $C$. Show that $Av = 0$.

(b) Use what you know about the equations for hyperplanes to show that there is a $1 \times r$ row vector $u = [a_1, a_2, \ldots, a_r]$ such that $uA$ is the $1 \times n$ incidence vector for the cocircuit $C^*$, i.e.,

$$uA = [\underbrace{1, \ldots, 1}_{k}, \underbrace{0, \ldots, 0}_{l}, \underbrace{1, \ldots, 1}_{m}, \underbrace{0, \ldots, 0}_{n-k-l-m}].$$

(c) Use the fact that $(uA)v = u(Av)$ to conclude $|C \cap C^*|$ is even.

(17) Suppose $M$ is a rank $r$ binary matroid on $n$ points, represented by the matrix $A$. Show the following three statements are equivalent.
(a) $M$ is affine, i.e., $M \subseteq AG(r, 2)$.
(b) All circuits in $M$ are even.
(c) The vector $\underbrace{(1, 1, \ldots, 1)}_{n}$ is in the row space of $A$.

(Hint: it may be useful to work through the vector representations of circuits and hyperplanes in Exercise 16.)

(18) This exercise is concerned with implication (2) $\Rightarrow$ (3) in Theorem 8.15, and is based on Example 8.17. Suppose $M$ is a matroid, *not* assumed to be binary, satisfying $|C \cap C^*|$ is even for any circuit $C$ and cocircuit $C^*$. Suppose $C_1, C_2, \ldots, C_k$ are circuits. We must show $C_1 \triangle C_2 \triangle \cdots \triangle C_k$ is a disjoint union of circuits.
(a) Let $D = C_1 \triangle C_2 \triangle \cdots \triangle C_k$. First show $D$ contains a circuit:
  (i) Suppose some $C_i$ is a one-element circuit, i.e., a loop. Then show $D$ contains a circuit.
  (ii) Now assume there are no loops, and suppose that $D$ contains no circuits, i.e., $D$ is independent. Then let $x \in D$ and extend $D$ to a basis $B$ and set $C^* = E - \overline{B - x}$. Show that $C^*$ is a cocircuit, and that $C^* \cap D = \{x\}$.

(iii) Show $|C^* \cap D|$ is even. Conclude from this contradiction that $D$ is a dependent set. (Hint: construct a matrix using incidence vectors, as in Example 8.17.)

(b) It remains to show $D$ is a disjoint union of circuits. Let $C'$ be a circuit with $C' \subseteq D$. If $C' = D$, then we are done, so assume $C' \subsetneq D$. Show

$$(C_1 \triangle C_2 \triangle \cdots \triangle C_k) - C' = C_1 \triangle C_2 \triangle \cdots \triangle C_k \triangle C'.$$

(c) Now show that the argument of part (a) can be repeated, showing

$$C_1 \triangle C_2 \triangle \cdots \triangle C_k \triangle C$$

contains a circuit. Since

$$|C_1 \triangle C_2 \triangle \cdots \triangle C_k \triangle C'| < |C_1 \triangle C_2 \triangle \cdots \triangle C_k|$$

and the universe is finite, this process eventually ends. Convince yourself (and whoever might be grading your work) that this eventually produces a disjoint union of circuits.

(19) This exercise guides you to a proof of the implication (3) $\Rightarrow$ (1) in Theorem 8.15 – see Example 8.18. Assume that $M$ is a matroid on $n$ points where the symmetric difference of any collection of circuits is always the disjoint union of circuits, or empty. We need to show $M$ is binary.

Let $\{C_1, C_2, \ldots, C_N\}$ be the collection of circuits of $M$, and let $v_1, v_2, \ldots, v_N$ be the corresponding incidence vectors in $\mathbb{F}_2^n$. Define $W$ to be the subspace of $\mathbb{F}_2^n$ generated by the incidence vectors.

(a) Let $Z = \{S \subseteq E \mid S \text{ is a disjoint union of circuits of } M\}$. Show the collection of all the incidence vectors in $Z$ is the subspace $W$.

(b) Find a (vector space) basis for $W^\perp$, say $\{v_1, v_2, \ldots, v_m\}$ and let $A$ be the $m \times n$ matrix whose rows are the $v_i$. Show the matroids $M$ and $M(A)$ (the binary matroid defined on the columns of $A$) are identical:

(i) Suppose $D$ is a dependent set in $M$, let $C$ be a circuit contained in $D$, and let $u \in W$ be the incidence vector for the circuit $C$. Use the fact that $Au = 0$ to show the columns corresponding to $D$ is also a dependent set in the binary matroid $M(A)$.

(ii) Now suppose $D$ is dependent in the binary matroid $M(A)$, let $C$ be a circuit contained in $D$, and let $v$ be the incidence vector for the circuit $C$. Use the fact that $(W^\perp)^\perp = W$ to show $v \in W$. Then use part (a) to conclude $D$ is dependent in $M$.

(iii) Then use parts (i) and (ii) to show $M$ and $M(A)$ are identical.

(20) This exercise outlines a proof that the symmetric difference of circuits is a disjoint union of circuits in a binary matroid. (This is the implication (1) $\Rightarrow$ (3) from Theorem 8.15.) Let $M$ be a binary matroid on $n$ ponts and let $A$ be a matrix representing $M$ over $\mathbb{F}_2$. Suppose $C_1, C_2, \ldots, C_k$ are a collection of circuits in $M$.

(a) Let $v_1, v_2, \ldots, v_k$ be the $n \times 1$ incidence vectors for the circuits $C_1, C_2, \ldots, C_k$, written as column vectors. Show that $A(v_1 + v_2 + \cdots + v_k) = 0$. Conclude that $C_1 \triangle C_2 \triangle \cdots \triangle C_k$ is dependent in $M$.

(b) Let $D$ be a circuit, with $D \subseteq C_1 \triangle C_2 \triangle \cdots \triangle C_k$. Show $(C_1 \triangle C_2 \triangle \cdots \triangle C_k) - D$ is dependent.[12]

(c) Use part (b) repeatedly, if needed, to conclude $C_1 \triangle C_2 \triangle \cdots \triangle C_k$ is a disjoint union of circuits.

## Section 8.3 – Excluded minors

(21) Show that the uniform matroids $U_{2,q+2}$ and $U_{q,q+2}$ are excluded minors for representability over $\mathbb{F}_q$.

(22) Show that the Fano plane $F_7$ and its dual $F_7^*$ are excluded minors for representability over $\mathbb{F}_q$ for any odd $q$.

(23) Suppose $\mathcal{M}$ is a hereditary class of matroids and $M$ is an excluded minor for $\mathcal{M}$. Show $M$ must be a connected matroid.

(24) Suppose $M$ is a rank $r$ simple matroid representable over the field $\mathbb{F}_q$. Show that $|E| \leq q^r$. Conclude that the number of excluded minors of rank at most $r$ for the class of matroids representable over $\mathbb{F}_q$ is finite.[13]

(25) Let $\mathcal{M}$ be the class of all *paving* matroids. (Recall: $M$ is a *paving* matroid if all its circuits have size $r(M) + 1$ or $r(M)$.)

(a) Show that the class $\mathcal{M}$ is hereditary, i.e., if $M$ is a paving matroid, then so is $M - x$ and $M/x$ whenever these are defined. (Hint: you might enjoy rereading Proposition 3.9.)

(b) Show that $\mathcal{M}$ is not closed under duality by finding an appropriate counterexample.

(c) Let $M = L \oplus I \oplus I$ be the matroid on three points consisting of two isthmuses and a loop. Show that $M$ is not paving, but any minor of $M$ is paving. Conclude $M$ is an excluded minor for the class $\mathcal{M}$.

(d) Show that $L \oplus I \oplus I$ is the *only* excluded minor for $\mathcal{M}$. (Hint: suppose $N \notin \mathcal{M}$. Find a small circuit $C$ in $N$ and extend $C - x$

---

[12] Hint: "nothing from nothing leaves nothing." Billy Preston.
[13] This is not quite as stupid as it sounds: there are an infinite number of simple, connected rank $r$ matroids for a fixed $r > 1$ – Exercise 43 of Chapter 3.

to a basis $B$ (for some $x \in C$). Then show $B \cup x$ contains $L \oplus I \oplus I$ as a minor.)

(e) Define a matroid $M$ to be *co-paving* if $M^*$ is paving. Find all the excluded minors for the class of co-paving matroids, and prove your answer.

(26) Call a matroid $M$ *2-paving* if all its circuits have size $r(M) - 1, r(M)$ or $r(M) + 1$. Let $\mathcal{M}$ be the class of all 2-paving matroids.

   (a) Show that the matroid on the left in Figure 8.14 is 2-paving, but not paving (see Exercise 25).

   (b) Show that the class $\mathcal{M}$ is hereditary.

   (c) Show that $\mathcal{M}$ contains a single excluded minor, find it, and prove your answer. (Hint: Exercise 25.)

(27) Show that the infinite family of matroids of Figure 8.11 are all representable over $\mathbb{Q}$, but no $M_i$ is a minor of $M_j$ for any pair $(M_i, M_j)$ of matroids in the family.

(28) Prove Theorem 8.28: Graphic matroids are regular. (Hint: modify the proof of Theorem 8.13 by orienting the edges of $G$, as in Figure 8.10. Then show the matrix $A$ obtained represents the graphic matroid $M(G)$ over any field by showing the circuits in the $M(G)$ and $M(A)$ are the same.)

(29) Figure 8.12 shows the containment relations among several classes of matroids. There are seven distinct regions in this figure. For each region, find an example of a matroid that resides in precisely the corresponding classes. (E.g., the Fano plane $F_7$ is binary, but not ternary, not regular, etc.)

# 9

# The Tutte polynomial

## 9.1 Motivation and history

It's not unusual for an open problem in mathematics to motivate significant research. One of the touchstone problems of modern combinatorics and graph theory was the Four Color Problem, now the Four Color Theorem after its resolution by Appel and Haken in 1976 [1]. This problem, which dates to a letter Francis Guthrie wrote to his brother Frederick in 1852, asks if it is always possible to color the regions of a map with four (or fewer) colors so that no two adjacent regions receive the same color. This is obviously important[1] when you are looking at a map – you don't want regions (countries or states) that share a border to receive the same color.

Appel and Haken's proof was noteworthy for two reasons:

- It resolved (in the affirmative) a 125 year-old conjecture.
- It was the first significant mathematical proof that made essential use of a computer.

In fact, Appel and Haken needed more than 1000 hours of computer time to complete the case checking involved in their proof. A more recent proof [28], modeled on the same approach, but streamlined, dramatically reduces the number of cases to check. But even this new proof uses a computer in an essential way to check cases.

Although the original problem is phrased in terms of coloring regions in a map, you can turn the entire enterprise into coloring *vertices* of a graph. First, the map is a planar graph itself; think of the edges as the boundaries between regions and the vertices as points where two (or more) boundaries meet. Then form the geometric dual of the graph (as in Section 3.4.1 of Chapter 3) and color the vertices of a graph instead of the regions of the map. Then a coloring is "proper" if no two adjacent     Proper coloring
vertices receive the same color.

---

[1] While it is important for adjacent countries to receive different colors, map makers were not the source of the problem and had little or no interest in determining if four colors were always enough to ensure a proper coloring.

What makes this problem interesting to us is a technique introduced by G. D. Birkhoff in 1912. He attacked the Four Color Problem by defining a function $\chi(G;\lambda)$ on a graph $G$ that simply counts the number of proper colorings of $G$ using $\lambda$ (or fewer) colors. Then it turns out $\chi(G;\lambda)$ is actually a polynomial in the variable $\lambda$. The Four Color Theorem can be rephrased in terms of this polynomial as follows:

**Theorem 9.1.** *Let $G$ be a planar graph. Then $\chi(G;4) > 0$.*

Then one might try to prove the Four Color Theorem by using analytic techniques to study the location of the roots of $\chi(G;\lambda)$ for planar graphs $G$. Unfortunately, no proof of the Four Color Theorem has arisen in this way. But the chromatic polynomial $\chi(G;\lambda)$ has been studied in a variety of other contexts, and it generalizes quite nicely to a two-variable polynomial defined on matroids, the *Tutte* polynomial.[2]

## 9.2 Definition and basic examples

Let $M$ be a matroid on ground set $E$. We define a two-variable polynomial $t(M;x, y)$ – the *Tutte* polynomial – as follows.

*Tutte polynomial*    **Definition 9.2.** The *Tutte polynomial* $t(M;x, y)$ of a matroid $M$ is defined recursively as follows:

(1) $t(M;x, y) = t(M - e;x, y) + t(M/e;x, y)$ if $e$ is neither an isthmus nor a loop;
(2) $t(M;x, y) = x \cdot t(M/e;x, y)$ if $e$ is an isthmus;
(3) $t(M;x, y) = y \cdot t(M - e;x, y)$ if $e$ is a loop;
(4) $t(M;x, y) = 1$ if $E = \emptyset$.

Let's play with some small matroids to get a feel for how this recursive definition works.

**Example 9.3.** There are four matroids with two points:

- $M_1 = U_{2,2}$ consists of two isthmuses;
- $M_2 = U_{1,1} \oplus U_{0,1}$ is the direct sum of a loop and an isthmus;
- $M_3 = U_{1,2}$ is a double point;
- $M_4 = U_{0,2}$ consists of two loops.

---

[2] William Tutte was born in England in 1917. When WWII broke out, Tutte's efforts were needed in the area of cryptanalysis; he became part of the famous Bletchley Park group, where he worked on decoding the FISH messages. When the War was over he returned to Cambridge to study for a doctorate in mathematics, although at this point he had no formal mathematics training. Tutte made fundamental contributions to the development of matroid theory, especially through his foundational work on excluded minors (see Chapter 8). He published some 168 papers and books, and the two-variable polynomial that bears his name is a fitting tribute to his impact on the field. An excellent, detailed reference for applications of the Tutte polynomial is [11].

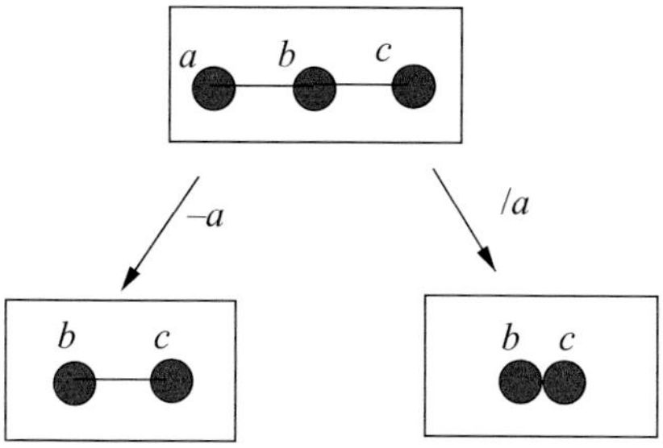

Figure 9.1. Delete and contract $a$ to compute $t(U_{2,3}; x, y)$.

Then we use the recursive definition to compute the Tutte polynomial for each of these matroids:

- For $M_1$, just apply part (2) of Definition 9.2 twice, contracting each isthmus to get $t(M_1; x, y) = x^2$.
- For $M_2$, apply part (2) to the isthmus and part (3) to the loop: $t(M_2; x, y) = xy$. (Note that $t(M_2; x, y) = t(U_{1,1}; x, y) \cdot t(U_{0,1}; x, y)$. The Tutte polynomial is *multiplicative* on direct sums – see Theorem 9.13.)
- $M_3$ requires part (1) of the definition. Let's say the ground set $E = \{a, b\}$, and we'll delete and contract $a$. Then $M_3/a$ is a loop, and $M_3 - a$ is an isthmus. So we get $t(M_3; x, y) = x + y$.
- Use part (3) of the definition twice: $t(M_4; x, y) = y^2$. (Note that $M_4 = M_1^*$, and $t(M_4; x, y) = t(M_1; y, x)$ – i.e., to get the polynomial for $M^*$, we just swap $x$ and $y$ in $t(M; x, y)$. See Theorem 9.12.)

One question concerning Definition 9.2 may occur to the alert reader:

> When computing $t(M; x, y)$, does the order in which we delete and contract matter?

If it did, then Definition 9.2 would be pretty silly, and you wouldn't be reading about it now. So we conclude the order *doesn't* matter.[3] We will prove the Tutte polynomial is well defined in Theorem 9.9. Before presenting that proof, we give two more examples.

**Example 9.4.** Figure 9.1 shows the calculation of the Tutte polynomial of the three-point line[4] $U_{2,3}$. Note that $M - a$ is the uniform matroid $U_{2,2}$ and $M/a$ is $U_{1,2}$. We know the Tutte polynomials of each of these matroids from Example 9.3. This gives

$$t(U_{2,3}; x, y) = t(U_{2,2}; x, y) + t(U_{1,2}; x, y) = x^2 + x + y.$$

---

[3] While this argument has some logical appeal, it should leave you unsatisfied.
[4] Some matroid theorists who take a graph-theoretic approach to the subject call $U_{2,3}$ a "triangle."

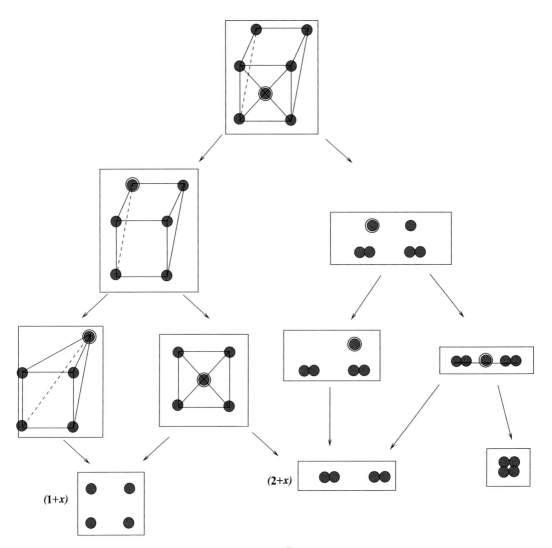

Figure 9.2. Computing the Tutte polynomial of a rank 4 matroid.

Applying the deletion–contraction recursion to $M = U_{2,4}$, we get

$$t(U_{2,4}; x, y) = t(U_{2,3}; x, y) + t(U_{1,3}; x, y)$$
$$= (x^2 + x + y) + (x + y + y^2) = x^2 + 2x + 2y + y^2.$$

It's not too hard to use induction to get a formula for the Tutte polynomial $t(U_{2,n}; x, y)$ for $n \geq 2$. See Exercise 2.

**Example 9.5.** Consider the matroid $M$ at the top of the diagram in Figure 9.2. As in Example 9.4, the box represents the Tutte polynomial of the matroid inside the box.[5] The circled elements represent those points deleted and contracted in the computation.

As a sample of how this computation is completed, note that the Tutte polynomial $t(U_{1,2} \oplus U_{1,2}; x, y) = \boxed{\bullet\bullet \quad \bullet\bullet}$ is multiplied by $(2 + x)$ in the last row of the computation. This represents the fact that $U_{1,2} \oplus U_{1,2}$ arises three times in the repeated deletion and contraction of points in $M$: twice via deletion and contraction of non-loops and non-isthmuses, and once via the contraction of an isthmus. The contraction of the isthmus contributes the factor of $x$.

The last row of the figure contains the matroids $U_{3,4}$, $U_{1,2} \oplus U_{1,2}$ and $U_{1,4}$. Then the Tutte polynomials of the matroids in the last row are:

- $t(U_{3,4}; x, y) = x^3 + x^2 + x + y$;
- $t(U_{1,4}; x, y) = x + y + y^2 + y^3$;
- $t(U_{1,2} \oplus U_{1,2}; x, y) = t(U_{1,2}; x, y)^2 = (x + y)^2$.

Putting these pieces together produces

$$t(M; x, y) = x^4 + 3x^3 + 4x^2 + 2x + 2x^2 y + 5xy + 2y + xy^2 + 3y^2 + y^3.$$

## 9.3 Corank–nullity polynomial

The most pressing issue for us to resolve is the proof that our recursive definition of the Tutte polynomial is meaningful; i.e., $t(M; x, y)$ is well-defined.[6] To do this, we introduce another two-variable polynomial, $s(M; u, v)$.

**Definition 9.6.** The *corank nullity polynomial* or *rank generating polynomial* $s(M; u, v)$ of a matroid $M$ defined on the ground set $E$ with rank function $r$ is defined as:

Corank–nullity polynomial

Rank generating polynomial

$$s(M; u, v) = \sum_{A \subseteq E} u^{r(E)-r(A)} v^{|A|-r(A)}.$$

The *corank* of $A \subseteq E$ is $r(E) - r(A)$, and the *nullity* is $|A| - r(A)$. These terms are borrowed from linear algebra.

**Example 9.7.** Consider the matroid in Figure 9.3. Table 9.1 gives the corank and nullity for various subsets of $E$.

---

[5] Homework problem: use this computation and the adage "Think outside the box" to construct a reasonable joke. Send your solution directly to the authors.
[6] If this were really the most pressing issue we faced, we should count ourselves as lucky, indeed.

Table 9.1.

| A | $r(A)$ | $\|A\|$ | $r(E) - r(A)$ | $\|A\| - r(A)$ | $u^{r(E)-r(A)}v^{\|A\|-r(A)}$ |
|---|---|---|---|---|---|
| $\emptyset$ | 0 | 0 | 3 | 0 | $u^3$ |
| $a$ | 1 | 1 | 2 | 0 | $u^2$ |
| $ab$ | 2 | 2 | 1 | 0 | $u$ |
| $abc$ | 2 | 3 | 1 | 0 | $uv$ |
| $abd$ | 3 | 3 | 0 | 0 | 1 |
| $E$ | 3 | 4 | 0 | 1 | $v$ |

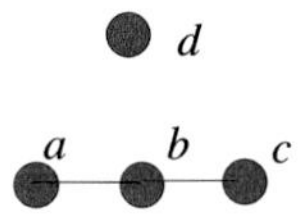

Figure 9.3. The matroid for
Example 9.7.

Adding up the terms in the right-most column gives us this polynomial:

$$s(M; u, v) = u^3 + 4u^2 + 6u + uv + 3 + v.$$

Matroid invariant    **Definition 9.8.**   A matroid *invariant* is a function $f : \mathcal{M} \to R$ from
the class of all matroids $\mathcal{M}$ to some commutative ring $R$ with the
property that isomorphic matroids given the same value: if $M_1 \cong M_2$,
then $f(M_1) = f(M_2)$.

Invariants are a very familiar and important idea in mathematics. For
instance, the number of bases is a matroid invariant, as is the corank–
nullity polynomial. We need to prove the Tutte polynomial is a matroid
invariant, too.

**Theorem 9.9.**   *For all matroids M, the Tutte polynomial is an evaluation
of the corank–nullity polynomial:*

$$t(M; x, y) = s(M; x - 1, y - 1).$$

*In particular, the Tutte polynomial $t(M; x, y)$ is a well-defined matroid
invariant.*

Before proving the theorem, let's check it for Example 9.7.
In that example, we have $M = U_{1,1} \oplus U_{2,3}$. Using the fact that
$t(M_1 \oplus M_2) = t(M_1) \cdot t(M_2)$ (see Theorem 9.13), we get $t(M; x, y) =
x \cdot (x^2 + x + y) = x^3 + x^2 + xy$. Then you can check this
equals $s(M; x - 1, y - 1) = (x - 1)^3 + 4(x - 1)^2 + 6(x - 1) +
(x - 1)(y - 1) + (y - 1) + 3.$

*Proof Theorem 9.9.* The proof is by induction[7] on $|E|$. While this proof may seem long, looks can be deceiving.[8] There are several steps, but the overall outline is rather straightforward.

When $|E| = 1$, then $M$ is either an isthmus or a loop. If $M$ is an isthmus, then $r(E) = 1$. We get

$$s(M; u, v) = u^{r(E)-r(E)}v^{|E|-r(E)} + u^{r(E)-r(\emptyset)}v^{|\emptyset|-r(\emptyset)} = u + 1.$$

But $t(M; x, y) = x$, so we have $s(M; x - 1, y - 1) = t(M; x, y)$.

If $|E| = 1$ and $M$ is a loop, then $r(E) = 0$, and we get

$$s(M; u, v) = u^{r(E)-r(E)}v^{|E|-r(E)} + u^{r(E)-r(\emptyset)}v^{|\emptyset|-r(\emptyset)} = v + 1.$$

But $t(M; x, y) = y$, so we again find $s(M; x - 1, y - 1) = t(M; x, y)$.

Now let $n \geq 2$, and assume the theorem is true for matroids with $|E| = n - 1$. We must show it remains true when $|E| = n$. Let $e \in E$. Then there are three cases to consider.

*Case 1. $e$ is neither an isthmus nor a loop.* Then the proof goes like this:

- First, show $s(M; u, v) = s(M - e; u, v) + s(M/e; u, v)$.
- Then, since the ground set for $M - e$ and for $M/e$ has $n - 1$ elements, use induction to claim $t(M - e; x, y) = s(M - e; x - 1, y - 1)$ and $t(M/e; x, y) = s(M/e; x - 1, y - 1)$.
- Finally, use the recursive definition of the Tutte polynomial (Definition 9.2(1)) to show $t(M; x, y) = s(M; x - 1, y - 1)$.

To accomplish this program, we need to compute the corank–nullity polynomials $s(M - e; u, v)$ and $s(M/e; u, v)$. To do this, we need the following information about the rank functions in the deletion $M - e$ and the contraction $M/e$. (Formulas for these rank functions were given in Proposition 3.9.)

Let $r$ be the rank function of $M$, $r'$ the rank function in $M - e$ and $r''$ the rank function in $M/e$. Then,

- Deletion: $r'(A) = r(A)$ for all $A \subseteq E - e$.
- Contraction: If $e \in A$, then $r''(A - e) = r(A) - 1$.

Now $s(M; u, v) = \sum_{A \subseteq E} u^{r(E)-r(A)}v^{|A|-r(A)}$. We break up the subsets of $E$ into two classes: let $S_1$ be the collection of all subsets of $E$ containing $e$, and $S_2$ be the collection of all subsets of $E$ avoiding $e$. (Note that $S_1$ and $S_2$ each contains $2^{n-1}$ subsets.)

---

[7] Proofs by induction should be the (more-or-less) default proof technique when using a recursive definition.

[8] As Mark Twain once observed about Richard Wagner's compositions, "Wagner's music is better than it sounds."

Then

$$s(M; u, v) = \sum_{A \in \mathcal{S}_1} u^{r(E)-r(A)} v^{|A|-r(A)} + \sum_{A \in \mathcal{S}_2} u^{r(E)-r(A)} v^{|A|-r(A)}. \quad (9.1)$$

(1) Now suppose $A \in \mathcal{S}_1$, so $e \in A$. Then, for each of these subsets, $r''(A - e) = r(A) - 1$, i.e., the rank drops by 1 in $M/e$. Then, for $A \in \mathcal{S}_1$, the corank of $A$, computed in $M$, equals the corank of $A - e$, computed in $M/e$:

$$r(E) - r(A) = r''(E - e) - r''(A - e).$$

For the nullity, both the rank of $A$ and the cardinality of $A$ drop by 1 when we remove $e$:

$$|A| - r(A) = |A - e| - r''(A - e).$$

Thus

$$\sum_{A \in \mathcal{S}_1} u^{r(E)-r(A)} v^{|A|-r(A)} = \sum_{A \in \mathcal{S}_1} u^{r''(E-e)-r''(A-e)} v^{|A-e|-r''(A-e)} \quad (9.2)$$

$$= \sum_{B \subseteq E-e} u^{r''(E-e)-r''(B)} v^{|B|-r''(B)} \quad (9.3)$$

$$= s(M/e; u, v). \quad (9.4)$$

(2) For $A \in \mathcal{S}_2$, the argument is a little easier. In this case, $e \notin A$, $r(E) = r'(E - e)$ and $r(A) = r'(A)$, so

$$\sum_{A \in \mathcal{S}_2} u^{r(E)-r(A)} v^{|A|-r(A)} = \sum_{A \in \mathcal{S}_2} u^{r'(E-e)-r'(A)} v^{|A|-r'(A)} \quad (9.5)$$

$$= s(M - e; u, v). \quad (9.6)$$

Putting Equations (9.1), (9.4) and (9.6) together gives

$$s(M; u, v) = s(M - e; u, v) + s(M/e; u, v).$$

This gives

$$t(M; x, y) = t(M - e; x, y) + t(M/e; x, y) \text{ (by Definition 9.2(1))}$$
$$= s(M - e; x - 1, y - 1)$$
$$\quad + s(M/e; x - 1, y - 1) \text{ (by induction)}$$
$$= s(M; x - 1, y - 1).$$

*Case 2. $e$ is an isthmus.* As in case 1, we partition $2^E$ into $\mathcal{S}_1$ and $\mathcal{S}_2$ according to whether the subset contains $e$ or not. Then, as above, for $\mathcal{S}_1$, the corank and nullity of $A$, computed in $M$, are the same as the corank and nullity of $A - e$, computed in $M/e$. This gives

$$\sum_{A \in \mathcal{S}_1} u^{r(E)-r(A)} v^{|A|-r(A)} = s(M/e; u, v),$$

as in case 1.

Things are a little different for $\mathcal{S}_2$ this time, though. Since $e$ is an isthmus, we compare $r(A)$ and $r''(A)$ in $M/e$ (instead of $r'(A)$ in $M - e$, which is not defined). If $e \notin A$, then $r(A) = r''(A)$ in $M/e$, because contracting an isthmus does not change the rank of sets not containing the isthmus. So our corank changes in $M/e$: $r(E) - r(A) = (r''(E - e) + 1) - r''(A)$. Our nullity is unchanged, though: $|A| - r(A) = |A| - r''(A)$, since $e \notin A$.

Since $r(E) - r(A) = (r''(E - e) + 1) - r''(A)$, we find

$$\sum_{A \in \mathcal{S}_2} u^{r(E)-r(A)} v^{|A|-r(A)} = \sum_{A \subseteq E-e} u^{(r''(E-e)+1)-r''(A)} v^{|A|-r''(A)}$$

$$= u \cdot \sum_{A \subseteq E-e} u^{r''(E-e)-r''(A)} v^{|A|-r''(A)}$$

$$= u \cdot s(M/e; u, v).$$

Then

$$s(M; u, v) = \sum_{A \in \mathcal{S}_1} u^{r(E)-r(A)} v^{|A|-r(A)} + \sum_{A \in \mathcal{S}_2} u^{r(E)-r(A)} v^{|A|-r(A)}$$

$$= s(M/e; u, v) + u \cdot s(M/e; u, v)$$

$$= (u+1)s(M/e; u, v).$$

Now set $u = x - 1$ and $v = y - 1$. Then $u + 1 = x$, so we get

$$s(M; x-1, y-1) = x \cdot s(M/e; x-1, y-1)$$

$$= x \cdot t(M/e; x, y) \text{ (by induction)}$$

$$= t(M; x, y) \text{ (by Definition 9.2(2))}.$$

*Case 3.* $e$ is a loop. This is like case 2, and we leave the details[9] to Exercise 10.    □

We conclude this section with some nice evaluations of the Tutte polynomial.

**Proposition 9.10.** *Let M be a matroid. Then:*

(1)  $t(M; 1, 1) = s(M; 0, 0)$ *equals the number of bases of M.*
(2)  $t(M; 2, 1) = s(M; 1, 0)$ *equals the number of independent sets of M.*
(3)  $t(M; 1, 2) = s(M; 0, 1)$ *equals the number of spanning sets of M.*
(4)  $t(M; 2, 2) = s(M; 1, 1) = 2^{|E|}$.

*Proof Proposition 9.10.* (1) We give two proofs; one proof uses the recursive Definition 9.2, and the other uses Theorem 9.9.

---

[9]  Math books, including this one, are filled with boring exercises that begin "Finish the proof of ...." This problem fits nicely into this genre, but it's worthwhile to go through the details of this case, and understanding those details is semi-rewarding.

- **Recursive proof** Let $b(M)$ be the number of bases of the matroid $M$. We proceed by induction. If $|E| = 1$, then $t(M) = x$ if $M$ is an isthmus and $t(M) = y$ if $M$ is a loop. In either case, we find $t(M; 1, 1) = 1 = b(M)$.

  Now suppose $|E| > 1$. Then, if $e$ is neither an isthmus nor a loop, we have $b(M) = b(M - e) + b(M/e)$ (this appears in Exercise 6 in Chapter 3, but it follows immediately from the definition of the independent sets in $M - e$ and $M/e$). Thus

  $$b(M) = b(M - e) + b(M/e)$$
  $$= t(M - e; 1, 1) + t(M/e; 1, 1) \text{ (by induction)}$$
  $$= t(M; 1, 1) \text{ (by Definition 9.2(1))}.$$

  What if $e$ is an isthmus? Then we have $b(M) = b(M/e)$ (since $e$ is in every basis of $M$), so

  $$b(M) = b(M/e)$$
  $$= t(M/e; 1, 1) \text{ (by induction)}$$
  $$= 1 \cdot t(M; 1, 1) \text{ (by Definition 9.2(2))}.$$

  If $e$ is a loop, then $b(M) = b(M - e)$ (since $e$ is in no basis of $M$), and the argument is analogous. We omit the details.

- **Corank–nullity proof**  We know $t(M; 1, 1) = s(M; 0, 0)$ (by Theorem 9.9). Now consider the evaluation of $s(M; u, v)$ at $u = v = 0$. Then a subset $B$ will contribute a non-zero term to this evaluation precisely when $r(B) = r(E)$ and $|B| = r(B)$. Now $r(B) = r(E)$ tells us $B$ is a spanning set, while $|B| = r(B)$ implies $B$ is an independent set. Thus each basis $B$ contributes 1 to $S(M; 0, 0)$ and each non-basis contributes 0. Hence $t(M; 1, 1) = s(M; 0, 0)$ equals the number of bases of $M$.

(2), (3) and (4): We leave the proofs to you[10] – see Exercise 11.   ∏

As a quick check, let's look at the matroid $M$ in Example 9.7. $M$ has three bases ($abd$, $acd$, $bcd$), 14 independent sets, four spanning sets, and there are $2^4 = 16$ subsets of $E$. We have $t(M; x, y) = x^3 + x^2 + xy$, and you can check $t(M; 1, 1) = 3$, $t(M, 2, 1) = 14$, $t(M; 1, 2) = 4$ and $t(2, 2) = 16$.

Can different (non-isomorphic) matroids have the same Tutte polynomial? The next example answers this question.

**Example 9.11.**  Let $M_1$ and $M_2$ be the two matroids at the top of Figure 9.4. Then you can check $M_1 - x \cong M_2 - y$ and $M_1/x \cong M_2/y$, as shown in the figure. Thus $t(M_1) = t(M_2)$, but it's clear $M_1$ and $M_2$ are not isomorphic matroids (since, for instance, $M_1$ has a four-point line, but $M_2$ does not).

[10] See previous footnote.

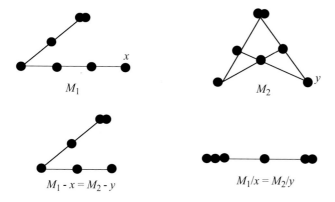

Figure 9.4. Two non-isomorphic matroids with the same Tutte polynomial.

$M_1$

$M_2$

$M_1 - x = M_2 - y$

$M_1/x = M_2/y$

One consequence of this example: any property one of these matroids has that the other matroid does not is a property the Tutte polynomial *cannot* determine. For instance, you can check $M_2$ is a graphic matroid (find the graph!), but $M_2$ is not even binary (it has a four-point line as a minor). Thus, you can't tell if $M$ is graphic (or binary) from its Tutte polynomial.

You can also see that $M_1$ and $M_2$ have different numbers of circuits, flats and hyperplanes. See Exercise 7.

## 9.4 Duality and direct sum

The Tutte polynomial has a very nice relationship with the matroid operations of duality and direct sum. The general relationship between the Tutte polynomials of a matroid and its dual is given in the next theorem.

**Theorem 9.12.** *If M is a matroid, then the Tutte polynomial of the dual matroid M\* is given by*

$$t(M^*; x, y) = t(M; y, x).$$

*Proof Theorem 9.12.* We use the corank–nullity polynomial formulation from Theorem 9.9. First, we need the formula (see Theorem 3.20) for the rank $r^*(A)$ in the dual matroid $M^*$:

$$r^*(A) = r(E - A) + |A| - r(E).$$

Then

$$r^*(E) - r^*(A) = |E - A| - r(E - A) \quad \text{and}$$
$$|A| - r^*(A) = r(E) - r(E - A).$$

Thus, the corank of $A$ in the dual matroid $M^*$ equals the nullity of $E - A$ in $M$, and the nullity of $A$ in $M^*$ is the corank of $E - A$ in $M$.

So

$$s(M^*; u, v) = \sum_{A \subseteq E} u^{r^*(E) - r^*(A)} v^{|A| - r^*(A)}$$

$$= \sum_{A \subseteq E} u^{|E-A| - r(E-A)} v^{r(E) - r(E-A)}$$

$$= \sum_{A' \subseteq E} u^{|A'| - r(A')} v^{r(E) - r(A')} \quad (\text{setting } A' = E - A)$$

$$= s(M; v, u).$$

The result now follows from Theorem 9.9. $\qquad\square$

It's worth proving $t(M^*; x, y) = t(M; y, x)$ directly from the recursive definition. The inductive proof structure is identical to our first proof of Proposition 9.10(1). See Exercise 16.

If $M = M_1 \oplus M_2$ is a direct sum of two matroids, it is natural to ask if $t(M)$ can be computed from $t(M_1)$ and $t(M_2)$. The next result tells us you can, and the formula is what you might guess.[11]

**Theorem 9.13.** *The Tutte polynomial of the direct sum is given by*

$$t(M_1 \oplus M_2; x, y) = t(M_1; x, y) t(M_2; x, y).$$

We will use the recursive definition of the Tutte polynomial to prove Theorem 9.13, but we need the following result on deletion and contraction in direct sums first.

**Proposition 9.14.** *Let $M_1, M_2$ be matroids defined on disjoint ground sets $E_1, E_2$ with $e \in E_1$. Then*

$$(M_1 \oplus M_2) - e = (M_1 - e) \oplus M_2$$

*and*

$$(M_1 \oplus M_2)/e = (M_1/e) \oplus M_2.$$

*A similar result holds for $e \in E_2$.*

The proof of Proposition 9.14 follows from Proposition 3.33 along with the definition of deletion and contraction. With this proposition in hand, we are ready to prove our theorem.

*Proof Theorem 9.13.* Let $M_1, M_2$ be matroids defined on disjoint ground sets $E_1, E_2$, and write $M = M_1 \oplus M_2$. We also write $t(M)$ instead of the more cumbersome $t(M; x, y)$ in formulas in the proof.

The proof is by induction on the size of $E_1$. Let $e \in E_1$. If $|E_1| = 1$, then $e$ is either a loop or an isthmus.

---

[11] This is an excellent opportunity to play the "cover up everything to the right of the equal sign" guessing game.

If $e$ is a loop, then $M - e = (M_1 - e) \oplus M_2 = M_2$, and $t(M_1) = y$, so

$$t(M) = y \cdot t(M - e) = y \cdot t(M_2) = t(M_1)t(M_2).$$

Similarly, if $e$ is an isthmus, then $M/e = (M_1/e) \oplus M_2 = M_2$, and $t(M_1) = x$, so, this time, we get

$$t(M) = x \cdot t(M/e) = x \cdot t(M_2) = t(M_1)t(M_2).$$

Now assume $|E_1| > 1$. If $e$ is not a loop or an isthmus, then

$$
\begin{aligned}
t(M) &= t(M - e) + t(M/e) \\
&= t(M_1 - e \oplus M_2) + t(M_1/e \oplus M_2) \\
&= t(M_1 - e)t(M_2) + t(M_1/e)t(M_2) \text{ (by induction)} \\
&= [t(M_1 - e) + t(M_1/e)]t(M_2) \\
&= t(M_1)t(M_2).
\end{aligned}
$$

If $e$ is a loop or an isthmus, a similar argument works. We leave those details to the reader. $\qquad\square$

One of the morals of our treatment of the Tutte polynomial is that we can prove everything twice;[12] either by induction (from Definition 9.2) or via a subset collection argument (from Theorem 9.9). In that spirit, Exercise 18 asks you to construct a proof that $t(M_1 \oplus M_2) = t(M_1)t(M_2)$ using the subset expansion of the corank–nullity polynomial.

If we write $t(M; x, y) = \sum b_{i,j} x^i y^j$, then the coefficient of the $x$ term in $t(M; x, y)$ is just $b_{1,0}$. This coefficient is important for a variety of applications, and it gets a special name.

**Definition 9.15.** If $M$ is a matroid with Tutte polynomial $t(M; x, y) = \sum b_{i,j} x^i y^j$, then we call $b_{1,0}$ the *beta invariant* of the matroid $M$, and we write $\beta(M)$.

Beta invariant

For instance, since $t(U_{2,4}; x, y) = x^2 + 2x + 2y + y^2$, we get $\beta(U_{2,4}) = 2$. Note that $\beta(M) \geq 0$ for all matroids $M$.

The proof of the next proposition follows immediately from our deletion–contraction definition of the Tutte polynomial.

**Proposition 9.16.** *Let $M$ be a matroid with $e$ neither an isthmus nor a loop. Then $\beta(M) = \beta(M/e) + \beta(M - e)$.*

We are interested in $\beta(M)$ because it tells us when the matroid $M$ is connected.

**Proposition 9.17.** *Suppose $M$ is a matroid and $M$ is not a loop. Then $M$ is a connected matroid if and only if $\beta(M) > 0$.*

---

[12] Why would you want two proofs of everything? You learn different things from different proofs, and you should end up with a deeper understanding of the results. At least, that's the hope.

*Proof Proposition 9.17.* First suppose $\beta(M) > 0$. We must show $M$ is connected. But, if not, then $M = M_1 \oplus M_2$ for non-trivial matroids $M_1$ and $M_2$, so $t(M) = t(M_1)t(M_2)$ by Theorem 9.13. Since all the coefficients of the Tutte polynomials $t(M_1)$ and $t(M_2)$ are non-negative, there is no way to produce the term $b_{1,0}x$ as a product of two terms from $t(M_1)$ and $t(M_2)$. (The only way this could happen would be if one of these two terms were a constant, but $b_{0,0} = 0$ for non-trivial matroids, so this can't happen.)

For the converse, suppose $M$ is a connected matroid. We use induction on $|E|$. If $|E| = 1$, then $M$ is an isthmus (we assumed $M$ is not a loop in the hypothesis). Then $t(M; x, y) = x$, so $\beta(M) = 1 > 0$, and the proposition is true. If $|E| = 2$, then $M$ is a double point (this is the only connected matroid on two points). Then $t(M; x, y) = x + y$, and, again, we find $\beta(M) = 1$.

Now assume $|E| > 2$ and $M$ is a connected matroid on the ground set $E$. Then, by Theorem 3.45, for all $e \in E$, either $M - e$ or $M/e$ (or both) is connected. Since $|E| > 2$, we know neither $M - e$ nor $M/e$ could be a single loop.[13] Now if $M/e$ is connected, then, by induction, $\beta(M/e) > 0$. Similarly, $\beta(M - e) > 0$ if $M - e$ is connected (also by the induction hypothesis). Thus, by Proposition 9.16, $\beta(M) = \beta(M/e) + \beta(M - e) > 0$, so we are done.    □

A dual version of Proposition 9.17 appears in Exercise 20. We also point out that if $L$ is the one-point matroid corresponding to a loop, then $\beta(L) = 0$ since $t(L; x, y) = y$. This matroid is connected, however; this explains the loop prohibition in the hypothesis of Proposition 9.17. But this trivial case is the only time $\beta(M) = 0$ for a connected matroid.

We conclude this section with a very nice, direct characterization of connected matroids via the Tutte polynomial. This theorem was proved by Merino, de Mier, and Noy in 2001 [24].

**Theorem 9.18.** *Let $M$ be a matroid with Tutte polynomial $t(M; x, y)$. Then $t(M; x, y)$ is an irreducible polynomial (in the polynomial ring $\mathbb{Z}[x, y]$) if and only if $M$ is a connected matroid.*

One direction of this proof follows immediately from Theorem 9.13: if $M$ is not connected, then $M = M_1 \oplus M_2$, so the Tutte polynomial $t(M) = t(M_1)t(M_2)$ factors. The converse is much harder, and involves a detailed analysis of the terms of the polynomial.

## 9.5 Tutte–Grothendieck invariants

The Tutte polynomial of a matroid gives lots of interesting information about the matroid. For instance, we can find the number of bases $b(M)$,

---

[13] This is why we included the $|E| = 2$ case as a base case for the induction.

independent sets $i(M)$ and spanning sets $sp(M)$ easily, using Proposition 9.10. We can't reconstruct the matroid itself, of course; that's the point of Example 9.11, where we saw $t(M_1) = t(M_2)$ is possible for non-isomorphic matroids $M_1$ and $M_2$.

The key property that each of $b(M), i(M)$ and $sp(M)$ satisfies is our deletion–contraction recursion (when $e$ is neither an isthmus nor a loop):

- $b(M) = b(M - e) + b(M/e)$,
- $i(M) = i(M - e) + i(M/e)$,
- $sp(M) = sp(M - e) + sp(M/e)$.

There is a vague, imprecise, moral.

> **Meta-theorem**[14] *Any matroid invariant $f(M)$ that satisfies*
>
> $$f(M) = f(M - e) + f(M/e)$$
>
> *can be obtained from the Tutte polynomial.*

We can give a more precise formulation of this idea.

**Theorem 9.19.** *[Brylawski [8]] Let $\mathcal{M}$ be the set of isomorphism classes of matroids, and let $f : \mathcal{M} \to \mathbb{Z}[x, y]$ satisfying*

(1) $f(M_1) = f(M_2)$ *for isomorphic matroids $M_1 \cong M_2$ (i.e., $f$ is a matroid invariant);*
(2) $f(M) = f(M - e) + f(M/e)$ *for $e$ not a loop or an isthmus;*
(3) $f(M) = f(I) \cdot f(M/e)$ *for $I$ an isthmus;*
(4) $f(M) = f(L) \cdot f(M - e)$ *for $L$ a loop.*

*Then $f(M) = t(M; f(I), f(L))$.*

An invariant satisfying (1)–(4) in Theorem 9.19 is called a *Tutte–Grothendieck* (T-G) invariant. For example, for the number of independent sets $i(M)$, we note

Tutte–Grothendieck invariant

- $i(M) = i(M - e) + i(M/e)$ when $e$ is not an isthmus or a loop;
- $i(M) = 2i(M/e)$ when $e$ is an isthmus;
- $i(M) = i(M - e)$ when $e$ is a loop.

Since $i(I) = 2$ and $i(L) = 1$ for an isthmus $I$ and a loop $L$, by the theorem, we must have $i(M) = t(M; 2, 1)$, i.e., we set $x = 2$ and $y = 1$ in $t(M; x, y)$ to find the number of independent sets. This agrees with Proposition 9.10(2).

We now give a proof of Theorem 9.19. The inductive technique should be familiar by now. First show the theorem holds for one-element

---

[14] A "meta-theorem" is a statement about collections of theorems.

matroids, then use the recursive definition of the Tutte polynomial along with properties (2), (3) and (4) of the invariant $f$ to complete the induction.

*Proof Theorem 9.19.* Suppose $M$ is a matroid on the ground set $E$. We use induction on $|E|$. If $|E| = 1$, then $M$ is either an isthmus or a loop. In either case, the result follows immediately from (3) or (4).

So now we suppose $|E| > 1$, and $e \in E$. As usual, we have three cases to consider.

*Case 1. $e$ is not an isthmus or a loop.* Then

$$f(M) = f(M - e) + f(M/e)$$
$$= t(M - e; f(I), f(L)) + t(M/e; f(I), f(L)) \text{ (by induction)}$$
$$= t(M; f(I), f(L)) \text{ (by Definition 9.2(1))}.$$

*Case 2. $e$ is an isthmus.* Then

$$f(M) = f(I) \cdot f(M/e)$$
$$= f(I) \cdot t(M/e; f(I), f(L)) \text{ (by induction)}$$
$$= t(M; f(I), f(L)) \text{ (by Definition 9.2(2))}.$$

*Case 3. $e$ is a loop.* Then the argument is entirely analogous to case 2:

$$f(M) = f(L) \cdot f(M - e)$$
$$= f(L) \cdot t(M - e; f(I), f(L)) \text{ (by induction)}$$
$$= t(M; f(I), f(L)) \text{ (by Definition 9.2(3))}.$$

$\square$

It's not too hard to generalize Theorem 9.19. One such generalization replaces the deletion–contraction recursion (2) in the theorem with the following: if $e$ is not a loop or an isthmus, then a *generalized T-G invariant* satisfies (1), (3), (4) and

*Generalized T-G invariant*

$$(2') \quad f(M) = a \cdot f(M - e) + b \cdot f(M/e),$$

where $a$ and $b$ are fixed constants. Then $f(M)$ can still be obtained from the Tutte polynomial:

$$f(M) = a^{|E|-r(E)} b^{r(E)} t \left( M; \frac{f(I)}{b}, \frac{f(L)}{a} \right).$$

See Exercise 21.

We conclude this section with a nice application of Theorem 9.19 to graphs. If $G$ is a graph, we can *orient* $G$ by assigning directions to all the edges of $G$. If an orientation of $G$ has no directed cycles, we say the orientation is *acyclic*.

**Example 9.20.** Let $C_3$ be the triangle of Figure 9.5. How many acyclic orientations does $C_3$ have? It's easier to count the *bad* orientations,

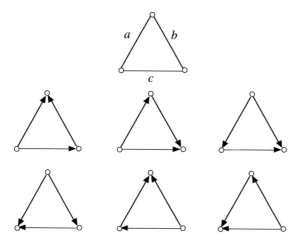

Figure 9.5. The six acyclic orientations of a triangle.

i.e., those that have cycles, and subtract the bad ones from the total number of orientations. But there are only two bad orientations; these are directed cycles themselves. Subtracting these two bad orientations from the $2^3 = 8$ total leaves six acyclic orientations. These six acyclic orientations are shown in Figure 9.5.

What about the Tutte polynomial? We know $M(C_3) = U_{2,3}$, so $t(M(C_3); x, y) = x^2 + x + y$. Evaluating the Tutte polynomial at $x = 2$, $y = 0$ gives $t(M(C_3), 2, 0) = 6$.

This always works.

**Theorem 9.21.**  *[Stanley [33]] Let $G$ be a graph and let $a(G)$ be the number of acyclic orientations of $G$. Then $a(G) = t(M(G); 2, 0)$.*

*Proof Theorem 9.21.* We first show $a(G)$ is a T-G invariant, then apply Theorem 9.19. Let $e$ be an edge of the graph $G$. Let's first dispense with loops and isthmuses.

If $L$ is a loop, then $L$ forms a cycle, and there is no way to orient $L$ without creating a directed cycle. This means $a(L) = 0$. Moreover, $a(G) = 0$ if $G$ contains a loop. We express this like so: $a(G) = 0 \cdot a(G - e)$ when $e$ is a loop.

If $I$ is an isthmus, then we can orient $I$ in two ways without creating a directed cycle: $a(I) = 2$. Further, if $e$ is an isthmus in the graph $G$, then $a(G) = 2 \cdot a(G/e)$. This is true because every acyclic orientation of $G$ gives an acyclic orientation of $G/e$, and each acyclic orientation of $G/e$ will appear twice: once for each of the two legal orientations of the isthmus $e$.

Now we assume $e$ is neither an isthmus nor a loop. We need to show $a(G) = a(G - e) + a(G/e)$. The theorem will then follow because $M(G - e) = M(G) - e$ and $M(G/e) = M(G)/e$, so we are using the standard matroid deletion and contraction operations.

Suppose the endpoints of $e$ are the vertices $u$ and $v$. Then we separate into four cases.

*Case 1.* There is no directed path from $u$ to $v$ or from $v$ to $u$ in $G - e$.
*Case 2.* There is a directed path from $u$ to $v$, but no directed path from $v$ to $u$ in $G - e$.
*Case 3.* There is a directed path from $v$ to $u$, but no directed path from $u$ to $v$ in $G - e$.
*Case 4.* There are directed paths from $u$ to $v$ and from $u$ to $v$ in $G - e$.

Let $a_1, a_2, a_3$ and $a_4$ be the number of acyclic orientations of $G - e$ in each of these cases. Then $a_4 = 0$; if there are directed paths from $u$ to $v$ and from $v$ to $u$, then we could piece these two paths together to get a directed cycle in $G - e$.

Thus, $a(G - e) = a_1 + a_2 + a_3$. What about $G/e$? Any acyclic orientation of $G - e$ in case 1 will give an acyclic orientation of $G/e$ (add the edge $e$, then contract it). Further, every acyclic orientation of $G/e$ arises in this way, so $a(G/e) = a_1$.

We still need to compute $a(G)$. Now every orientation in cases 2 and 3 gives rise to a unique acyclic orientation of $G$. The orientations in case 1 give rise to *two* acyclic orientations of $G$: we can orient $e$ in two ways without creating a directed cycle. Thus $a(G) = 2a_1 + a_2 + a_3$.

Putting this all together gives us $a(G) = a(G - e) + a(G/e)$, and that completes the proof that $a(G)$ is a T-G invariant. Since $a(I) = 2$ and $a(L) = 0$, we have $a(G) = t(M(G); 2, 0)$ by Theorem 9.19.    □

One consequence of Theorem 9.21 is that the number of acyclic orientations only depends on the matroid structure of $M(G)$. So if $G_1$ and $G_2$ are different graphs with $M(G_1) = M(G_2)$, then $a(G_1) = a(G_2)$. Recall from Section 4.4 that two graphs give the same matroid precisely when we can obtain $G_1$ from $G_2$ by a sequence of vertex gluing, vertex splitting and twisting operations. Thus, these operations do not change the number of acyclic orientations of the graph.

A more striking application is given in Exercise 27: let $G$ be a graph and let $v$ be a vertex of $G$. Let $a_v(G)$ be the number of acyclic orientations of a graph $G$ with $v$ as the unique source (so every edge incident to $v$ is directed away from $v$). Then $a_v(G) = t(M(G); 1, 0)$. So this number depends only on the graph, not the vertex you chose as your source.

## 9.6 The chromatic polynomial

Much of Tutte's original motivation for introducing a two-variable polynomial for graphs is based on the chromatic polynomial. In fact, Tutte (modestly) called his two-variable generalization a *dichromatic* polynomial. Crapo [12] generalized Tutte's work to matroids, and

Brylawski [8] proved many of the foundational matroid Tutte polynomial results.

In this section, we show how the Tutte polynomial of a cycle matroid of a graph is related to the chromatic polynomial of the graph. Let $G = (V, E)$ be a graph and $M(G)$ the corresponding cycle matroid of $G$.

## 9.6.1 Basics of graph coloring

**Definition 9.22.** A vertex coloring of a graph $G$ with vertex set $V$ is a map $c : V \to \mathbb{Z}^+$, where $\mathbb{Z}^+$ denotes the positive integers. The coloring is *proper* if $c(u) \neq c(v)$ whenever the vertices $u$ and $v$ are joined by an edge.

*Vertex coloring*

So a coloring is proper if and only if adjacent vertices receive different colors. The smallest number of colors needed to achieve a proper coloring is the *chromatic number* of $G$. Then it is easy to see that $G$ has chromatic number 1 if and only if $G$ has no edges. It is also easy to check the chromatic number of the $n$ cycle $C_n$ is 2 or 3, depending on whether $n$ is even or odd, respectively.

A slightly more interesting class of graphs are those with chromatic number equal to 2. You can see that any bipartite graph can be 2-colored; simply color all the vertices on one side red, and color the rest of the vertices blue. Conversely, if you can color the vertices of $G$ using two colors, then your graph is bipartite: place all the red vertices on one side and all the blue vertices on the other.

$G$ is bipartite if and only if $G$ has chromatic number (at most) 2.

Returning to polynomial invariants, we define the chromatic polynomial $\chi_G(\lambda)$.

**Definition 9.23.** Let $\lambda$ be a non-negative integer. The *chromatic polynomial* $\chi_G(\lambda)$ of a graph $G$ is the number of proper colorings of $G$ using at most $\lambda$ colors.

*Chromatic polynomial*

So $\chi_G(\lambda) > 0$ if and only if $G$ has chromatic number at most $\lambda$, where $\lambda \in \mathbb{Z}^+$. Let's look at a few small examples.

**Example 9.24.** Consider the three graphs in Figure 9.6. Then $G_1$ is the complete graph $K_4$, and it's pretty clear we can color the vertices "greedily:" first, we have $\lambda$ choices for the color we use for vertex 1. Since vertex 2 must receive a different color from vertex 1, we have $\lambda - 1$ choices for the color of vertex 2. Continuing in this way, we have $\lambda - 2$ choices for vertex 3 and $\lambda - 3$ choices for vertex 4. This gives $\chi_{G_1}(\lambda) = \lambda(\lambda - 1)(\lambda - 2)(\lambda - 3)$. So, for instance, there are $\chi_{G_1}(4) = 24$ ways to properly 4-color the vertices, but $\chi_{G_1}(3) = 0$, so it is impossible to properly color $K_4$ using three colors.

Figure 9.6. Three graphs for
Example 9.24.

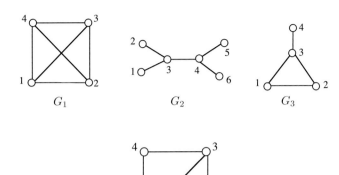

Figure 9.7. Graph $G$ for
Example 9.25.

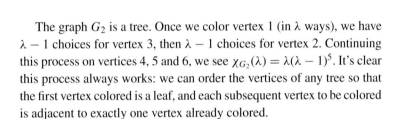

The graph $G_2$ is a tree. Once we color vertex 1 (in $\lambda$ ways), we have
$\lambda - 1$ choices for vertex 3, then $\lambda - 1$ choices for vertex 2. Continuing
this process on vertices 4, 5 and 6, we see $\chi_{G_2}(\lambda) = \lambda(\lambda - 1)^5$. It's clear
this process always works: we can order the vertices of any tree so that
the first vertex colored is a leaf, and each subsequent vertex to be colored
is adjacent to exactly one vertex already colored.

Finally, let's compute $\chi_{G_3}(\lambda)$. First, color vertices 1, 2 and 3 in
$\lambda(\lambda - 1)(\lambda - 2)$ ways. Then vertex 4 has $\lambda - 1$ colors available, so
$\chi_{G_3}(\lambda) = \lambda(\lambda - 1)^2(\lambda - 2)$.

The argument used to compute $\chi_{G_2}(\lambda)$ for the tree $G_2$ generalizes.
$\chi_T(\lambda) = \lambda(\lambda - 1)^{n-1}$ for any tree $T$ with $n$ vertices. Then $\chi_T(2) > 0$, so
$T$ can be 2-colored. This shows that trees are bipartite graphs.

We also note[15] that the three graphs of Example 9.24 are deceivingly
easy to color. In general, we can't just order the vertices, color them in
some number of colors, then multiply to get $\chi_G(\lambda)$.

**Example 9.25.**   For a more typical example, consider the graph $G$ in
Figure 9.7.

For this graph, we can color vertex 1 in $\lambda$ ways, vertex 2 in $\lambda - 1$
ways, and vertex 3 in $\lambda - 2$ ways. But in order to color vertex 4, we
need to consider two cases:

- vertex 4 gets the same color as vertex 2;
- vertices 2 and 4 receive different colors.

In the first case, there is one choice for vertex 4 (its color is already
determined). In the second case, we have $\lambda - 3$ choices. Thus

$$\chi_G(\lambda) = \lambda(\lambda - 1)(\lambda - 2) + \lambda(\lambda - 1)(\lambda - 2)(\lambda - 3)$$
$$= \lambda(\lambda - 1)(\lambda - 2)^2.$$

---

[15] Somewhat sheepishly.

Although we've called $\chi_G(\lambda)$ the chromatic *polynomial*, it's not immediately obvious $\chi_G(\lambda)$ is always a polynomial in $\lambda$. Let's prove that our terminology is valid. First, we will need the following recursive formula for $\chi_G(\lambda)$.

**Proposition 9.26.** *Suppose G is a simple graph (no loops or multiple edges) and e is an edge. Then*

$$\chi_G(\lambda) = \chi_{G-e}(\lambda) - \chi_{G/e}(\lambda).$$

*Proof Proposition 9.26.* For each proper coloring of $G - e$, either the endpoints $u$ and $v$ of $e$ are colored the same or different. If they are different, then the coloring of $G - e$ is also a proper coloring of $G$. Further, every proper coloring of $G$ corresponds to a proper coloring of $G - e$ in which $u$ and $v$ are colored differently.

On the other hand, if $u$ and $v$ receive the same color in $G - e$, then we get a proper coloring of $G/e$, and every proper coloring of $G/e$ arises in this way. Thus, $\chi_{G-e}(\lambda) = \chi_G(\lambda) + \chi_{G/e}(\lambda)$. □

As a quick check, consider the graph $G$ from Figure 9.7. Let $e$ be the edge joining vertices 1 and 2. Then $G - e$ is isomorphic to the graph $G_3$ from Figure 9.6, and $G/e$ is the 3-cycle $C_3$ with one pair of parallel edges. We can eliminate one of the parallel edges without changing the chromatic polynomial, so we get:

$$\begin{aligned}
\chi_G(\lambda) &= \chi_{G-e}(\lambda) - \chi_{G/e}(\lambda) \\
&= \chi_{G_3}(\lambda) - \chi_{C_3}(\lambda) \\
&= \lambda(\lambda - 1)^2(\lambda - 2) - \lambda(\lambda - 1)(\lambda - 2) \\
&= \lambda(\lambda - 1)(\lambda - 2)^2,
\end{aligned}$$

which agrees with our earlier calculation.

Now we can use Proposition 9.26 and induction to prove $\chi_G(\lambda)$ is always a polynomial in $\lambda$. (If every edge of $G$ is an isthmus or a loop, then we can't apply Proposition 9.26, but it's easy to compute $\chi_G(\lambda)$ in those cases.)

**Corollary 9.27.** $\chi_G(\lambda)$ *is a polynomial in* $\lambda$.

### 9.6.2 Connection to Tutte polynomial

Think about this: the cycle matroid $M(G)$ of a graph is defined on the edges of $G$, but the chromatic polynomial of a graph is defined on the vertices of $G$. Thus, any connection between the Tutte polynomial $t(M(G))$ and $\chi_G(\lambda)$ will need an adjustment involving the vertices of $G$. Here is that connection.

**Theorem 9.28.** *Let G be a connected graph with vertex-set V. Then*

$$\chi_G(\lambda) = (-1)^{|V|-1}\lambda \cdot t(M(G); 1 - \lambda, 0).$$

Figure 9.8. Coloring a graph
with an isthmus.

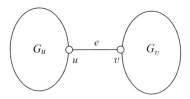

Before the proof, a quick example, why not?

**Example 9.29.** Let's verify Theorem 9.28 for the 3-cycle $C_3$. We computed the Tutte polynomial in Example 9.4: $t(C_3; x, y) = x^2 + x + y$. So, according to Theorem 9.28, we should have

$$\chi_{C_3}(\lambda) = (-1)^{3-1}\lambda \cdot \left((1 - \lambda)^2 + (1 - \lambda)\right) = \lambda(\lambda - 1)(\lambda - 2).$$

This agrees with our computation for complete graphs in Example 9.24.

*Proof Theorem 9.28.* We would like to use Theorem 9.19 by cooking up an appropriate T-G invariant, based on $\chi_G(\lambda)$. Before doing that, however, we need to understand loops and isthmuses for $\chi_G(\lambda)$.

If $I$ is an isthmus and $L$ is a loop, then $\chi_I(\lambda) = \lambda(\lambda - 1)$ and $\chi_L(\lambda) = 0$. Moreover, it is impossible to properly color any graph that contains a loop; $\chi_G(\lambda) = 0$ (as a polynomial) in this case. Thus, $\chi_G(\lambda) = \chi_L(\lambda) \cdot \chi_{G-e}(\lambda)$ when $e$ is a loop.

We need to relate $\chi_{G/e}(\lambda)$ and $\chi_G(\lambda)$ when $e$ is an isthmus. In this case, let $u$ and $v$ be the two vertices incident to the isthmus $e$, and let $G_u$ and $G_v$ be the two components of $G - e$, as in Figure 9.8.

Then it's easy to see $\chi_{G-e}(\lambda) = \chi_{G_u}(\lambda) \cdot \chi_{G_v}(\lambda)$. By Proposition 9.26, we also know $\chi_{G-e}(\lambda) = \chi_G(\lambda) + \chi_{G/e}(\lambda)$, so

$$\chi_G(\lambda) + \chi_{G/e}(\lambda) = \chi_{G_u}(\lambda) \cdot \chi_{G_v}(\lambda).$$

Now if $u$ and $v$ receive the same color, we get a proper coloring of $G/e$. The number of such proper colorings of $G$ is $\dfrac{\chi_{G_u}(\lambda) \cdot \chi_{G_v}(\lambda)}{\lambda}$. To see this, first color $G_u$ in $\chi_{G_u}(\lambda)$ ways, then note that we can partition the $\chi_{G_v}(\lambda)$ colorings of $G_v$ into $\lambda$ classes, each with size $\dfrac{\chi_{G_v}(\lambda)}{\lambda}$, based on the color of $v$. Only one of these classes gives a proper coloring of $G/e$, the class where the color of $v$ matches the color of $u$.

Then we get

$$\chi_{G/e}(\lambda) = \frac{\chi_{G_u}(\lambda) \cdot \chi_{G_v}(\lambda)}{\lambda} = \frac{\chi_G(\lambda)}{\lambda} + \frac{\chi_{G/e}(\lambda)}{\lambda}.$$

So, when $e$ is an isthmus, we have

$$\chi_G(\lambda) = (\lambda - 1)\chi_{G/e}(\lambda).$$

We are now ready to define a T-G invariant based on $\chi_G(\lambda)$. We let

$$f_G(\lambda) = \frac{(-1)^{|V|-1}\chi_G(\lambda)}{\lambda}.$$

We will show $f_G(\lambda)$ is a matroid T-G invariant, and then we'll apply Theorem 9.19.

- Loops: If $L$ is a loop, then $f_L(\lambda) = 0$. Furthermore, if $e$ is a loop in the graph $G$, we also have $f_G(\lambda) = f_L(\lambda) \cdot f_{G-e}(\lambda) = 0$.
- Isthmuses: If $I$ is an isthmus, then $f_I(\lambda) = (-1)\lambda(\lambda - 1)/\lambda = (1 - \lambda)$. Now if $e$ is an isthmus in the graph $G$, from above, we have $(\lambda - 1)\chi_{G/e}(\lambda) = \chi_G(\lambda)$, so

$$f_G(\lambda) = (-1)^{|V|-1} \frac{\chi_G(\lambda)}{\lambda}$$

$$= (-1)^{|V|-1} \frac{(\lambda - 1)\chi_{G/e}(\lambda)}{\lambda}$$

$$= (1 - \lambda) \cdot (-1)^{|V|-2} \frac{\chi_{G/e}(\lambda)}{\lambda}$$

$$= (1 - \lambda)f_{G/e}(\lambda),$$

since $G/e$ has $|V| - 1$ vertices.

- If $e$ is neither an isthmus nor a loop, then

$$f_G(\lambda) = (-1)^{|V|-1} \frac{\chi_G(\lambda)}{\lambda}$$

$$= (-1)^{|V|-1} \frac{\chi_{G-e}(\lambda) - \chi_{G/e}(\lambda)}{\lambda} \quad \text{(by Prop. 9.26)}$$

$$= (-1)^{|V|-1} \frac{\chi_{G-e}(\lambda)}{\lambda} + (-1)^{|V|-2} \frac{\chi_{G/e}(\lambda)}{\lambda}$$

$$= f_{G-e}(\lambda) + f_{G/e}(\lambda).$$

So $f_G(\lambda)$ meets all the requirements of a T-G invariant, so, by Theorem 9.19, $f_G(\lambda) = t(M(G); 1 - \lambda, 0)$. But $f_G(\lambda) = (-1)^{|V|-1}\chi_G(\lambda)/\lambda$, so

$$\chi_G(\lambda) = (-1)^{|V|-1}\lambda \cdot t(M(G); 1 - \lambda, 0).$$

$\square$

A few comments about the proof are in order:

- The key to the proof is the construction of $f_G(\lambda)$. Why didn't $\chi_G(\lambda)$ work as a T-G invariant? There are two problems here:
  (1) The recursion $\chi_G(\lambda) = \chi_{G-e}(\lambda) - \chi_{G/e}(\lambda)$ has a $-$ instead of a $+$.
  (2) Even if we fixed this, by defining $g_G(\lambda) = (-1)^{|V|-1}\chi_G(\lambda)$, say, it still fails. The problem is that $g_I(\lambda) = -\lambda(\lambda - 1)$, but $g_G(\lambda) = (1 - \lambda)g_{G/e}(\lambda)$, so part (3) of Theorem 9.19 is not satisfied.
- More generally, this difficulty points out one of the problems in using induction as a proof tool: *you have to know the answer before you start.* If you don't, you might be able to fix things up with a little cleverness,

as we were able to do here. But it's also easy to go astray.[16] The moral: be careful with induction!

If $G$ is not connected, then it is easy to modify the evaluation in Theorem 9.28. If $G$ has $\kappa(G)$ connected components, then

$$\chi_G(\lambda) = \lambda^{\kappa(G)}(-1)^{|V|-\kappa(G)}t(M(G); 1 - \lambda, 0).$$

See Exercise 25.

The evaluation $t(M; \lambda - 1, 0)$ makes sense even when $M$ is not a graphic matroid. In this case, we can *define* a new, one-variable invariant, the *characteristic polynomial*

Characteristic polynomial

$$p(M; \lambda) = (-1)^{r(E)}t(M; 1 - \lambda, 0).$$

This polynomial is also an interesting invariant. We can use Theorem 9.9 to rewrite it as follows (Exercise 24):

**Proposition 9.30.** *Let $M$ be a matroid on the ground set $E$. Then*

$$p(M; \lambda) = \sum_{A \subseteq E}(-1)^{|A|}\lambda^{r(E)-r(A)}.$$

Since $t(M^*; x, y) = t(M; y, x)$, it makes sense to ask if the Tutte polynomial evaluation $t(M(G); 0, 1 - \lambda)$ gives you any meaningful information for a graph $G$. The answer is yes, twice!

(1) When $G$ is planar, we know the matroid dual $M(G)^*$ is the cycle matroid of the geometric dual $G^*$, i.e., $M(G)^* = M(G^*)$ (this is Theorem 3.27). Thus, the evaluation $(-1)^{|V|-1}\lambda \cdot t(M(G); 0, 1 - \lambda)$ is the chromatic polynomial for the dual graph $G^*$.
(2) There is another, surprising interpretation for $t(M(G); 0, 1 - k)$ that doesn't depend on whether $G$ is planar. This connection involves *flows* in graphs, and we give a few details now.

### 9.6.3 Nowhere-zero flows

Let $G$ be a graph and $H$ a finite abelian group. We can orient the edges of $G$ to get a directed graph $G_{\mathcal{O}}$, where $\mathcal{O}$ is the orientation. Write $H$ as an additive group, with identity element 0, and assign an element of $H$ to each directed edge of $G_{\mathcal{O}}$. Call the element $h$ assigned to the edge $e$ the *weight* of $e$, and denote it $w(e)$.

H-flow

An *H-flow* is an assignment of elements of $H$ to the edges of the graph such that Kirchhoff's current law is satisfied: at each vertex of $G_{\mathcal{O}}$, the sum of the weights of the edges directed in equals the sum of

---

[16]  Polya designed an inductive proof that "All horses are the same color." He gave his "proof" as an exercise to his students, asking them to find the flaw in his reasoning.

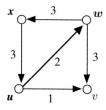

Figure 9.9. The oriented graph has a nowhere-zero flow with $H = \mathbb{Z}_4$.

the weights directed out. Finally, an $H$-flow is said to be *nowhere-zero* if the 0-element of the group is not assigned to any edge, i.e., $w(e) \neq 0$ for all edges of $G_\mathcal{O}$.

For example, let $G$ be the graph of Figure 9.9, with the given orientation. Then you can check the edge weights from $\mathbb{Z}_4$ given in the figure give a nowhere-zero flow. For example, at vertex $w$, the weight of the edge directed in is 2, and the total weight of the two edges directed out is 6, but $6 \equiv 2 \pmod 4$.

Given a graph $G$, an orientation of the edges of $G$, and a finite abelian group $H$, how many nowhere-zero flows are there? The answer to that question has three surprises.[17]

### First surprise

**Proposition 9.31.** *For a given graph $G$ and finite abelian group $H$, the number of nowhere-zero flows is independent of the orientation $\mathcal{O}$.*

*Proof Proposition 9.31.* First, you can change a given orientation $\mathcal{O}$ into any other orientation $\mathcal{O}'$ be switching the direction of some of the edges. Let $e$ be an edge of the graph. Now if we have a nowhere-zero flow using a given orientation $\mathcal{O}$, we can get a nowhere-zero flow on the orientation $\mathcal{O}'$ with the direction of $e$ reversed if we replace the weight $w(e)$ by its inverse $-w(e)$ in the group $H$. Then we get a bijection between nowhere-zero flows using $\mathcal{O}$ and those using $\mathcal{O}'$. That does it. □

As an example of how the proof of Proposition 9.31 works, consider the two graphs $G_1$ and $G_2$ in Figure 9.10. We reversed edge $\overline{wx}$ to change $G_1$ to $G_2$, and, consequently, we also changed the weight of that edge from 3 in $G_1$ to 1 in $G_2$, the additive inverse of 3 in $\mathbb{Z}_4$.

### Second surprise

**Theorem 9.32.** *Let $G$ be a graph and $H$ a finite abelian group. Then the number of nowhere-zero flows depends only on $|H|$ (the number of elements of $H$).*

---

[17] Surprises in life can be good or bad. Mathematical surprises are good, but usually not very exciting.

Figure 9.10. The edge $\overline{wx}$ is reversed, and its weight $w(e) = 3$ is replaced by $-w(e) = 1$ in $\mathbb{Z}_4$. This operation preserves the property that the flow is nowhere-zero.

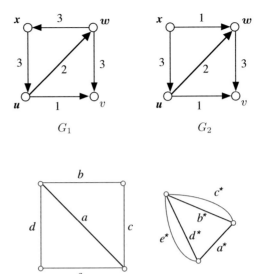

Figure 9.11. The graph $G$ and its dual, $G^*$.

So, a consequence of Theorem 9.32 is that the structure of the abelian group $H$ doesn't matter.[18] For the rest of this section, we just use $\mathbb{Z}_k$ any time we need an abelian group of order $k$.

The proof of Theorem 9.32 will follow from the third surprise. In light of the first two surprises, we define $\chi_G^*(k)$ to be the number of nowhere-zero flows in the graph $G$ with weights from the abelian group *Flow polynomial* $H$, where $|H| = k$. We call this the *flow polynomial* for $G$.

### Third surprise

**Theorem 9.33.** *Let $G$ be a connected graph and $H$ a finite abelian group. Then*

$$\chi_G^*(k) = (-1)^{|V|+|E|+1} t(M(G); 0, 1-k).$$

The only contribution of the group $H$ to this calculation occurs when you set $y = 1 - k$ in the Tutte polynomial, where $k = |H|$. So Theorem 9.32 follows from Theorem 9.33.

Why is this a surprise? Well, at the very least, the relation between $\chi_G^*(k)$ and the chromatic polynomial of the dual graph is somewhat unexpected. As an example, in Figure 9.11, let's first find the flow polynomial $\chi_G^*(k)$ for the graph $G$. First, the Tutte polynomial (see Exercise 14):

$$t(M(G); x, y) = x^3 + 2x^2 + 2xy + x + y + y^2.$$

---

[18] If you've spent lots of time learning about the structure of finite abelian groups, forget it. For now.

By Theorem 9.33, we have

$$\chi_G^*(k) = (-1)^{10}t(M(G); 0, 1-k) = (1-k) + (1-k)^2$$
$$= (k-1)(k-2).$$

This tells us there are no nowhere-zero flows if $|H| < 3$. The number of nowhere-zero flows for $\mathbb{Z}_3$ is 6; you might enjoy listing the six distinct nowhere zero flows.[19]

How is this related to the chromatic polynomial of the dual graph $G^*$? We get $\chi_{G^*}(\lambda) = \lambda(\lambda-1)(\lambda-2)$, since the multiple edges don't affect the chromatic polynomial. Thus, $\chi_{G^*}(\lambda) = \lambda \cdot \chi_G^*(\lambda)$. More prosaically, the chromatic polynomial of the dual graph $G^*$ differs from the flow polynomial of $G$ by a factor of $\lambda$.[20] We state the connection as a corollary.

**Corollary 9.34.** *Suppose G is a connected planar graph with connected planar dual $G^*$. Then* $\chi_G^*(k) = k \cdot \chi_{G^*}(k)$.

Here's something worth chewing over: for the chromatic polynomial, the Tutte evaluation $\chi_G(\lambda) = (-1)^{|V|-1}\lambda \cdot t(M(G); 1-\lambda, 0)$ has $y = 0$. This corresponds to the fact that graphs with loops cannot be properly colored, as we noted above. For the flow polynomial, the Tutte evaluation $\chi_G^*(k) = (-1)^{|V|+|E|+1}t(M(G); 0, 1-k)$ has $x = 0$. The implication is that graphs with isthmuses have no nowhere-zero flows – see Exercise 30.

Coming full circle, when $G$ is a planar graph, we can interpret the Four Color Theorem (Theorem 9.1) in terms of nowhere-zero flows:

**Corollary 9.35.** *Suppose G is a planar graph. Then G has a nowhere-zero $\mathbb{Z}_4$ flow.*

The proof follows from applying Theorem 9.1 to the planar graph $G^*$ to show $\chi_{G^*}(4) > 0$, then using Corollary 9.34 to show $\chi_G^*(4) > 0$. In theory, this approach could be used to *prove* the Four-Color Theorem.[21]

Finally, how do we prove Theorem 9.33? The proof technique is exactly the same as the proof of Theorem 9.28. We leave the details to Exercise 31.

## Exercises

### Section 9.2 – The Tutte polynomial via deletion and contraction
(1) Show the matroids $M_1$ and $M_2$ in Figure 9.12 have the same Tutte polynomial by finding $x$ and $y$ with $M_1 - x \cong M_2 - y$ and

---

[19] At this point in the text, there's no telling what pleases you.
[20] We are, indeed, surprised.
[21] We emphasize the "in theory" clause – no one has been able to successfully prove the Four-Color Theorem via flows, flow polynomials or chromatic polynomials.

Figure 9.12. Two matroids with the same Tutte polynomial – see Exercise 1.

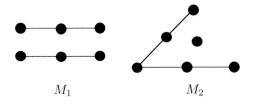

$$M_1 \qquad\qquad\qquad M_2$$

$M_1/x \cong M_2/y$. Then find the polynomial. (This is the smallest such example.)

(2) (a) Show $t(U_{n,n}; x, y) = x^n$ and $t(U_{0,n}; x, y) = y^n$.

    (b) Now find formulas for $t(U_{n-1,n}; x, y)$ and $t(U_{1,n}; x, y)$.

    (c) Show that the Tutte polynomial of $U_{2,n}$ is given by the following formula:

$$t(U_{2,n}; x, y) = x^2 + (n-2)x + (n-2)y + (n-3)y^2 + \cdots$$
$$+ (n-k-1)y^k + \cdots y^{n-2}.$$

(3) Show that the Tutte polynomial for a uniform matroid $U_{r,n}$ satisfies the recursion

$$t(U_{r,n}; x, y) = t(U_{r-1,n}; x, y) + t(U_{r-1,n-1}; x, y),$$

where $1 \le r \le n - 1$.

(4) Suppose $t(M; x, y) = x^a y^b$ for some $a, b \ge 0$. Show that $M = U_{a,a} \oplus U_{0,b}$, i.e., $M$ is the direct sum of $a$ isthmuses and $b$ loops.

(5) (a) Show that $t(U_{r,n}; x, y)$ has no terms of the form $x^i y^j$ for $i$, $j > 0$.

    (b) (Converse of (a)) Show that if $t(M; x, y)$ has no terms of the form $x^i y^j$ for $i, j > 0$, then $M$ is a uniform matroid. (It's worth observing the connection between this exercise and Exercise 3 of Chapter 8, where we asked you to show a matroid is a uniform matroid if and only if it does not contain $U_{1,1} \oplus U_{0,1}$ as a minor.)

(6) Let $M$ be a non-empty matroid. Show that $t(M; 0, 0) = 0$.

(7) Let $M_1$ and $M_2$ be the two matroids from Figure 9.4. Show that $M_1$ and $M_2$ have different numbers of circuits, flats and hyperplanes. Conclude that these three matroid invariants (the number of circuits, flats and hyperplanes) are not evaluations of the Tutte polynomial.

## Section 9.3 – The corank–nullity polynomial

(8) Use the corank–nullity polynomial to find a formula for $t(U_{r,n}; x, y)$.

(9) Example 9.11 shows that non-isomorphic matroids can have the same Tutte polynomials. Show that this does not happen for uniform matroids: show that if $t(M; x, y) = t(U_{r,n}; x, y)$ for some

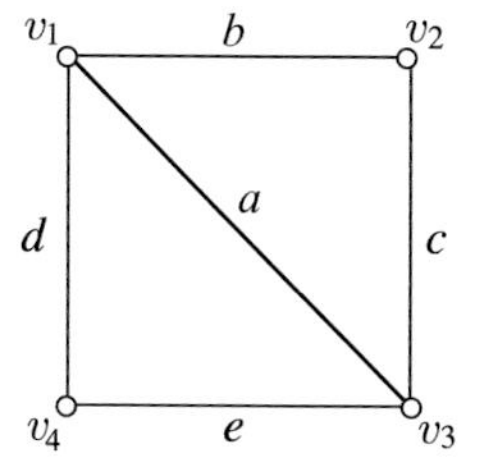

Figure 9.13. A graph for Exercises 14 and 27.

uniform matroid $U_{r,n}$, then $M \cong U_{r,n}$. (Hint: show that a rank $r$ matroid on $n$ points has at most $\binom{n}{r}$ bases, and this maximum is achieved if and only if $M = U_{r,n}$.)

(10) Complete the proof of Theorem 9.9 by proving that $s(M;u, v) = (v + 1)s(M - e;u, v)$ when $e$ is a loop.

(11) Complete the proofs of parts (2), (3) and (4) in Proposition 9.10.

(12) Let $M$ be a matroid with Tutte polynomial $t(M;x, y)$. Show that the highest power of $x$ that appears in $t(M;x, y)$ is $r(M)$, and the highest power of $y$ is $|E| - r(M)$.

(13) Suppose $M$ is a matroid. This exercise shows that the Tutte polynomial determines the size of the smallest circuit of $M$.

    (a) Show that if $C$ is a circuit, then $C$ contributes the term $u^{r(E)-|C|+1}v$ to the corank–nullity polynomial $s(M;u, v)$.

    (b) Suppose the term $u^k v$ appears in the corank–nullity polynomial $s(M;u, v)$, and $k$ is as large as possible (meaning that if $u^m v$ also appears as a term, then $m \leq k$). Show that there is a circuit $C$ of $M$ with $|C| = r(M) - k + 1$, and this is the size of the smallest circuit in $M$.

    (c) Use part (b) to conclude the Tutte polynomial determines the size of the smallest circuit in $M$. (A counterexample found in 1991 by Schwärzler [30] shows that the Tutte polynomial does *not* determine the size of the *largest* circuit.)

## Section 9.4 – Duality and direct sums

(14) Let $G$ be the graph of Figure 9.13.

    (a) Show that $t(M(G);x, y) = x^3 + 2x^2 + 2xy + x + y + y^2$.

    (b) Draw the dual graph $G^*$ and compute $t(M(G^*);x, y)$ via deletion–contraction or Theorem 9.9.

    (c) Check your answer to part (b) by using Theorem 9.12.

(15) Show that $t(U_{k,2k};x, y)$ is a symmetric polynomial in $x$ and $y$, i.e., $t(U_{k,2k};x, y) = t(U_{k,2k};y, x)$.

(16) Prove Theorem 9.12 that $t(M^*;x, y) = t(M;y, x)$ by induction directly from the recursive Definition 9.2.

(17) Find the chromatic polynomial for the $n$-cycle $C_n$. (You can do this directly, by induction, or by using the evaluation of Theorem 9.28 with $t(U_{n-1,n};x, y)$. Or something else. See also Exercise 26.)

(18) Prove Theorem 9.13 from the subset expansion of the corank–nullity polynomial $s(u, v)$, i.e., prove

$$s(M_1 \oplus M_2) = s(M_1)s(M_2).$$

(Hint: you will need the description of the rank function of $M_1 \oplus M_2$ in terms of the rank functions $r_1$ of $M_1$ and $r_2$ of $M_2$. See Proposition 3.33.)

(19) (a) Suppose a matroid $M$ has exactly $k$ isthmuses and $m$ loops. Show that $t(M; x, y) = x^k y^m t(M'; x, y)$, where $M'$ is the minor of $M$ obtained by contracting the $k$ isthmuses and deleting the $m$ loops of $M$. (Hint: see Theorem 9.13.)

(b) Now suppose the matroid $M$ has Tutte polynomial $t(M; x, y) = x^k y^m f(x, y)$, where $k$ and $m$ are as large as possible and $f(x, y)$ is some polynomial in $\mathbb{Z}[x, y]$. Show that $M$ has exactly $k$ isthmuses and $m$ loops, and $f(x, y) = t(M'; x, y)$ for the minor $M'$ of $M$ obtained by contracting the isthmuses and deleting the loops of $M$. (This is the converse of (a).)

(20) Let $M$ be a matroid with Tutte polynomial $\sum b_{i,j} x^i y^j$, and define $\beta^*(M)$ to be the coefficient of $y$ in $t(M; x, y)$, i.e., $\beta^*(M) = b_{0,1}$.

(a) Show that $\beta^*(M) = \beta(M^*)$ for all matroids $M$.

(b) Show that if $M$ is not an isthmus, then $M$ is a connected matroid if and only if $\beta^*(M) > 0$. (This is a dual version of Proposition 9.17.)

(c) In fact, more is true: if $|E| \geq 2$, then $\beta(M) = \beta^*(M)$. Prove this fact, and then use it to provide an alternate proof of (b).

(Hint: for part (a), use Theorem 9.12. For part (b), either use Corollary 3.47 and duality, or modify the proof of Proposition 9.17.)

## Section 9.5 – Tutte–Grothendieck invariants

Generalized T-G invariants (21) This exercise presents a generalization of Theorem 9.19 on Tutte–Grothendieck invariants. Let $\mathcal{M}$ be the set of isomorphism classes of matroids, and let $f : \mathcal{M} \to \mathbb{Z}[x, y]$ satisfying

(a) $f(M_1) = f(M_2)$ for isomorphic matroids $M_1 \cong M_2$,

(b) $f(M) = a \cdot f(M - e) + b \cdot f(M/e)$ for some fixed integers $a$ and $b$, and for $e$ not a loop or an isthmus,

(c) $f(M) = f(I) \cdot f(M/e)$ for $I$ an isthmus,

(d) $f(M) = f(L) \cdot f(M - e)$ for $L$ a loop.

Show that $f(M) = a^{|E|-r(E)} b^{r(E)} t\left(M; \dfrac{f(I)}{b}, \dfrac{f(L)}{a}\right)$. (Hint: modify the inductive proof of Theorem 9.19.)

(22) Let $\mathcal{A}$ be a collection of hyperplanes in $\mathbb{R}^n$ (see the discussion in Section 7.4). Assume the arrangement is non-central (so the hyperplanes do not all intersect at the origin), but they define a matroid $M(\mathcal{A})$ (so there are no parallel intersections).

(a) Let $c(\mathcal{A})$ be the number of regions determined by $\mathcal{A}$. Show that $c(\mathcal{A}) = t(M(\mathcal{A}); 2, 0) - t(M(\mathcal{A}); 1, 0)$. (Recall Theorem 7.27: the number of regions satisfies the deletion–contraction recursion for T-G invariants: $c(\mathcal{A}) = c(\mathcal{A}') + c(\mathcal{A}'')$, where $H \in \mathcal{A}$ is a hyperplane, $\mathcal{A}'$ is the arrangement with $H$ deleted and $\mathcal{A}''$ is the arrangement with $H$ contracted.)

(b) Let $b(\mathcal{A})$ be the number of *bounded* regions determined by $\mathcal{A}$. Show that $b(\mathcal{A}) = t(M(\mathcal{A}); 1, 0)$.

## Section 9.6 – Chromatic and flow polynomials

(23) Show that $G$ is bipartite if and only if $G$ has no odd cycles. (Hint: start 2-coloring the vertices, and show that you get stuck precisely when there is an odd cycle.)

(24) Let $p(M; \lambda) = (-1)^{r(E)} t(M; 1 - \lambda, 0)$ be the characteristic polynomial of a matroid $M$.

(a) Prove Proposition 9.30: $p(M; \lambda) = \sum_{A \subseteq E} (-1)^{|A|} \lambda^{r(E)-r(A)}$.

(b) Show that if $e$ is not a loop or an isthmus, then $p(M; \lambda) = p(M - e; \lambda) - p(M/e; \lambda)$.

(c) If we write $p(M; \lambda) = \sum_{i=0}^{m} a_i \lambda^i$, where $m = r(E)$, then show that the coefficients $a_0, a_1, a_2 \ldots$ alternate in sign:

$$a_m > 0, \quad a_{m-1} < 0, \quad a_{m-2} > 0, \quad a_{m-3} < 0, \ldots.$$

Conclude the same thing is true for the chromatic polynomial $\chi_G(\lambda)$ of a graph.

(25) Prove the generalization of Theorem 9.28 when $G$ is not a connected graph: If $G$ has $\kappa(G)$ connected components, then

$$\chi_G(\lambda) = \lambda^{\kappa(G)} (-1)^{|V|-\kappa(G)} t(M(G); 1 - \lambda, 0).$$

(Hint: use Theorem 9.28 and the fact that the chromatic polynomial and the Tutte polynomial are multiplicative on connected components.)

(26) Let $C_n$ be the graph consisting of a cycle with $n$ edges.

(a) Observe that the cycle matroid $M(C_n) = U_{n-1,n}$, and show $t(M(C_n); x, y) = x^{n-1} + x^{n-2} + \cdots + x + y$. (Note: see part(b) of Exercise 2 or Exercise 8.)

(b) Now use Theorem 9.28 to find the chromatic polynomial $\chi_{C_n}(\lambda)$. Conclude that the chromatic number of $C_n$ is 2 if $n$ is even and 3 if $n$ is odd.

(c) This time, use Theorem 9.21 to show the number of acyclic orientations of $C_n$ is $2^n - 2$. Then give a direct argument (not using the Tutte polynomial) to show the same thing.

(27) Let $G$ be a graph and let $v$ be a vertex of $G$. Let $a_v(G)$ be the number of acyclic orientations of a graph $G$ with $v$ as the unique source (so every edge incident to $v$ is directed away from $v$).

Figure 9.14. The wheel $W_4$ – see Exercise 29.

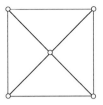

(a) Show $a_v(G) = 4$ for any $v$ for the graph of Figure 9.13. (There are two vertices to check, up to symmetry.)

(b) Show $a_v(G) = t(M(G); 1, 0)$. Conclude that $a_v(G)$ is independent of the vertex $v$. Really.

(c) Check your answer to part (b), using $t(M(G); x, y) = x^3 + 2x^2 + 2xy + x + y + y^2$.

(28) Let $C_n$ be the $n$-cycle.

(a) Show $\chi^*_{C_n}(k) = (k - 1)$ directly by finding all the nowhere-zero flows for (some orientation of) $C_n$.

(b) Check your answer to part (a) using Theorem 9.33 and the Tutte polynomial of $C_n$ (given in part (b) of Exercise 26).

(c) Now redo part (a) using the connection between $\chi^*_G(k)$ and $\chi_{G^*}(k)$ for planar graphs given in Corollary 9.34.

(29) Let $W_n$ be the $n$-spoked wheel, as in Figure 9.14.

(a) Show that $\chi_{W_n}(\lambda) = \lambda \cdot \chi_{C_n}(\lambda - 1)$, where $C_n$ is the $n$-cycle.

(b) Use part (a) and the fact that $W_n$ is planar and self-dual (Exercise 30 of Chapter 3) to find the flow polynomial $\chi^*_{W_n}(k)$.

(30) Suppose $G$ is a connected graph and $H$ is an abelian group with $|H| = k$. Suppose we have a nowhere-zero flow for some orientation of $G$ using elements of $H$.

(a) Let $C^*$ be a cocircuit in $M(G)$, i.e., a minimal collection of edges whose removal disconnects the graph. Let $G_1$ and $G_2$ be the two components obtained when the edges of $C^*$ are removed. Show that the total weight of the edges of $C^*$ directed from $G_1$ to $G_2$ equals the total weight of the edges directed from $G_2$ to $G_1$.

(b) Use part (a) to show that if $G$ has an isthmus $e$, then $\chi^*_G(k) = 0$ for all $k$, i.e., $G$ has no nowhere-zero flows.

(31) Prove Theorem 9.33: Let $G$ be a connected graph and $H$ a finite abelian group. Then the flow polynomial $\chi^*_G(k) = (-1)^{|V|+|E|+1} t(M(G); 0, 1 - k)$. (Hint: first show $\chi^*_G(k) = \chi^*_{G/e}(k) - \chi^*_{G-e}(k)$, then show $f(G; k) = (-1)^{|V|+|E|+1} \chi^*_G(k)$ is a T-G invariant.)

# Projects

## P.1 The number of matroids

How many rank $r$ matroids are there on $n$ elements? In general, there is no simple formula in terms of $n$ and $r$. In this project, we'll look at three cases: $r = 1, 2$ and $3$. Throughout, assume the ground set $E$ has $n$ elements for $n > 0$, and let $f(n, r)$ be the number of distinct (non-isomorphic) matroids of rank $r$ on the ground set $E$.

(1) Show $f(n, 1) = n$.
(2) Let $p(n)$ be the number of partitions of the integer $n$ into one or more parts, without regard to the order of the summands. For instance, $p(4) = 5$ because we can partition 4 as follows: $4, 3 + 1, 2 + 2, 2 + 1 + 1$, or $1 + 1 + 1 + 1$. Show

$$f(n, 2) = \left( \sum_{k=1}^{n} p(k) \right) - n.$$

For example, $f(4, 2) = p(1) + p(2) + p(3) + p(4) - 4 = 1 + 2 + 3 + 5 - 4 = 7$. See Figure P.1. (Using an asymptotic estimate for $p(k)$ gives $f(n, 2) \sim e^{\sqrt{n}}$.)

(3) Let $M$ be a rank 3 matroid on $E$ with bases $\{B_1, B_2, \ldots\}$. In Chapter 2, we showed that a matroid $M$ is uniquely determined by its bases. Show that $f(n, 3) \leq 2^{\binom{n}{3}} < 2^{n^3/6}$.

(4) (Brylawski 1982) We can sharpen the estimate given for $f(n, 3)$. Write $E = \{f_1, f_2, \ldots, f_n\}$; we assume the matroid has no loops. Define a function $\phi$ from the collection of subsets of $E$ of size 2 to the set $E$ as follows. For $i < j$,

$$\phi(f_i, f_j) = \begin{cases} f_j & \text{if } \{f_i, f_j\} \text{ is a multiple point,} \\ f_k & \text{if } \{f_i, f_j, f_k\} \text{ is collinear} \\ & \text{where } k > j \text{ is minimal,} \\ f_i & \text{if } \{f_i, f_j, f_k\} \text{ is not collinear for any } k > j. \end{cases}$$

Figure P.1. The seven rank 2
matroids on four elements.
Loops are depicted as hollow
points inside clouds.

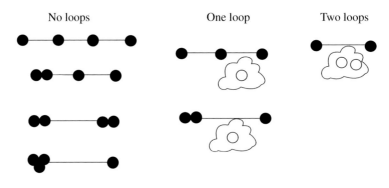

Figure P.2.

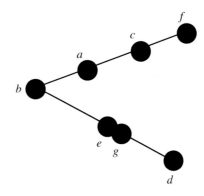

For example, if we let $M$ be the matroid in Figure P.2, then the
function $\phi$ is given by:

$$ab \mapsto c \quad ac \mapsto f \quad ad \mapsto a \quad ae \mapsto g \quad af \mapsto a \quad ag \mapsto a \quad bc \mapsto f$$
$$bd \mapsto e \quad be \mapsto g \quad bf \mapsto b \quad bg \mapsto b \quad cd \mapsto c \quad ce \mapsto g \quad cf \mapsto c$$
$$cg \mapsto c \quad de \mapsto g \quad df \mapsto d \quad dg \mapsto d \quad ef \mapsto e \quad eg \mapsto g \quad fg \mapsto f.$$

(a) Let $M$ be a rank 3 matroid with no loops. Show that the function
$\phi$ uniquely determines all the multiple points in the rank 3
matroid.

(b) Show the matroid $M$ can be uniquely reconstructed from the
function $\phi$.

(c) Show the number of possible functions $\phi$ is bounded above by
$n^{\binom{n}{2}}$.

(d) Show that $n^{\binom{n}{2}} < 2^{(n^2 \log n)/2}$. Conclude $f(n, 3) < 2^{(n^2 \log n)/2}$.

Note that different labelings of the same matroid $M$ can produce
different functions $\phi$, so the argument in part (d) overestimates
$f(n, 3)$. Brylawski also obtains a lower bound $2^{n^2/6} < f(n, 3)$ by
considering *Steiner triples*.

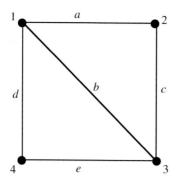

## P.2 Matrix-Tree Theorem

This theorem gives the number of spanning trees in a graph as a determinant of a matrix. Here's the set-up. Let $G$ be a graph, and let $A$ be the *Laplacian* matrix for $G$.

Here's how to create the matrix $A$: if $G$ has $m$ vertices $\{v_1, v_2, \ldots, v_m\}$, then let $d_i$ be the degree of the vertex $v_i$, and let $e_{i,j}$ be the number of edges joining vertices $i$ and $j$. Then $A_G$ is an $m \times m$ matrix:

$$a_{i,j} = \begin{cases} d_i & \text{if } i = j, \\ -e_{i,j} & \text{otherwise.} \end{cases}$$

Recall that a *principal minor* $A_i$ of an $m \times m$ matrix $A$ is the $(m-1) \times (m-1)$ matrix you get by deleting row $i$ and column $i$ from $A$.

**Theorem P.1.** *Let $G$ be a graph and form the Laplacian matrix $A$. Then the determinant of any principal minor of $A$ equals the number of spanning trees for $G$.*

We'll work through an example carefully; the example will also suggest a slightly more general theorem. Let $G$ be the graph in Figure P.3.

The generalization keeps track of all the edge *labels*. For a graph $G$, let $T(G)$ be the sum of all the monomials corresponding to all the spanning trees, where we identify a labeled edge with its label, and we think of the label as a variable. For this example, we have

$$T(G) = abd + abe + acd + ace + ade + bcd + bce + cde.$$

The proof relies on deletion and contraction (and induction). So let's delete and contract the edge $a$: we now have three matrices:

$$A_G = \begin{array}{c} \\ 1 \\ 2 \\ 3 \\ 4 \end{array} \begin{array}{cccc} 1 & 2 & 3 & 4 \\ \begin{bmatrix} a+b+d & -a & -b & -d \\ -a & a+c & -c & 0 \\ -b & -c & b+c+e & -e \\ -d & 0 & -e & d+e \end{bmatrix} \end{array}$$

$$A_{G-a} = \begin{array}{c} \\ 1 \\ 2 \\ 3 \\ 4 \end{array} \begin{array}{cccc} 1 & 2 & 3 & 4 \\ \left[ \begin{array}{cccc} b+d & 0 & -b & -d \\ 0 & c & -c & 0 \\ -b & -c & b+c+e & -e \\ -d & 0 & -e & d+e \end{array} \right] \end{array}$$

$$A_{G/a} = \begin{array}{c} \\ 1=2 \\ 3 \\ 4 \end{array} \begin{array}{ccc} 1=2 & 3 & 4 \\ \left[ \begin{array}{ccc} b+c+d & -b-c & -d \\ -b-c & b+c+e & -e \\ -d & -e & d+e \end{array} \right] \end{array}.$$

Note these three matrices are generalizations of the Laplacians, where we keep the edge labels in the matrix. (So you can get the Laplacians by setting all variables equal to 1.)

We need principal minors of each of these matrices. We'll cross out the first row and the first column of each matrix, then compute the determinants:

$$D_1 \qquad\qquad D_2 \qquad\qquad D_3$$

$$\begin{vmatrix} a+c & -c & 0 \\ -c & b+c+e & -e \\ 0 & -e & d+e \end{vmatrix} \quad \begin{vmatrix} c & -c & 0 \\ -c & b+c+e & -e \\ 0 & -e & d+e \end{vmatrix} \quad \begin{vmatrix} b+c+e & -e \\ -e & d+e \end{vmatrix}.$$

Our goal is to show $D_1 = T(G)$. (Of course, you could do a direct calculation here, but that won't help when you need a general proof.) The proof depends on a nice deletion–contraction formula (you are asked to prove this below):

$$T(G) = T(G-a) + aT(G/a).$$

Assuming this formula for now, to show $D_1 = T(G)$, we will show $D_1 = D_2 + aD_3$. Then, since we know $D_2 = T(G-a)$ and $D_3 = T(G/a)$ (by induction), we get $D_1 = T(G-a) + aT(G/a) = T(G)$, which is what we want.

Use a computer algebra system (or your brain) to show $D_2 = bcd + bce + cde$ and $D_3 = bd + be + cd + ce + de$. Then this finishes the example: $D_1 = D_2 + aD_3$.

Here's your handy outline that should lead you to a proof of Theorem P.1.

(1) First, prove that, for any non-loop edge $a$,

$$T(G) = T(G-a) + aT(G/a).$$

(2) To compute the determinant $D_1$, use the permutation definition of determinant:

$$\det(A) = \sum_{\sigma \in S_n} sgn(\sigma) a_{1,\sigma(1)} a_{2,\sigma(2)} \cdots a_{n,\sigma(n)}.$$

This definition sums over all permutations of $\{1, 2, \ldots, n\}$. You can think of it as selecting one element from each row and each column of $A$ (that's the permutation $\sigma$), multiplying those $n$ elements together, then multiplying this product by the *sign* of the permutation, which is 1 or $-1$, depending on the whether $\sigma$ is even or odd.

(3) Then break up the terms in $D_1$ into those that use the entry $A_{1,1}$ and those that don't. For the terms that avoid using the $A_{1,1}$ entry, note that all of them appear in the determinant $D_2$ (which only differs from $D_1$ in the $A_{1,1}$ entry).

(4) For terms that use $A_{1,1}$ in the computation of $D_1$, note that $A_{1,1} = a + x$, where $x$ is the $A_{1,1}$ entry of $D_2$. Thus, when computing the determinant $D_1$, we'll get $(a + x)D_3$.

(5) Now $x D_3$ is exactly what you get from the remaining terms of $D_2$, i.e., those terms of $D_2$ that use the $A_{1,1}$ entry.

(6) What's left? Precisely $a D_3$.

In summary, you should get $D_1 = D_2' + D_2'' + a D_3$, where $D_2'$ consists of those terms of $D_2$ *not* using the first entry in the matrix for $D_2$ and $D_2''$ are the terms that do use that first entry of the matrix.

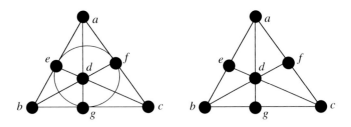

## P.3  Relaxing a hyperplane

It is frequently the case that two matroids on the same ground set are closely related. For instance, the Fano plane and non-Fano plane have identical ground sets, and almost identical sets of bases (see Figure P.4). The difference between these two matroids is that the Fano plane has one extra dependence. That extra "line" in the Fano plane, $efg$, is both a circuit and a hyperplane.

This "always" works in the following sense:

> Let $M$ be a matroid with bases $\mathcal{B}$ and suppose $C$ is both a circuit and a hyperplane in $M$. Let $\mathcal{B}'$ be the collection of bases of $M$ together with $C$, i.e., $\mathcal{B}' = \mathcal{B} \cup C$. Then the family $\mathcal{B}'$ is the collection of bases of a matroid.

We denote the matroid obtained in this way $M'$, and we say $M'$ is obtained from $M$ by *relaxing*[1] the circuit-hyperplane $C$. The next theorem shows that this always gives a matroid.

**Theorem P.2.**  *If $C$ is a circuit and a hyperplane of a matroid $M$, then the relaxation $M'$ is also a matroid.*

### Problem 1

Prove this theorem. Suggestion: verify the independent set axioms. The only interesting thing to do is to verify the augmentation axiom (I3) is true when the larger of the sets is $C$.

Here are a few more relaxing examples, some of whom you have already met.

**Example P.3.**  For each of the following matroid pairs, one of the matroids is a relaxation of the the other.

(1) The two matroids $M$ and $M'$ in Figure P.5.
(2) The Pappus and non-Pappus configurations (Example 6.20 and Theorem 6.21). See Figure P.6.
(3) The representable cube and the Vamos cube (Example 6.30). Let $M$ be the rank 4 matroid on the left in Figure P.7. Then $|E| = 8$ and there are eight four-point circuit-hyperplanes in $M$: the six "faces" of the cube and the two planes $acfh$ and $bdeg$. Relaxing $bdeg$

---

[1] We hope this term induces a calm, content demeanor for this project.

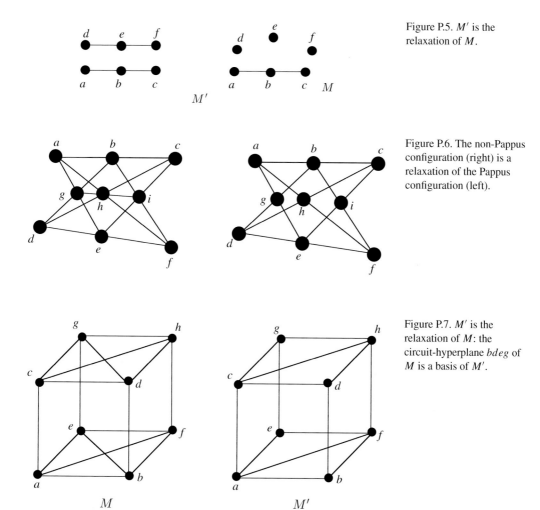

Figure P.5. $M'$ is the relaxation of $M$.

Figure P.6. The non-Pappus configuration (right) is a relaxation of the Pappus configuration (left).

Figure P.7. $M'$ is the relaxation of $M$: the circuit-hyperplane $bdeg$ of $M$ is a basis of $M'$.

produces the matroid $M'$ on the right in the figure, the so-called "Vamos cube."
(4) The Desargues and non-Desargues configurations (Exercise 26 of Chapter 6).
(5) Wheels and whirls (Exercise 30 from Chapter 3). The *wheel* $W_n$ is formed adding one vertex to the $n$-cycle and then joining this new vertex to each vertex in the cycle (see Figure 3.37 for a picture of $W_4$). So $M(W_n)$ is a rank $n$ matroid on $2n$ points. The *whirl* $\mathcal{W}_n$ is obtained from the wheel $W_n$ by relaxing the circuit corresponding to the rim of the wheel.

Finally, you could continue this process, relaxing another circuit-hyperplane of $M'$ to produce $M''$, and so on.

It is often the case that $M$ and $M'$ are representable over different fields. For instance, $F_7$ is only representable over fields of characteristic 2, while $F_7'$ is representable only over fields of characteristic not equal to 2: In terms of characteristic sets (Definition 6.24), we have $\chi(F_7) = \{2\}$ and $\chi(F_7') = \{0, 3, 5, 7, 11, \ldots\}$ ($=$ everything except 2). See Propositions 6.16 and 6.17.

But $M$ and $M'$ can sometimes be represented over the same field:

## Problem 2

Show that the matroids $M$ and $M'$ in Figure P.5 are both representable over all fields with at least four elements. (It is easy to find representing matrices, if you've gone through Chapter 6.) On the other hand, show that $M$ is representable over $\mathbb{F}_3$, but $M'$ is not.

You can also find a relationship between the Tutte polynomials $t(M)$ and $t(M')$.

## Problem 3

If $M'$ is the relaxation of the matroid $M$, then use the corank–nullity formulation of the Tutte polynomial to show

$$t(M'; x, y) = t(M; x, y) - xy + x + y.$$

Which affine and projective planes have circuit-hyperplanes? We saw the Fano plane $PG(2, 2)$ can be relaxed to form the non-Fano plane. Moreover, the binary cube $AG(3, 2)$ also has a circuit-hyperplane. But this is the entire story:

## Problem 4

Suppose $C$ is both a circuit and a hyperplane in the matroid $M$, where $M$ is either the projective space $PG(n, q)$ or the affine space $AG(n, q)$. Prove that $M = PG(2, 2)$, the Fano plane, or $M = AG(3, 2)$, the binary cube.

## Problem 5

Show that the Vamos cube can be obtained from $AG(3, 2)$ by repeatedly relaxing circuit-hyperplanes. How many relaxations are needed to get from $AG(3, 2)$ to the Vamos cube?

# P.4 Bases and probability in affine and projective space

Here are three related problems:

(1) Suppose you are given an $n \times n$ matrix, with entries in the finite field $\mathbb{F}_q$. What is the probability your matrix is invertible?

(2) Suppose you randomly choose a set $B$ of $n + 1$ points from the rank $n + 1$ affine geometry $AG(n, q)$. What is the probability $B$ is a basis?

(3) Same question as (2), but for the rank $n + 1$ projective space $PG(n, q)$.

The goal of this project is to find formulas for each of these three probabilities, and then to establish a connection between them. In fact, for a fixed $q$, all three probabilities approach the same value as $n \to \infty$![2]

## A. First, the formulas

For problem (1), let $f_{n,q}$ be the probability that an $n \times n$ matrix over $\mathbb{F}_q$ is invertible. For this problem, for a given $a \in \mathbb{F}_q$, assume the probability that a given entry $m_{i,j} = a$ of the $n \times n$ matrix equals $1/q$, i.e., each of the $q^{n^2}$ matrices is equally probable.

(a) As a warm up, show $f_{2,2} = \frac{3}{8}$ by listing all 16 $2 \times 2$ matrices over $\mathbb{F}_2$.

(b) Show

$$f_{n,q} = \frac{(q - 1)}{q} \cdot \frac{(q^2 - 1)}{q^2} \cdots \frac{(q^n - 1)}{q^n}.$$

(Hint: use the idea from Theorem 5.28 to count the number of ordered bases of $PG(n - 1, q)$.)

For (2), suppose a subset $B$ of $n + 1$ distinct points is picked at random from the affine geometry $AG(n, q)$, where each subset has an equal probability of being selected. Let $g_{n,q}$ be the probability $B$ is a basis of $AG(n, q)$ (assume $n \geq 2$).

(c) This time, for warm-ups, show $g_{2,2} = 1$ and $g_{2,3} = \frac{6}{7}$. (Hint: for the affine plane $AG(2, q)$, a set of three points is a basis if and only if the three points are not collinear.)

(d) Now do the general case: use the formula for the number of independent sets of size $k$ given in Exercise 25(a) of Chapter 5 for $a_{n,k,q}$ with $k = n + 1$ to show that

$$g_{n,q} = \frac{(q^n - 1)(q^n - q)(q^n - q^2) \cdots (q^n - q^{n-1})}{(q^n - 1)(q^n - 2)(q^n - 3) \cdots (q^n - n)}.$$

---

[2] We are expressing excitement here; we are not invoking a factorial.

(Hint: the total number of ways to select $n + 1$ points is the binomial coefficient $\binom{q^n}{n+1}$. Then divide $a_{n,n+1,q}$ by this binomial coefficient to get $g_{n,q}$.)

For (3), suppose $B$ is an $(n + 1)$-point subset picked at random from the projective geometry $PG(n, q)$. Let $h_{n,q}$ be the probability $B$ is a basis of $PG(n, q)$ (assume $n \geq 1$).

(e) Show $h_{2,2} = \frac{4}{5}$ and $h_{2,3} = \frac{9}{11}$.
(f) The general case: use the formula from Exercise 25(b) of Chapter 5 for $b_{n,k,q}$ with $k = n + 1$ to show that $h_{n,q}$ is given by:

$$\frac{(q^{n+1} - 1)(q^{n+1} - q)(q^{n+1} - q^2)\cdots(q^{n+1} - q^n)}{(q^{n+1} - 1)(q^{n+1} - q)(q^{n+1} - 2q + 1)(q^{n+1} - 3q + 2)\cdots(q^{n+1} - nq + n - 1)}.$$

## B. Next, some analysis

We are interested in three limits:

$$\lim_{n\to\infty} f_{n,q} \qquad \lim_{n\to\infty} g_{n,q} \qquad \lim_{n\to\infty} h_{n,q}.$$

(a) Fix $q$. Let's first verify all three limits exist. Use the three formulas you just verified in A to show that $f_{n,q}, g_{n,q}$ and $h_{n,q}$ are bounded, monotonically decreasing sequences: hence all three limits exist. (This is a famous property of the real numbers, and it depends on the fact that every bounded set has a greatest lower bound and a least upper bound. This is the *completeness* property of $\mathbb{R}$.)

(b) To compare the three limits, we need to do some more work. Use the formulas from part A to check the following inequalities.
   (i) Show $g_{n+1,q} \leq h_{n,q} \leq g_{n,q}$ for all $n \geq 1$.
   (ii) Show $f_{n,q} \leq g_{n,q}$ for all $n \geq 1$.
   This shows $\lim_{n\to\infty} g_{n,q} = \lim_{n\to\infty} h_{n,q}$ and $\lim_{n\to\infty} f_{n,q} \leq \lim_{n\to\infty} g_{n,q}$. It remains to show $\lim_{n\to\infty} f_{n,q} = \lim_{n\to\infty} g_{n,q}$.
   (iii) To show $\lim_{n\to\infty} g_{n,q} = \lim_{n\to\infty} f_{n,q}$, try proving $\frac{1}{f_{n,q}} - \frac{1}{g_{n,q}} \to 0$ as $n \to \infty$ directly by getting common denominators and noting the degree of the numerator is smaller than the degree of the denominator.
   (Global hint: all of this is a bit easier if you rewrite $f_{n,q}, g_{n-1,q}$ and $h_{n,q}$ so they all have the same *numerators*.[3] Then all the inequalities can be obtained by examining the denominators.)

Thus, for fixed $q$, we now know

$$\lim_{n\to\infty} f_{n,q} = \lim_{n\to\infty} g_{n,q} = \lim_{n\to\infty} h_{n,q}.$$

[3] We know your high school teacher drilled you on finding common *denominators*, but they didn't write this text, did they?

(c) It's easiest to get asymptotic bounds on this common limit by examining $f_{n,q}$ (among $f_{n,q}$, $g_{n,q}$ and $h_{n,q}$, our formula for $f_{n,q}$ is the easiest formula to digest). Show

$$1 - \frac{1}{q} - \frac{1}{q^2} < \lim_{n\to\infty} f_{n,q} < 1 - \frac{1}{q}.$$

Hint: the upper bound is trivial. For the lower bound, try this:

(i) Let $r_n = \prod_{k=1}^{n}\left(\frac{q^k - 1}{q^k}\right)$. Then show that $r_n$ satisfies the relation

$$\frac{r_{n-1}}{q^n} = r_{n-1} - r_n \text{ for } n \geq 2.$$

(ii) Sum both sides of the above relation to get a telescoping series:

$$r_n = r_1 - \sum_{k=1}^{n-1} \frac{r_k}{q^{k+1}}.$$

(iii) Use a geometric series and the fact that $r_1 > r_2 > \cdots$ to show

$$\sum_{k=1}^{n-1} \frac{r_k}{q^{k+1}} \leq \frac{1}{q^2}.$$

Conclude that the determinant of an $n \times n$ matrix over $\mathbb{F}_q$ is (slightly) more likely to equal zero than any fixed non-zero $a \in \mathbb{F}_q$.

As an example of the above bounds for $n = 2$, we get:

$$\frac{1}{4} < f_{n,2} = \left(\frac{1}{2} \cdot \frac{3}{4} \cdot \frac{7}{8} \cdots \frac{2^n - 1}{2^n}\right) < \frac{1}{2}.$$

As an aside, it is an interesting (if somewhat ill-defined) exercise to figure out how rapidly your fractions need to approach 1 in order to bound the limit away from 0. More precisely, if $0 < r_1 < r_2 \cdots$, with $r_n \to 1$ as $n \to \infty$, then $1 - r_n = \frac{1}{q^n}$ guarantees $r_1 \cdot r_2 \cdot r_3 \cdots > 0$ (provided $q > 1$).

On the other hand, if $1 - r_n = \frac{1}{2n}$, then the infinite product $r_1 \cdot r_2 \cdot r_3 \cdots = 0$. You can check this for yourself:

$$\lim_{n\to\infty} \left(\frac{1}{2} \cdot \frac{3}{4} \cdot \frac{5}{6} \cdots \frac{2n-1}{2n}\right) = 0.$$

Let's return briefly to the common $\lim_{n\to\infty} f_{n,q} = \lim_{n\to\infty} g_{n,q} = \lim_{n\to\infty} h_{n,q}$. Here is a chart of some "limiting" values for the three limits (computed by evaluating $f_{n,q}$ at $n = 100$) for some small values of $q$. Note the accuracy of the lower bound in the table.

| $q$ | $1 - \frac{1}{q} - \frac{1}{q^2}$ | $f_{100,q}$ | $1 - \frac{1}{q}$ |
|---|---|---|---|
| 2 | 0.25000000 | 0.28878810 | 0.50000000 |
| 3 | 0.55555556 | 0.56012608 | 0.66666667 |
| 4 | 0.68750000 | 0.68853754 | 0.75000000 |
| 5 | 0.76000000 | 0.76033280 | 0.80000000 |
| 7 | 0.83673469 | 0.83679541 | 0.85714286 |
| 8 | 0.85937500 | 0.85940599 | 0.87500000 |
| 9 | 0.87654321 | 0.87656035 | 0.88888889 |
| 11 | 0.90082645 | 0.90083271 | 0.90909091 |
| 13 | 0.91715976 | 0.91716247 | 0.92307692 |

When $q$ is large, the lower bound converges roughly twice as quickly as the upper bound in the following sense: when $n = 100$ and $q = 2^{10} = 1024$, we find the lower bound is accurate to 16 decimals, while the upper bound is accurate to around seven decimals.

### C. A surprise

There is an even closer connection between affine and projective space in the binary case. Show that

$$g_{n+1,2} = h_{n,2}$$

for all $n \geq 1$.

For instance, there are 28 bases in the Fano plane $PG(2, 2)$, so the probability a three-point subset is a basis is $h_{2,2} = \frac{28}{35} = \frac{4}{5}$. For the affine cube $AG(3, 2)$, there are 14 four-point planes and 70 subsets of size 4, so there are $70 - 14 = 56$ bases. Thus, you still get a basis with probability $g_{3,2} = \frac{56}{70} = \frac{4}{5}$.

*Hint.* this is easy from the formulas for $g_{n,2}$ and $h_{n,2}$, but try to construct a combinatorial/incidence-count proof that uses the fact that $AG(n + 1, 2)/x = PG(n, 2)$ for any point $x \in AG(n + 1, 2)$.

## P.5 Representing affine space – the characteristic set of $AG(n, q)$

The projective plane $PG(2, p)$ can only be represented as a matroid over fields of characteristic $p$, where $p$ is a given prime number. This is the content of Theorem 6.27, and its proof was based on finding a subset (a matroid) $M$ of the projective plane $PG(2, p)$ with $\chi(M) = \{p\}$. (Recall: $\chi(M)$ is the collection of all field characteristics that $M$ can be represented over.)

We can rephrase Theorem 6.27 in terms of characteristic sets: $\chi(PG(2, p)) = \{p\}$. Our goal in this project is to compute $\chi(AG(2, p))$.

Here is the idea, introduced by way of an example. (It might be a good idea to reread Example 6.18 before getting started – the technique we use here is very similar.) Let $p = 13$ be our prime; we will show $AG(2, 13)$ can only be represented over fields of characteristic 13, i.e., $\chi(AG(2, 13)) = \{13\}$.

First, let $A_{13}$ be the matrix:

$$
A_{13} = \begin{array}{c} \\ \\ \\ \\ \end{array}
\begin{array}{ccccccccccccc}
a & b & c & d & e & f & g & h & i & j & k & l & m \\
\end{array}
$$

$$
A_{13} = \begin{bmatrix}
1 & 0 & 0 & 1 & 1 & 1 & 0 & 1 & 1 & 1 & 0 & 1 & 0 \\
0 & 1 & 0 & 1 & 1 & 0 & 1 & 2 & 2 & 2 & 1 & 2 & 1 \\
0 & 0 & 1 & 1 & 0 & 1 & 1 & 1 & 0 & 3 & 3 & 6 & 6
\end{bmatrix}.
$$

The columns have been ordered so that, after the first four columns, each column corresponds to a point in the matroid that is on the intersection of two lines already determined. In this case, you should check the following intersections determine the coordinates (up to projective uniqueness).

$$
h = \overline{bd} \cap \overline{eg}; \quad i = \overline{ab} \cap \overline{ch}; \quad j = \overline{fg} \cap \overline{ch}; \quad k = \overline{bc} \cap \overline{ej};
$$
$$
l = \overline{ak} \cap \overline{ch}; \quad m = \overline{bc} \cap \overline{el}.
$$

For example, since the point $h$ is on $\overline{bd}$, we must have the first and last coordinates of $h$ the same, so, projectively, we may assume both coordinates equal 1. So, $h$ looks like $\begin{bmatrix} 1 \\ x \\ 1 \end{bmatrix}$ for some $x$. But since $h$ is also on the line $\overline{eg}$, we know the determinant

$$
\begin{vmatrix}
1 & 0 & 1 \\
1 & 1 & x \\
0 & 1 & 1
\end{vmatrix} = 0.
$$

This forces $x = 2$.

Figure P.8. The matroid $M[A_{13}]$ is affine, and $\chi(M[A_{13}]) = \{13\}$.

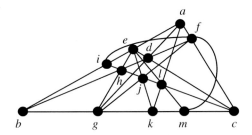

Now add the dependence $\overline{fim}$. This forces those three columns to be dependent. But the determinant

$$\begin{vmatrix} 1 & 1 & 0 \\ 0 & 2 & 1 \\ 1 & 0 & 6 \end{vmatrix} = 13.$$

So we conclude the matroid $M[A_{13}]$ can only be represented over fields of characteristic 13. See Figure P.8 for a drawing of this matroid.

Finally, we come to the punch line of this example. The matroid $M[A_{13}]$ is affine over $\mathbb{F}_{13}$, i.e., $M[A_{13}] \subseteq AG(2, 13)$. One way to see this is to find a hyperplane ($=$ line) in $PG(2, 13)$ that misses all the points of the matroid. In this case, you can check no point of the matroid is on the line $x + y + z = 12$.

Since the matroid is only representable over fields of characteristic 13, we conclude $AG(2, 13)$ is, too: $\chi(AG(2, 13)) = \{13\}$.

The reason this worked is that we were able to create a determinant of 13 efficiently – no line in the matroid has more than five points. This is based on the last coordinates of the points $g$, $j$ and $k$: $1 \mapsto 3 \mapsto 6$.

The general idea for $AG(2, p)$ goes like this (assume $p > 3$):

(1) First, compute the numbers you need to efficiently generate the prime $p$. These are generated by dividing $p$ by powers of 2, and rounding down:

$$\left\lfloor \frac{p}{2} \right\rfloor, \quad \left\lfloor \frac{p}{4} \right\rfloor, \quad \left\lfloor \frac{p}{4} \right\rfloor, \quad \cdots \quad \left\lfloor \frac{p}{2^s} \right\rfloor = 1,$$

where $s = \lfloor \log_2 p \rfloor$. In the example, we had $\lfloor \frac{13}{2} \rfloor = 6, \lfloor \frac{13}{4} \rfloor = 3, \lfloor \frac{13}{8} \rfloor = 1$. (Note: these are the integers you would use to create the base 2 expansion of $p$.)

(2) Let's get some notation for these numbers. Let $b_1 = \lfloor \frac{p}{2^s} \rfloor = 1, b_2 = \lfloor \frac{p}{2^{s-1}} \rfloor = 2$ or $3, \ldots, b_s = \lfloor \frac{p}{2} \rfloor$. Then show $b_{k+1} = 2b_k$ or $2b_k + 1$ for all $k$, where $1 \le k \le s - 1$.

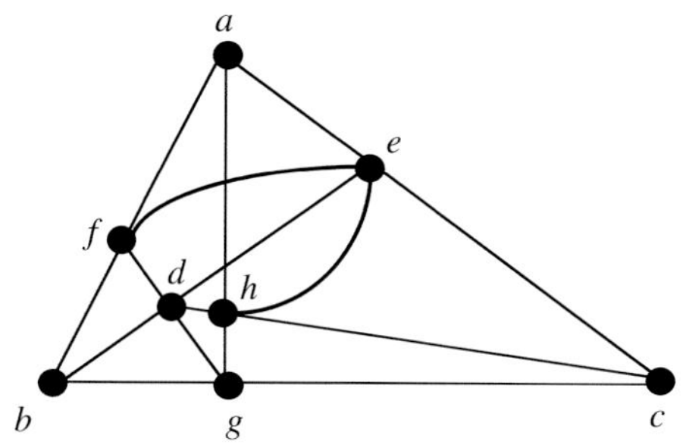

Figure P.9. The matroid
$M_{\sqrt{-3}}$ from Example 6.11.

(3) Use the $b_k$ to form the $3 \times (2\lfloor \log_2 p \rfloor + 7)$ matrix $A_p$:

$$
A_p = 
\begin{array}{c}
\phantom{A_p=}\begin{array}{ccccccccccccccccc} a & b & c & d & e & f & g & h & i & j_2 & k_2 & j_3 & k_3 & \cdots & j_s & k_s \end{array}\\
\left[
\begin{array}{ccccccccccccccccc}
1 & 0 & 0 & 1 & 1 & 1 & 0 & 1 & 1 & 1 & 0 & 1 & 0 & \cdots & 1 & 0 \\
0 & 1 & 0 & 1 & 1 & 0 & 1 & 2 & 2 & 2 & 1 & 2 & 1 & \cdots & 2 & 1 \\
0 & 0 & 1 & 1 & 0 & 1 & 1 & 1 & 0 & b_2 & b_2 & b_3 & b_3 & \cdots & b_s & b_s
\end{array}
\right].
\end{array}
$$

(4) Show that the coordinates of each point, after the first four points, are uniquely determined by showing each such point is the intersection of two lines that have already been determined. (Hint: if $b_{r+1} = 2b_r$, the point corresponding to column $j_{r+1}$ is $\overline{bc} \cap \overline{ak_r}$. If $b_{r+1} = 2b_r + 1$, then this point is $\overline{bc} \cap \overline{fk_r}$.)

(5) Add the dependence $\overline{fik_s}$ to show the matroid $M[A_p]$ is representable only over characteristic $p$.

(6) Finally, show that this matroid is affine by finding a line of $PG(2, p)$ that misses it.

Here's what you just proved:

**Theorem P.4.** *Let $p$ be a prime number with $p > 3$. Then the affine plane $AG(2, p)$ is representable only over characteristic $p$, i.e., $\chi(AG(2, p)) = \{p\}$.*

What's up with the primes 2 and 3? The affine plane $AG(2, 2)$ has four points, no three on a line. This configuration is (obviously) representable over all fields (in fact, the existence of this configuration in any projective plane is forced by the axioms for projective planes).

For $p = 3$, the construction above works, i.e., the matrix $A_3$ is defined. But the resulting matroid, which has the nine points $a, b, \ldots, i$, is not affine over $\mathbb{F}_3$. (This follows because the points $abei$ are collinear, but every line in $AG(2, 3)$ has three points.)

To get the characteristic set for the affine plane $AG(2, 3)$, we use the matroid $M_{\sqrt{-3}}$ from Example 6.11 (Figure P.9). In Exercise 29 of Chapter 6, you were asked to show $M_{\sqrt{-3}} \subseteq AG(2, 3)$.

This matroid is represented by the matrix

$$
\begin{array}{c}
\begin{array}{cccccccc} a & b & c & d & e & f & g & h \end{array} \\
A = \left[\begin{array}{cccccccc}
1 & 0 & 0 & 1 & 1 & 1 & 0 & 1 \\
0 & 1 & 0 & 1 & 0 & x & 1 & 1 \\
0 & 0 & 1 & 1 & 1 & 0 & \frac{1}{1-x} & \frac{1}{1-x}
\end{array}\right],
\end{array}
$$

where $x$ satisfies the polynomial $x^2 - x + 1 = 0$. Now add the point $i$ as the point of intersection of the four lines $\overline{ad}, \overline{bh}, \overline{cf}$ and $\overline{eg}$. Show that this forces

$$
i \leftrightarrow \left[\begin{array}{c} 1 \\ x \\ x \end{array}\right],
$$

but does not force $x$ to satisfy any new relations. Conclude that adding the point $i$ in this way gives the affine plane $AG(2, 3)$.

Here's what we/you have proven about $AG(2, 2)$ and $AG(2, 3)$.

**Theorem P.5.** *$AG(2, 2)$ is representable over any field. $AG(2, 3)$ is representable over a filed $\mathbb{F}$ if and only if $\mathbb{F}$ contains a root of the equation $x^2 - x + 1$. In particular, $AG(2, 3)$ is representable over all characteristics.*

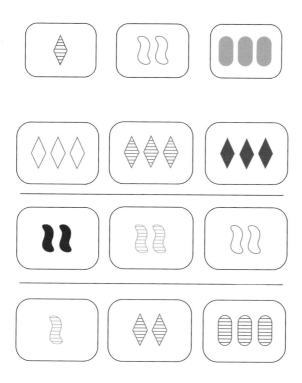

Figure P.10. Typical cards in the game of SET®. These cards form a *SET* because they differ in all four attributes: Number, color, shading and shape.

Figure P.11. Top: a *SET* formed by cards that differ in one attribute (shading) and are the same in the other three (color = red, shape = diamond, number = 3). Middle: a *SET* formed by cards that differ in two attributes. Bottom: a *SET* formed by cards that differ in three attributes.

# P.6 The card game SET® and affine geometry[4]

## P.6.1 The game

The game of SET® is played with a deck of 81 cards. Each card has one, two or three identical symbols, and each symbol has a color (red, green or purple), shading (solid, empty or striped) and shape (typically diamonds, ovals or "squiggles" that look like fat, backwards integral signs). A collection of three cards forms a *SET* if the three cards are either all the same or all different for each of the four attributes (color, shading, shape and number). For instance, the three cards in Figure P.10 form a *SET* since they differ in each attribute.

The game is played by two or more people. Initially, a dealer places 12 cards face-up on a table, and the first person who sees a *SET* calls "*SET!*"[5] and takes the three cards that form a *SET*. The dealer then replaces the cards taken by dealing out three new cards from the deck, again face-up, and the game continues.

Three cards can form a *SET* in four different ways: all four attributes are different among the three cards, as in Figure P.10, or three attributes differ while one is the same, or two differ and two are the same, or one differs and three are the same. See Figure P.11 for examples of these last three kinds of *SET*s.

---

[4] Cards and game © 1988, 1991 Cannei, LLC. All rights reserved. SET® and all associated logos and taglines are registered trademarks of Cannei, LLC. Figures P.10, P.11, P.13, P.14 and P.15 are used with permission from Set Enterprises, Inc.

[5] Often quite loudly, in our experience.

When playing the game, it's possible for the 12 cards to contain no *SET*s. When this happens, the dealer deals out three more cards, face-up. Then, typically, someone finds a *SET*, and the game then proceeds with 12 cards, unless no *SET* is found, in which case the dealer deals another three cards. It's possible to have 20 cards with no *SET*s, but, in practice, it's rare for 15 cards to contain no *SET*s.

This would be an excellent time for you to stop reading, get a deck of cards (or go to the SET® website: http://www.setgame.com to play the daily puzzle).

## P.6.2  Connections to affine geometry

Since each of the four attributes has three possible values, there are $3^4$ cards. Furthermore, we can associate an ordered 4-tuple with each card. A card will be expressed as *(number, color, shading, shape)*, where the value of each attribute is 0, 1 or 2, according to the following table:

| Number | $3 \mapsto 0$ | $1 \mapsto 1$ | $2 \mapsto 2$ |
|---|---|---|---|
| Color | purple $\mapsto 0$ | red $\mapsto 1$ | green $\mapsto 2$ |
| Shading | empty $\mapsto 0$ | striped $\mapsto 1$ | solid $\mapsto 2$ |
| Shape | oval $\mapsto 0$ | diamond $\mapsto 1$ | squiggle $\mapsto 2$ |

Thus, for example, the card consisting of one purple, striped, squiggle would correspond to the 4-tuple $(1, 0, 1, 2)$. Then we can think of the 81 cards as the *points* in the affine geometry $AG(4, 3)$, the four-dimensional affine geometry built over the field $\mathbb{F}_3 = \{0, 1, 2\}$. All the lines in this affine geometry contain three points, every plane has nine points, and the hyperplanes have 27 points each. (Recall the points of $AG(4, 3)$ are all possible ordered 4-tuples $(a_1, a_2, a_3, a_4)$, where each $a_i = 0, 1$ or 2. See Chapter 5 for a review.)

When do three cards form a *SET* in $AG(4, 3)$? The answer is simple and quite satisfying:

**Proposition P.6.**  *Three cards form a SET if and only if the corresponding three points in $AG(4, 3)$ form a line.*

The proposition is easiest to prove if we first show that the lines in $AG(n, 3)$ all have a very special property:

**Lemma P.7.**  *Three points $P = (a_1, a_2, \ldots, a_n)$, $Q = (b_1, b_2, \ldots, b_n)$ and $R = (c_1, c_2, \ldots, c_n)$ form a line in $AG(n, 3)$ if and only if $a_i + b_i + c_i \equiv 0 \pmod 3$ for all $1 \leq i \leq n$.*

## Problem 1
Prove Lemma P.7. One way to attack this problem is to express the lines in $AG(n, 3)$ through vectors, as depicted in Figure P.12. Then, if $P$ and

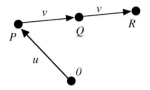

Figure P.12. The line
$\overline{PQR}$ in $AG(n, 3)$. The
coordinates for the point $R$
on the line through $P$ and $Q$
is uniquely determined as
$R = 2P + 2Q$.

$Q$ are two given points, the coordinates for the unique third point $R$ on $\overline{PQ}$ is determined as follows:

$$R = \overline{0P} + 2\overline{PQ} = P + 2(Q - P) = 2P + 2Q.$$

## Problem 2

Now use Lemma P.7 to prove Proposition P.6.

Thus, given any two cards, there is a unique card that completes a *SET* with the two given cards. This result depends on Lemma P.7 in a crucial way, and tells us the field $\mathbb{F}_3$ is special.

For instance, over the field $\mathbb{F}_5$, every line has five points. While it's still true that each of the coordinates of the five points on a line sum to zero (mod 5), the converse is false: you can find collections of five points whose coordinates sum to 0, but are not lines. For a quick, non-numbered exercise, prove that the five points on a line in $AG(2, 5)$ sum to the zero vector, and then find an example of five non-collinear points that also sum to $\overline{0}$.

What do affine planes look like in SET^®?

SET^® aficionados use the term *magic square*[6] to refer to a collection of cards arranged as in Figure P.13.

## P.6.3 Counting problems

Because the cards in the deck are points in $AG(4, 3)$ and the *SET*s are the lines in that geometry, we can use our discussion on $q$-binomial coefficients (see Section 5.5) to answer all sorts of counting questions. Here are several for you to check:

## Problem 3

Show the following (it may be useful to refer to Theorem 5.30):

(1) The number of *SET*s in the deck is 1080, and each card is in exactly 40 *SET*s.
(2) The 1080 *SET*s are partitioned into the four kinds as follows:
  (a) All four attributes different: 216 (exactly 20% of all *SET*s).
  (b) Three attributes different, one attribute the same: 432 (exactly 40% of all *SET*s).

---

[6] This terminology is a bit unfortunate; mathematicians reserve the term *magic square* for an $n \times n$ array of integers, usually composed of the numbers 1 through $n^2$, whose rows, columns and two main diagonals have the same sum.

Figure P.13. The affine plane
$AG(2, 3)$ in SET cards. The
configuration of nine cards
contains 12 sets, the 12 lines of
$AG(2, 3)$.

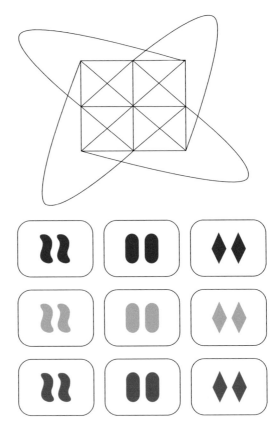

(c) Two attributes different, two attributes the same: 324 (exactly 30% of all *SET*s).

(d) One attribute different, three attributes the same: 108 (exactly 10% of all *SET*s).

(3) The number of nine-card affine planes is 1170, and each card is in exactly 130 planes.

(4) The number of 27-card affine hyperplanes is 120, and each card is in exactly 40 hyperplanes. (See Figure P.14 for a picture of one hyperplane.)

(5) The expected number of *SET*s that the first 12 cards contain is $220/79 \approx 2.78481 \ldots$

## Problem 4

Generalize as many of the counts from Problem 3 as you can to the general case $AG(n, 3)$, for $n \geq 4$. In particular, are the percentages of "*SET*s" of the different kinds always "nice"?

## P.6.4 Caps

We mentioned above that it is possible for 12 cards to contain no *SET*s. What is the largest number of *SET*-free cards you can find? Such a

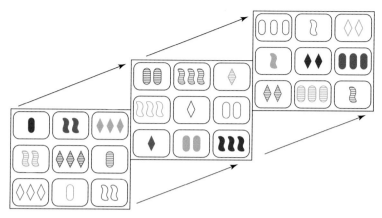

Figure P.14. A hyperplane of 27 cards is isomorphic to $AG(3, 3)$.

collection is called a *cap* by geometers, and we are asking for the maximum size of a cap in $AG(4, 3)$.

It's not too hard to find 16 cards with no *SET*s: take all cards with $a_i = 1$ or 2 for $1 \leq i \leq 4$. Using our coding, that means all cards that are red or green, have one or two symbols, are striped or solid, and use diamonds or squiggles. There are $2^4 = 16$ such cards, and it's clear from Lemma P.7 there are no *SET*s contained among these cards.

### Problem 5

Show that the size of the largest cap in $AG(n, 3)$ is at least $2^n$.

It's interesting to determine the exact value for the size of the maximum cap for small values of $n$. Then you can check the following:

| $n$ | 1 | 2 | 3 | 4 | 5 |
|---|---|---|---|---|---|
| Maximum cap size | 2 | 4 | 9 | 20 | 45 |

A maximum size cap of size 20 in $AG(4, 3)$ is shown in Figure P.15. The 20 points can be paired off to form *SET*s with the card set off to the

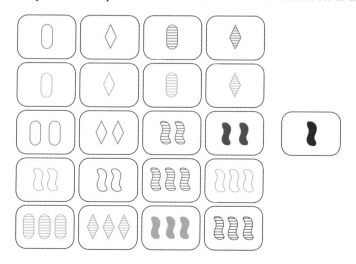

Figure P.15. A maximum size cap has the structure of a pencil of 10 lines through a point $P$, with $P$ removed.

side. This means this collection of 21 points (the 20 points in the cap and the additional card) forms a pencil of 10 lines through a point.

We conclude this project by stating a theorem about caps in $AG(4, 3)$.

**Theorem P.8.**  *Let C be a maximum size cap in $AG(4, 3)$. Then $|C| = 20$, and C is composed of a pencil of 10 lines through a point P, with P removed.*

As a general research question, it would be of interest to determine when the maximum caps have this structure for $n > 4$. (Note that for $n = 3$ and $n = 5$, this does not hold: a pencil of lines, with one point removed, will always have an even number of points, but the maximum size caps in these two cases are odd.)

## P.7 More matroid constructions – truncation

Given a matroid $M$, how can we define the matroid $M'$, derived from $M$? This project explores a few ways we can construct new matroids from existing ones. In Chapter 3, we introduced three important operations, deletion, contraction and duality. In each case, we produced a new matroid from a given matroid. Now we define a new operation on a matroid $M$, *truncation*.

**Definition P.9.** Let $M$ be a rank $r$ matroid on the ground set $E$. Let $\mathcal{I}_k$ be the collection of all independent sets of cardinality at most $k$. Then define the *truncation* $T(M)$ to be the matroid with independent sets $\mathcal{I}_{r-1}$.

Then it's obvious $\mathcal{I}_{r-1}$ satisfies (I1), (I2) and (I3), so $T(M)$ is a matroid. For some practice, you should check $T(F_7) = T(F_7^-) = U_{2,7}$, where $F_7$ is the Fano plane and $F_7^-$ is the non-Fano plane. In fact, you can show more:

(1) (a) If $M$ is a rank 3 simple matroid on $n$ points, show that $T(M) = U_{2,n}$.
    (b) Show $T(U_{r,n}) = U_{r-1,n}$

     Generally, it is straightforward to describe the bases, circuits, rank function, flats and geometric lattice for $T(M)$:

(2) Let $M$ be a matroid with truncation $T(M)$. Show the following:
    (a) Bases $\mathcal{B}(T(M))$: $B$ is a basis in $T(M)$ if and only if $B$ is independent in $M$ and $|B| = r - 1$. Equivalently, $\mathcal{B}(T(M)) = \{B - x \mid B \in \mathcal{B}(M) \text{ and } x \in B\}$.
    (b) Circuits $\mathcal{C}(T(M))$: $C$ is a circuit in $T(M)$ if and only if $C$ is a circuit of $M$ or $C$ is a basis of $M$, i.e., $\mathcal{C}(T(M)) = \mathcal{C}(M) \cup \mathcal{B}(M)$.
    (c) Rank $r_{T(M)}$: $r_{T(M)}(A) = \min\{r_M(A), r - 1\}$, i.e., $r_{T(M)}(A) = r_M(A)$ if $r_M(A) \leq r - 1$, and $r_{T(M)}(A) = r - 1$ if $A$ is a spanning set in $M$.
    (d) Flats $\mathcal{F}(T(M))$: $F$ is a flat of $T(M)$ if and only if $F$ is a non-hyperplane flat of $M$, i.e., $\mathcal{F}(T(M)) = \mathcal{F}(M) - \mathcal{H}(M)$, where $\mathcal{H}(M)$ is the collection of hyperplanes of $M$.
    (e) Geometric lattice $\mathcal{L}(T(M))$: the geometric lattice of flats of $T(M)$ is obtained by removing the hyperplanes from the lattice of flats of $M$.
    (f) Spanning sets $\mathcal{S}(T(M))$: $S$ spans $T(M)$ if $S$ was a spanning set for $M$, or if $r(S) = r - 1$.
    That $\mathcal{C} \cup \mathcal{B}$ satisfies the circuit axioms for any matroid is the point of Exercise 26 in Chapter 2.

When $r(M) = 4$, it's especially easy to draw a picture of $T(M)$ from a picture of $M$. See Figure P.16.

Figure P.16. Left: a rank 4 matroid $M$. Right: the rank 3 truncation $T(M)$. A picture of the truncation $T(M)$ is a projection of the picture for $M$ – just remove the two-point lines drawn for perspective in $M$.

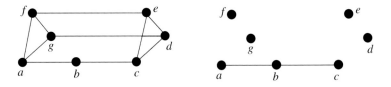

(3) Suppose $G$ is a graph with associated cycle matroid $M(G)$. Give an example to show $T(M(G))$ need not be a graphic matroid. (See if you can find an example where $T(M(G))$ is not binary.)

(4) Suppose $M$ is a paving matroid, i.e., all the circuits of $M$ have size $r(M)$ or $r(M) + 1$. Show that $T(M)$ is paving, too.

(5) Show that truncation commutes with deletion and contraction: if $M$ is a matroid on the ground set $E$ with $e \in E$, then

(a) If $e$ is not an isthmus, then $T(M - e) = T(M) - e$.

(b) If $e$ is not a loop, then $T(M/e) = T(M)/e$.

You can also describe the bases for the dual matroid $T(M)^*$:

(6) Dual $T(M)^*$: let $M$ be a rank $r$ matroid on the ground set $E$. Then $B$ is a basis of $T(M)^*$ if $B = (E - B) \cup x$ for some basis $B$ of $M$ with $x \notin B$, i.e., $\mathcal{B}(T(M)^*) = \{B \cup x \mid B \in \mathcal{B}(M^*), x \notin B\}$.

Note that $r(T(M)^*) = r(M^*) + 1$, so this process increases rank. In fact, you can define this operation without using duality at all:

(7) Let $M$ be a rank $r$ matroid with bases $\mathcal{B}(M)$, and let $\mathcal{B}' = \{B \cup x \mid B \in \mathcal{B}(M) \text{ and } x \notin B\}$. Then show $\mathcal{B}'$ is the collection of bases of a matroid, the *elongation* of $M$. We write $\mathcal{E}(M)$ for this matroid.

In terms of duality, we can now write

$$T(M)^* = \mathcal{E}(M^*).$$

(8) Give an example to show that truncation and elongation are *not* inverse operations, in general:

$$T(\mathcal{E}(M)) \neq \mathcal{E}(T(M)).$$

We can define the $i$th truncation $T^i(M)$ inductively by $T^i(M) = T(T^{i-1}(M))$, provided $i \leq r(M)$. Then it's clear that $r(T^i(M)) = r(M) - i$. How many times do we need to truncate until we get a uniform matroid? The answer is given in another truncation exercise. First, one more definition:

**Definition P.10.** If the smallest circuit of a matroid $M$ has $k$ elements, then we say $M$ has *girth* $k$. (If $M$ has no circuits, then $M$ is a Boolean algebra, and we set the girth to be $\infty$.)

Girth

(9) Let $M$ be a rank $r$ matroid with $n$ points and girth $k$, and assume $r < n$ (so $M$ is not the Boolean algebra). Show that $T^{r-k+1}(M) =$

$U_{k-1,n}$, and $T^i(M)$ is not a uniform matroid for any $i < r - k + 1$. (If $k = r + 1$, then this gives $T^0(M)$. You should interpret $T^0(M)$ as $M$.)

It's possible to get the Tutte polynomial $t(T(M); x, y)$ from the Tutte polynomial $t(M; x, y)$. First, an example.

**Example P.11.** Let $M$ be the uniform matroid $U_{3,4}$. Then $M$ is a circuit, and we know $t(M; x, y) = x^3 + x^2 + x + y$, and the truncation $T(M) = U_{2,4}$, with $t(T(M); x, y) = x^2 + 2x + 2y + y^2$ (see Examples 9.4 and 9.5).

Let's rewrite these polynomials using the corank–nullity formulation of Section 9.3. Now $s(M; u, v)$ is related to the Tutte polynomial by the simple change of variable: $x = u + 1$ and $y = v + 1$. Then we get

$$s(U_{3,4}; u, v) = u^3 + 4u^2 + 6u + v + 4 \quad \text{and}$$
$$s(U_{2,4}; u, v) = u^2 + 4u + v^2 + 4v + 6.$$

But $s(M; u, v) = \sum_{A \subseteq E} u^{r(M)-r(A)} v^{|A|-r(A)}$. We concentrate on two families of subsets:

- $A$ is spanning in $M = U_{3,4}$, i.e., $r(A) = 3$. The spanning sets $A$ are the subsets of $E$ of cardinality at least 3, and these give the following terms in $s(U_{3,4}; u, v)$:

  *Computation of terms in $s(U_{3,4}; u, v)$.*

  | $A$ | abc | abd | acd | bcd | abcd |
  |---|---|---|---|---|---|
  | $u^{r(M)-r(A)}v^{|A|-r(A)}$ | 1 | 1 | 1 | 1 | $v$ |

  So the total contribution to $s(U_{3,4}; u, v)$ from these subsets is $v + 4$. Now in the truncation $T(U_{3,4}) = U_{2,4}$, these subsets are still spanning, but their rank has dropped by 1, and so has the rank of the matroid. So the *corank* $r(M) - r(A)$ is unchanged (it's still 0), but the *nullity* is increased by 1:

  *Computation of terms in $s(U_{2,4}; u, v)$.*

  | $A$ | abc | abd | acd | bcd | abcd |
  |---|---|---|---|---|---|
  | $u^{r(M)-r(A)}v^{|A|-r(A)}$ | $v$ | $v$ | $v$ | $v$ | $v^2$ |

- $A$ is not a spanning set in $M = U_{3,4}$, i.e., $r(A) < 3$. This time, we are looking at subsets $A$ with $|A| < 3$. For each such subset, the rank remains the same in the truncation, so the corank is decreased by 1 and the nullity is unchanged:

| A | Ø | a | b | c | d | ab | ac | ad | bc | bd | cd |
|---|---|---|---|---|---|----|----|----|----|----|----|
| Term in $U_{3,4}$ | $u^3$ | $u^2$ | $u^2$ | $u^2$ | $u^2$ | $u$ | $u$ | $u$ | $u$ | $u$ | $u$ |
| Term in $U_{2,4}$ | $u^2$ | $u$ | $u$ | $u$ | $u$ | 1 | 1 | 1 | 1 | 1 | 1 |

So each term in this case has seen its $u$-exponent drop by 1, and its $v$-exponent stay the same.

The main point of Example P.11 is that we can determine the corank–nullity polynomial $s(T(M))$ from $s(M)$, so we can do the same thing for the Tutte polynomial.

Last exercise:

(1) Show that the corank–nullity polynomial for the $T(M)$ can be computed from the corank–nullity polynomial of $M$:

$$s(T(M); u, v) = (v - u^{-1})s(M; 0, v) + u^{-1}s(M; u, v).$$

(2) Use the change of variables $x = u + 1$ and $y = v + 1$ to do the same thing for the Tutte polynomials:

$$t(T(M); x, y) = (x - 1)^{-1} ((xy - x - y)t(M; 1, y) + t(M; x, y)).$$

# Appendix

## Matroid axiom systems

### A.1 Axiom lists

Axiom systems for a matroid $M$ defined on a ground set $E$.

**Independent sets $\mathcal{I}$**

(I1)  $\mathcal{I} \neq \emptyset$.

(I1')  $\emptyset \in \mathcal{I}$.

(I2)  If $J \in \mathcal{I}$ and $I \subseteq J$, then $I \in \mathcal{I}$.

(I3)  If $I, J \in \mathcal{I}$ with $|I| < |J|$, then there is some element $x \in J - I$ with $I \cup \{x\} \in \mathcal{I}$.

(I3')  If $I, J \in \mathcal{I}$ with $|J| = |I| + 1$, then there is some element $x \in J - I$ with $I \cup \{x\} \in \mathcal{I}$.

**Bases $\mathcal{B}$**

(B1)  $\mathcal{B} \neq \emptyset$.

(B2)  If $B_1, B_2 \in \mathcal{B}$ and $B_1 \subseteq B_2$, then $B_1 = B_2$.

(B2')  If $B_1, B_2 \in \mathcal{B}$, then $|B_1| = |B_2|$.

(B3)  If $B_1, B_2 \in \mathcal{B}$ and $x \in B_1 - B_2$, then there is an element $y \in B_2 - B_1$ so that $B_1 - x \cup \{y\} \in \mathcal{B}$.

(B3'')  If $B_1, B_2 \in \mathcal{B}$ with $x \in B_1 - B_2$, then there is an $y \in B_2 - B_1$, so that $B_1 - x \cup y$ and $B_2 - y \cup x$ are bases.

**Circuits $\mathcal{C}$**

(C1)  $\emptyset \notin \mathcal{C}$.

(C2)  If $C_1, C_2 \in \mathcal{C}$ and $C_1 \subseteq C_2$, then $C_1 = C_2$.

(C3)  If $C_1, C_2 \in \mathcal{C}$ with $C_1 \neq C_2$, and $x \in C_1 \cap C_2$, then $C_3 \subseteq C_1 \cup C_2 - x$ for some $C_3 \in \mathcal{C}$.

(C3')  If $C_1, C_2 \in \mathcal{C}$ with $C_1 \neq C_2$, $x \in C_1 \cap C_2$, and $y \in C_1 - C_2$ then $y \in C_3 \subseteq C_1 \cup C_2 - x$ for some $C_3 \in \mathcal{C}$.

**Rank function $r$** For any $A \subseteq E$:

(r1)  $0 \leq r(A) \leq |A|$.

(r1')  $r(\emptyset) = 0$.

(r2)    If $A \subseteq B$, then $r(A) \leq r(B)$.

(r2')    $r(A \cup \{x\}) = r(A)$ or $r(A) + 1$.

(r3)    $r(A \cup B) + r(A \cap B) \leq r(A) + r(B)$.

(r3')    If $r(A) = r(A \cup x) = r(A \cup y)$, then $r(A \cup \{x, y\}) = r(A)$.

(r3'')    If $r(A) = r(A \cup x_1) = r(A \cup x_2) = \cdots = r(A \cup x_n)$, then

$$r(A \cup \{x_1, x_2, \ldots, x_n\}) = r(A).$$

**Flats $\mathcal{F}$**

(F1)    $E \in \mathcal{F}$.

(F2)    If $A, B \in \mathcal{F}$, then $A \cap B \in \mathcal{F}$.

(F3)    If $F \in \mathcal{F}$ and $\{F_1, F_2, \ldots F_k\}$ is the set of flats that cover $F$, then $\{F_1 - F, F_2 - F, \ldots, F_k - F\}$ partition $E - F$.

**Hyperplanes $\mathcal{H}$**

(H1)    $E \notin \mathcal{H}$.

(H2)    If $H_1, H_2 \in \mathcal{H}$ and $H_1 \subseteq H_2$, then $H_1 = H_2$.

(H3)    For all distinct $H_1, H_2 \in \mathcal{H}$ and for all $x \in E$, there exists $H \in \mathcal{H}$ with $(H_1 \cap H_2) \cup x \subseteq H$.

(H3')    For all distinct $H_1, H_2 \in \mathcal{H}$, $x \notin H_1 \cup H_2$ and $y \in H_1 - H_2$, there exists $H \in \mathcal{H}$ with $(H_1 \cap H_2) \cup x \subseteq H$ and $y \notin H$.

# A.2  Axiom tables

## Graphic matroids

Let $M = M(G)$ denote a graphic matroid for $G$ a connected graph.

Table A.1. *Matroid terms interpreted for graphs.*

| Matroid term | Symbol | Graph interpretation |
|---|---|---|
| Independent sets | $\mathcal{I}$ | Edges of forests. |
| Bases | $\mathcal{B}$ | Edges of spanning forests. (Spanning trees if $G$ is connected.) |
| Circuits | $\mathcal{C}$ | Edges of cycles. |
| Rank | $r$ | $r(A)$ is the number of edges of a spanning forest in $A$. |
| Flats | $\mathcal{F}$ | Edges $F$ for which there is a partition $\Pi$ of the vertices so that $e \in F$ whenever $e$ joins two vertices of the same block of $\Pi$. |
| Cocircuits | $\mathcal{C}^*$ | Minimal edge cut-sets. |
| Hyperplanes | $\mathcal{H}$ | Cocircuit complements or maximal flats. |
| Closure | $^-$ | $\bar{A}$ contains all edges whose endpoints are connected by a path in $A$. |

## Representable matroids

Let $M$ be a matroid represented by a matrix over a field.

Table A.2. *Matroid terms interpreted for matrices.*

| Matroid term | Symbol | Matrix description |
|---|---|---|
| Independent sets | $\mathcal{I}$ | Linearly independent sets of columns. |
| Bases | $\mathcal{B}$ | Maximal independent sets. |
| Circuits | $\mathcal{C}$ | Minimal linearly dependent sets. |
| Rank | $r$ | Rank of the corresponding submatrix. |
| Flats | $\mathcal{F}$ | Sets equal to their linear span. |
| Hyperplanes | $\mathcal{H}$ | Corank 1 flats. |
| Closure | – | Linear span. |
| Cocircuits | $\mathcal{C}^*$ | Complements of hyperplanes. |
| Spanning sets | $\mathcal{S}$ | Subsets of columns whose linear span contains all the columns of the matrix. |

## Transversal matroids

We consider the transversal matroid $M$ associated with the bipartite graph with bipartition $E \cup X$.

Table A.3. *Matroid terms interpreted for transversal matroids.*

| Matroid term | Symbol | Description |
|---|---|---|
| Independent sets | $\mathcal{I}$ | Matchings |
| Bases | $\mathcal{B}$ | Maximal matchings |
| Circuits | $\mathcal{C}$ | Minimal sets without a matching |
| Rank | $r$ | $r(A)$ is size of a matching contained in $A$ |

Table A.4. *Matroid interpretations for affine geometry.*

| Matroid concept | Interpretation in $AG(n, q)$ |
|---|---|
| Points | Ordered $n$-tuples in $\mathbb{F}_q^n$. |
| Independent sets | $\{P_1, P_2, \ldots, P_n\}$ is affinely independent if there are no non-trivial scalars $a_i$ such that $\sum_{i=1}^{k} a_i = 0$ and $\sum_{i=1}^{k} a_i P_i = \bar{0}$. |
| Closure | Affine closure: for $S = \{P_1, P_2, \ldots, P_n\}$, $Q \in \bar{S}$ if there are non-trivial scalars $a_i$ such that $\sum_{i=1}^{n} a_i = 1$ and $Q = \sum_{i=1}^{n} a_i P_i$. |
| Flats | Affine flats: translates of subspaces of $\mathbb{F}_q^n$. Equivalently, all $x \in \mathbb{F}_q^n$ satisfying $Ax = Au$ for some $(n-r) \times n$ matrix $A$ and $u \in \mathbb{F}_q^n$. |
| Hyperplanes | Maximal flats: $\left\{ (x_1, x_2, \ldots, x_n) \mid \sum_{i=1}^{n} a_i x_i = b \text{ for } a_i, b \in \mathbb{F}_q \right\}$. |
| Rank | $r(AG(n, q)) = n + 1$, and $r(S) = \dim(\bar{S}) + 1$. |

Table A.5. *Matroid interpretations for the projective geometry* $PG(n, q)$.

| Matroid concept | Interpretation in $PG(n, q)$ |
|---|---|
| Points | Lines through $(0, 0, \ldots, 0)$ in $AG(n + 1, q)$. Equivalently, equivalence classes $[v]$ for $\bar{0} \neq v \in \mathbb{F}_q^{n+1}$, where $v \sim v'$ if $v = k \cdot v'$ for a non-zero scalar $k$. |
| Independent sets | Linear independence: $\{P_1, P_2, \ldots, P_n\}$ is independent if there are no non-trivial scalars $a_i$ such that $\sum_{i=1}^{k} a_i P_i = \bar{0}$. |
| Closure | Linear closure: for $S = \{P_1, P_2, \ldots, P_n\}$, $Q \in \bar{S}$ if there are non-trivial scalars $a_i$ such that $Q = \sum_{i=1}^{n} a_i P_i$. |
| Flats | The points of $F$ correspond to a subspace of $\mathbb{F}_q^{n+1}$. Equivalently, a $k$-flat consists of all vectors $v$ satisfying $Av = \bar{0}$ for some $(n - k) \times n$ matrix $A$. |
| Hyperplanes | Maximal flats: $\left\{ (x_0, x_1, \ldots, x_n) \mid \sum_{i=0}^{n} a_i x_i = 0 \text{ for } a_i \in \mathbb{F}_q \right\}$. |
| Rank | $r(S)$ is the matrix rank of the matrix whose column vectors are the points of $S$. $r(PG(n, q)) = n + 1$. |

# Bibliography

[1] K. Appel and W. Haken. Every planar map is four colorable. *Bull. Amer. Math. Soc.*, 82:711–712, 1976.

> This paper is really just an announcement of the proof of the Four-Color Theorem, a problem that had been open for more than 100 years at the time of its resolution. The authors made fundamental use of a computer program to check nearly 2000 cases, and this generated some mathematical controversy at the time. A streamlined proof (that also uses a computer) appears in [28].

[2] L. Babai, C. Pomerance and P. Vèrtesi. The mathematics of Paul Erdös. *Notices of the AMS*, 45(1):19–31, 1998.

> This article is composed of three essays by the three authors, each of whom knew Erdös well and collaborated with him. They survey his contributions to many fields, and his influence on the development of those fields. His impact was enormous.

[3] G. Birkhoff. Abstract linear dependence in lattices. *Amer. J. Math*, 57:800–804, 1935.

> An early, foundational paper linking matroids to *geometric* lattices.

[4] G. Birkhoff. *Lattice Theory*, third edition, volume XXV. American Mathematical Society, Providence, RI, 1967.

> This text treats geometric lattices within the broader context of lattice theory. This is where the term "crypto-isomorphism" first appears.

[5] R. Bixby. On Reid's characterization of the ternary matroids. *J. Combin. Theory Ser. B*, 26:174–204, 1979.

> Bixby's proof models Reid's unpublished proof. See the comment concerning Seymour's paper [31].

[6] A. Björner, M. Las Vergnas, B. Sturmfels, N. White and G. Ziegler. *Oriented Matroids, Encyclopedia of Mathematics and Its Applications*, volume 46. Cambridge, New York, 1993.

> This text develops the theory of oriented matroids in an informal style. The authors emphasize the connections to topology and geometry, and include applications to linear programming. Electronic updates have been published in the journal *Electronic J. Combinatorics* since the publication of this text.

[7] A. Bondy and U. S. R. Murty. *Graph Theory (Graduate Texts in Mathematics)*, volume 244. Springer, New York, 2008.

> This is an updated version of a classic text on graph theory. One especially nice feature is the inclusion of several applications of graph theory at the end of the chapters. The text is very clearly written and well motivated. The 1976 edition can be downloaded from the website of http://book.huihoo.com/pdf/graph-theory-With-applications/

[8] T. Brylawski. A decomposition for combinatorial geometries. *Trans. Amer. Math. Soc.*, 171:235–282, 1972.

> This paper develops much of the theory of the Tutte polynomial for matroids, extending the foundational work of Tutte [35] and Crapo [12]. The paper includes several examples to illustrate the theory.

[9] T. Brylawski. Appendix of matroid cryptomorphisms. *Encyclopedia Math. Appl. – Theory of Matroids*, 26(1):298–312, 1986.

> This is a novel presentation of 13 different axiomatizations of matroids, with many variants of axiom systems presented. It also includes specializations of the axiom systems for special classes, including representable, affine, transversal, graphic and binary matroids.

[10] T. Brylawski and D. Kelly. *Matroids and Combinatorial Geometries, Carolina Lecture Series*, volume 8. Department of Mathematics, University of North Carolina, Chapel Hill, NC, 1980.

> This short text is designed for self-study, with very short 2–3 page chapters followed by 3–5 pages of exercises. Many of these exercises remain quite valuable to work through.

[11] T. Brylawski and J. Oxley. The Tutte polynomial and its applications. *Encyclopedia Math. Appl. – Matroid Applications*, 40:123–225, 1992.

> This excellent article appears as a chapter in *Matroid Applications* [41]. It gives a detailed introduction to the Tutte polynomial, including many applications to graphs. It also includes 97 exercises of various levels.

[12] H. Crapo. The Tutte polynomial. *Aequationes Math.*, 3:211–229, 1969.

> This paper extends Tutte's work on his *dichromatic* polynomial from graphs to matroids.

[13] P. Davis and R. Hersh. *The Mathematical Experience*. Birkhäuser, Boston, 1981.

> This text is written for the non-mathematician, but includes some interesting philosophical discussions about the nature of mathematical proof.

[14] J. Geelen, A. Gerards and A. Kapoor. The excluded minors for GF(4)-representable matroids. *J. Combin. Theory Ser. B*, 79:247–299, 2000.

> This paper proves there are precisely seven excluded minors for the class of matroids representable over the four-element field $GF(4)$. The proof is quite technical.

[15] R. Graham and P. Hell. On the history of the minimum spanning tree problem. *IEEE Ann. Hist. Comput.*, 7:43–57, 1985.

> The minimum weight spanning tree problem has an interesting history, and this article traces various greedy solutions to several different researchers working in Europe in the early part of the twentieth century. In particular, they point out that Kruskal's discovery of the greedy procedure was a rediscovery of work done in Czechoslovakia, Poland and France.

[16] R. L. Graham, M. Grötschel and L. Lovász, editors. *Handbook of Combinatorics*, volume 1. Elsevier, Amsterdam, 1995.

> This comprehensive volume includes three articles devoted to matroids. The chapter "Matroid minors" by Seymour is an advanced introduction to the subject that gives a good overview of the role connectivity and structure plays in proving excluded minor theorems.

[17] D. Hughes and F. Piper. *Projective Planes, Graduate Texts in Mathematics*, volume 6. Springer, New York, 1973.

> This graduate level text contains a proof of the Bruck–Ryser Theorem (Theorem 5.34).

[18] A. W. Ingleton. A note on independence functions and rank. *J. London Math. Soc.*, 34:4–56, 1959.

> The author proves that the non-Pappus configuration is not representable over commutative fields, but is representable over non-commutative division rings. He then modifies the configuration by adding an additional three-point dependence to produce a matroid not representable over division rings. See Exercise 27 of Chapter 6.

[19] D. Kelly and G.-C. Rota. Some problems in combinatorial geometry. *Proc. Internat. Sympos., Colorado State Univ., Fort Collins, Colo., 1971*, 57:309–312, 1973.

> This short paper includes several problems in matroid theory, but is included here for its characterization of the term "matroid."

[20] J. Kruskal. On the shortest spanning tree of a graph and the traveling salesman problem. *Proc. Amer. Math. Soc.*, 7:48–50, 1956.

Kruskal presents the algorithm that is now known as "Kruskal's algorithm" to solve the minimum weight spanning tree problem in a graph. See the comment for [15].

[21]  C. W. H. Lam, L. Thiel and S. Swiercz. The non-existence of finite projective planers of order 10. *Canad. J. Math.*, XLI:1117–1123, 1989.

This paper reports on the computer search that eventually verified that there are no projective planes of order 10. As was the case with the Four Color Theorem [1], this proof relies on the use of a computer in an essential way.

[22]  J. Lipton. *An Exaltation of Larks*. Penguin Books, New York, 1968, 1977, 1991.

This charming book gives the proper term for many collective nouns. Examples range from the mathematical ("a pencil of lines"), to the biblical ("a host of angels"), to the unexpected ("an exaltation of larks").

[23]  S. MacLane. Some interpretations of abstract linear dependence in terms of projective geometry. *Amer. J. Math*, 58:236–240, 1936.

This foundational paper connects matroids to projective geometry.

[24]  C. Merino, A. de Mier and M. Noy. Irreducibility of the Tutte polynomial of a connected matroid. *J. Combin. Theory Ser. B*, 83:298–304, 2001.

This paper proves that the Tutte polynomial of a connected matroid is irreducible over the polynomial ring $\mathbb{Z}[x, y]$. See Theorem 9.18.

[25]  J. Oxley. What is a matroid? *Cubo Math. Educ.*, 5(3):179–218, 2003.

This is a gentle introduction to matroids that concentrates on graphic and representable matroids, with connections to finite geometry. It is available from the author's website: https://www.math.lsu.edu/~oxley/survey4.pdf

[26]  J. Oxley. *Matroid Theory*, second edition. Oxford University Press, New York, 2011.

This text is the industry standard for the subject. The first edition was a necessity for graduate students and researchers, and the new edition includes many of the important results of the last 20 years. This very clearly written text is designed for graduate students and

researchers, and its value is enhanced by the fact that it is self-contained, and includes the proofs of several very important theorems.

[27]  R. Rado. A theorem on independence relations. *Quart. J. Math*, 13:83–89, 1942.

Transversal matroids are introduced here; the author shows that matchings in a bipartite graph satisfy the abstract notion of independence characterizing matroids.

[28]  N. Robertson, D. Sanders, P. Seymour and R. Thomas. A new proof of the four-colour theorem. *Electron. Res. Announc. Amer. Math. Soc.*, 2:17–25, 1996.

The authors provide a new proof of the Four-Color Theorem for planar graphs. The proof is modeled after the original proof of Appel and Haken [1], and, like that proof, still requires a computer to check several hundred cases. But this is a streamlined version of the original proof, and offers the hope that a non-computer based proof might still be possible (albeit with humans checking several cases by hand).

[29]  G. C. Rota. On the foundations of combinatorial theory 1. Theory of Möbius functions. *Z. Wahrscheinlichkeits theorie u. Verw. Gebiete*, 2:340–368, 1964.

This article won the prestigious Steele prize in 1988. The paper gives an algebraic formulation that sets results from combinatorics alongside more established fields, like number theory. One measure of the success of this paper is that it (almost) instantly changed the way this material is taught. As we noted in the text, the commendation called this ". . . the single paper most responsible for the revolution that incorporated combinatorics into the mainstream of modern mathematics."

[30]  W. Schwärzler. Being Hamiltonian is not a Tutte invariant. *Discrete Math.*, 91:87–89, 1991.

This paper shows that the Tutte polynomial does not determine the size of the largest circuit in a matroid by producing a counterexample. In contrast, the polynomial does determine the size of the *smallest* circuit. See Exercise 13 in Chapter 9.

[31]  P. Seymour. Matroid representation over GF(3). *J. Combin. Theory Ser. B*, 26:159–173, 1979.

This paper, along with Bixby's [5], gives two different proofs of Reid's characterization of the excluded minors for ternary matroids. The two papers appeared consecutively in *J. Combin. Theory*, and Bixby reviewed this paper in *Math. Rev.*, where he summarizes Seymour's proof.

[32] R. Stanley. *An Introduction to Hyperplane Arrangements, IAS/Park City Mathematical Series*, volume 14. American Mathematical Society, Providence, RI, 2004.

This collection of notes is based on a series of lectures the author gave in Park City, Utah. The notes are self-contained, and they are written at an advanced undergraduate/graduate level. Matroids are treated here, although they are not the main focus. The notes are available from the following website: http://www.math.umn.edu/~ezra/PCMI2004/stanley.jcp.pdf

[33] R. P. Stanley. Acyclic orientations of graphs. *Discrete Math.*, 5:17–178, 1973.

Stanley showed that the number of acyclic orientations of a graph are counted by the evaluation $t(M(G); 2, 0)$. See Theorem 9.21.

[34] F. Stevenson. *Projective Planes*. Freeman, San Francisco, 1972.

This text is still a good resource for classical results on projective planes, including substantial information about Pappus' and Desargues' Theorems.

[35] W. T. Tutte. A homotopy theorem for matroids I, II. *Trans. Amer. Math. Soc.*, 88:144–174, 1958.

Tutte proved the first two excluded minor theorems, characterizing regular matroids and binary matroids.

[36] J. H. van Lint and R. M. Wilson. *A Course in Combinatorics*. Cambridge University Press, Cambridge, 1992.

This is an excellent advanced undergraduate/graduate level text that includes an interesting chapter on $q$-binomial coefficients.

[37] D. J. A. Welsh. *Matroid Theory. L. M. S. Monographs*, volume 8. Academic Press, London, 1976.

This was the first "real" matroid theory book. Although it went out of print, it has been reissued by the publisher *Dover*. The text covers the basics and some advanced topics in an informal, but thorough style. It concentrates extensively on applications and connections between matroids and graphs.

[38] D. West. *Introduction to Graph Theory*, second edition. Prentice Hall, 2001.

This is a standard text on graph theory that includes a chapter on matroids, with many quick proofs of various cryptomorphic descriptions.

[39] N. White. *Theory of Matroids, Encyclopedia of Mathematics and Its Applications*, volume 26. Cambridge University Press, New York, 1986.

[40] N. White. *Combinatorial Geometries, Encyclopedia of Mathematics and Its Applications*, volume 29. Cambridge University Press, New York, 1987.

[41] N. White. *Matroid Applications, Encyclopedia of Mathematics and Its Applications*, volume 40. Cambridge University Press, New York, 1992.

The three books *Theory of Matroids, Combinatorial Geometries* and *Matroid Applications* appearing in Cambridge's *Encyclopedia of Mathematics and Its Applications* series are graduate level texts all edited by Neil White. These books contain chapters written by a variety of researchers on a wide range of topics within matroid theory (and a few chapters on closely related topics). One chapter of [41] is the standard Tutte polynomial reference [11], and Brylawski's "Appendix of Matroid Cryptomorphisms" [9] appears at the end of [39].

[42] H. Whitney. On the abstract properties of linear dependence. *Amer. J. Math*, 57:509–533, 1935.

This is the foundational paper that introduced matroids, including the term "matroid." Google Scholar lists 630 citations for this paper. The paper is remarkable for many reasons, and it still rewards a careful read. Whitney defined matroids via independent set axioms, but also introduced bases, circuits and the rank function. He found axiomatizations of all four of these concepts, and proved they were equivalent (the first *cryptomorphisms*). Whitney then defines duality, and interprets it for representable matroids (proving the relation between the matrix representing a matroid and one representing the dual) and graphs. He also introduces connectivity, and he uses fundamental circuits to get a special axiomatization for binary matroids. He concludes with a proof that the Fano plane is not representable over fields of characteristic 0.

[43] T. Zaslavsky. *Facing up to Arrangements: Face-count Formulas for Partitions of Space by Hyperplanes*, volume 154. American Mathematical Society, 1975.

This is a systematic study of how to count the regions in various hyperplane arrangements.

[44] G. Ziegler. *Lectures on Polytopes, Graduate Texts in Mathematics*, volume 152. Springer, New York, 1995.

This graduate level text is also an excellent reference for historical information. It includes information on Steinitz's Theorem on 3-connected graphs and polytopes.

# Index

Printed in the United States
By Bookmasters